Physics for Medical Imaging Applications

NATO Science Series

A Series presenting the results of scientific meetings supported under the NATO Science Programme.

The Series is published by IOS Press, Amsterdam, and Springer in conjunction with the NATO Public Diplomacy Division

Sub-Series

I. Life and Behavioural Sciences	IOS Press
II. Mathematics, Physics and Chemistry	Springer
III. Computer and Systems Science	IOS Press
IV. Earth and Environmental Sciences	Springer

The NATO Science Series continues the series of books published formerly as the NATO ASI Series.

The NATO Science Programme offers support for collaboration in civil science between scientists of countries of the Euro-Atlantic Partnership Council. The types of scientific meeting generally supported are "Advanced Study Institutes" and "Advanced Research Workshops", and the NATO Science Series collects together the results of these meetings. The meetings are co-organized bij scientists from NATO countries and scientists from NATO's Partner countries – countries of the CIS and Central and Eastern Europe.

Advanced Study Institutes are high-level tutorial courses offering in-depth study of latest advances in a field.
Advanced Research Workshops are expert meetings aimed at critical assessment of a field, and identification of directions for future action.

As a consequence of the restructuring of the NATO Science Programme in 1999, the NATO Science Series was re-organised to the four sub-series noted above. Please consult the following web sites for information on previous volumes published in the Series.

http://www.nato.int/science
http://www.springer.com
http://www.iospress.nl

Series II: Mathematics, Physics and Chemistry – Vol. 240

Physics for Medical Imaging Applications

edited by

Yves Lemoigne
European Scientific Institute, Site d'Archamps,
Archamps, France

Alessandra Caner
European Scientific Institute, Site d'Archamps,
Archamps, France

and

Ghita Rahal
European Scientific Institute, Site d'Archamps,
Archamps, France

Springer

Published in cooperation with NATO Public Diplomacy Division

Proceedings of the NATO Advanced Study Institute on
Optimising Detectors, Imaging and Computing Technologies from Nuclear Physics in General and Security Applications
Archamps, France
27 October – 8 November 2005

A C.I.P. Catalogue record for this book is available from the Library of Congress.

ISBN-10 1-4020-5650-8 (PB)
ISBN-13 978-1-4020-5650-5 (PB)
ISBN-10 1-4020-5649-4 (HB)
ISBN-13 978-1-4020-5649-9 (HB)
ISBN-10 1-4020-5653-2 (e-book)
ISBN-13 978-1-4020-5653-6 (e-book)

Published by Springer,
P.O. Box 17, 3300 AA Dordrecht, The Netherlands.

www.springer.com

Printed on acid-free paper

CONTENTS

PREFACE

The Nato Advanced Study Institute ASI-981171 was organised at the European Scientific Institute of Archamps (ESI, Archamps - France) from October 26 to November 8, 2005. Every year in autumn ESI organises the European School of Medical Physics, which covers a large spectrum of topics, ranging from Medical Imaging to Radiotherapy, over a period of five weeks. Thanks to the Cooperative Science and Technology sub-programme of the NATO Science Division, the first two weeks were replaced this year by the ASI course dedicated to "Optimising Detectors, Imaging & Computing Technologies from Nuclear Physics in General to Security Applications". This allowed the participation of experts and students from 20 different countries, with diverse cultural background and professional experience (Africa, America, Asia, and Europe). A further positive outcome of NATO ASI participation is the publication of this book, which contains the lectures series contributed by speakers during the ASI weeks. We hope it will be a reference book in Medical Imaging, addressing an audience of young medical physicists everywhere in the world, who are wishing to review the physics foundations of the relevant technologies, catch up with the state of the art and look ahead into future developments in their field.

The ASI course was structured in two parts, defined by the use or absence of ionising radiation in the medical imaging techniques of interest. The section on medical imaging started with a refresher on fundamental aspects of digital imaging given by Karl-Freidrich Kamm (Germany).

The courses on imaging techniques not based on ionising radiation began with Magnetic Resonance Imaging and Ultrasound. For Magnetic Resonance Imaging, Martin Leach (UK) reviewed the physics principles and illustrated how they lead to imaging while Emile Hiltbrand (CH) explained the main parameters in hospital MRI. François Lazeyras discussed security and quality assurance issues related to such complex and powerful equipment. The recent developments and future applications of MRI were addressed by Jeff Alger (USA), who introduced as well the novel and very promising field of Magnetic Resonance Spectroscopy (MRS). Alessandro Alimenti (CH) concluded the Magnetic Resonance Imaging series with a presentation on clinical aspects of the technology.

The fundamental principles of Ultrasound were reviewed by Teresa Robinson (UK). This contribution was complemented by Franco Bertora (IT) with a lecture on design principles and technical aspects of ultrasound transducers. Piero Tortoli (IT) reviewed the Doppler techniques in ultrasounds. The topic of image formation and processing was treated by Johan Thijssen

(NL), who also developed the important subject of quality assurance in Ultrasound. A review of contrast agents was presented by Christian Cachard (Fr) and novel applications of Ultrasound, such as elastography and ultrasound therapeutical applications were illustrated by Remi Souchon (FR) and Cyril Lafon (FR), respectively. A key presentation on quality assurance of ultrasound equipment was given by Johan Thijssen (NL). Image post processing was treated by Olivier Basset (Fr). Ultrasound lectures were concluded by Nico de Jong (NL), with a presentation on ultrasound intravascular imaging and a review of recent and future applications of echocardiography.

The program devoted to techniques based on ionising radiation includes two parts: Radiology and Nuclear Medicine. Some knowledge about electronic signal processing and evaluation of image quality is required in these disciplines. Electronics signals were addressed by Renato Turchetta (UK) while Peter Sharp covered the image quality aspects (UK). For Radiology, Isabel Castellano (UK) explained the principles of computed tomography, David Dance (UK) developed the principles of digital X-ray imaging and their application to mammography while Alain Noel (FR) shared his important professional know-how by discussing mammography quality control. In the context of security considerations, Alain Noel (FR) explained as well the patient dose regulations.

The Nuclear Medicine part began with Robert OTT (UK), who reviewed in detail the working principles of PET and SPECT cameras. Next, Gerd Beyer (CH) explained the production and use of radioisotopes, essential for this type of detectors. Data acquisition in radioisotope imaging was also illustrated by Robert Ott (UK) whereas PET applications to drug development were reviewed by Piero Salvadori (IT). Image reconstruction algorithms were discussed by Claude Comtat (FR) whereas David Bandon (CH) explained the PACS and DICOM standards for medical images management. To conclude these lecture series, Andrew Todd-Pokropek (UK) illustrated the latest advances in PET and their applications.

Lectures were not the only activity proposed to participants of the ASI: the possibility was offered to visit the equipments at the neighbouring Geneva hospital, hosting MRI devices and PET/SPECT cameras. Production of radioisotopes was explained in-situ at a new private firm called "Advanced Accelerators Applications" (AAA) in St-Genis, France. To add to their scientific education students were invited to CERN, located only 13 Km away from Archamps. At the renowned European experimental research centre, they were shown and taught about a huge crystal electromagnetic calorimeter, i.e. a detector representing a very large scale version of the main constituent of modern PET cameras. Note that for the students who wished to practise their skills in dealing with experimental equipment for beta and gamma radiation detection, lab courses were also organised at Geneva University.

We wish to thank all the participants, who allowed the ASI at Archamps to be a success within an excellent international atmosphere: lecturers, students (who participated actively) and all the ESI team (Manfred Buhler-Broglin, Alessandra Caner, Severine Guilland, Haidi Palenque and Ghita Rahal).

Many thanks to The Physics Department of the Geneva University which allowed us to propose pactical exercises in its laboratory directed by Jean Divic Rapin. We deeply thank also the CERN team in Geneva for openning the laboratory and the "Hôpital Cantonal de Genève" which welcomed us twice for equipment study (MRI, PET and SPECT camera).

Finally, we wish to thank and express the gratitude of all participants to the Cooperative Science and Technology sub-programme of the NATO Science Division, lead by Prof. Fausto Pedrazzini, without whom this Advance Study Institute could not have been possible.

Yves Lemoigne[1],
Co-Director of ASI-981171

Vitali Kaftanov[2]
Co-Director of ASI-981171

[1] European School of Medical Physics, European Scientific Institute, Bâtiment Le. Salève, Site d'Archamps F-74166 Archamps (France);

[2] ITEP, B. Cheremuskinskaya 25, Moscow, RU-117259 (Russia)

LIST OF PARTICIPANTS

Lecturers and assistants are listed in bold characters.

Surname	Name	Origin	City / Affiliation	E-mail
1. **Mr Alger**	Jeff	USA	Los Angeles / David Geffen School of Medicine, UCLA	jralger@ucla.edu
2. **Mr Alimenti**	Alessandro	Italian	Geneva / Hôpitaux Universitaires de Genève	alessandro.alimenti@hcuge.ch
3. Ms Andrei	Daniela Julia	Romanian	Bucharest/ National Comission for Nuclear Activities Control	dana_iulia_andrei@yahoo.com
4. **Mr Bandon**	David	Swiss	Geneva / Hôpitaux Universitaires de Genève	david.bandon@hcuge.ch
5. **Mr Basset**	Olivier	French	Lyon / Université Claude Bernard Inserm U630	olivier.basset@creatis.insa-lyon.fr
6. Ms Baytosum	Filiz	Turkish	Izmir/ Ege University	filiz_zilif@yahoo.com
7. **Mr Bertora**	Franco	Italian	Genoa/ Easote	bertora@iol.it
8. Mr Beyer	Gerd-Jurguen	German	Geneva / ESI	gerd.beyer@cern.ch
9. **Mr Cachard**	Christian	French	Lyon / Université Claude Bernard Inserm U630	christian.cachard@creatis.insa-lyon.fr
10. Mr Çakir	Aydin	Turkish	Istanbul / Istanbul University	cakiraydin@yahoo.com
11. Ms Camara	estela	Spain	Barcelona / University of Barcelona	ecamara@ub.edu
12. **Ms Caner**	Alessandra	Italian	Archamps / ESI	alessandracaner@freesurf.fr
13. Mr Carlsohn	Matthias	German	Bremen / Engineering & Consultancy Dr. Carlsohn	matthias.carlsohn@t-online.de
14. **Ms Castellano**	Isabel	Spanish	London / Royal Marsden Hospital	elly.castellano@rmh.nhs.uk

15. Ms Cho	Jennifer	USA	Los Angeles / UCLA	jscho@mednet.ucla.edu
16. Ms Cojanu	Elena-Camelia	Romanian	Bucharest / National Institute for Laser Plasma and Radiation Physics	k_meli_ec@yahoo.com
17. **Mr Comtat**	Claude	French	Orsay / CEA-SHFJ	comtat@cea.fr
19. **Mr Dance**	David	British	London / Joint Department of Physics, Institute of Cancer Research and The Royal Marsden NHS FT	drdance@f2s.com
20. Ms Daníèková	Katerina	Czech	Prague / University Hospital Motol	katerina.danickova@fnmotol.cz
21. **Mr de Jong**	Nico	Deutch	Rotterdam / Erasmus University Rotterdam	n.dejong@erasmusmc.nl
22. Mr Di Leo	Giovanni	Italian	Genova / San Martino Hospital	dileo@na.infn.it
24. Ms Donchenko	Lyudmyla	Ukraine	Donetsk / Institute for Traumatology & Orthopedic of Donetsk State Medicine University	donch@rambler.ru or shishkov@kinetic.ac.donetsk.ua
25. Ms Doyeux	Kaya	French	Saclay / CEA Saclay	kdoyeux@cea.fr
26. Mr El deeb	Ahmed	Egyptian	Cairo / Physics dept AL Azhar University	ahmedeldeeb2003@yahoo.com
27. Mr El Hamadi	LakbirR	Morocco	Raba t/ Nationale Institute of Oncology	l lakbirelhamidi@yahoo.fr
28. Mr Emam	Ismail	Egyptian.	Cairo / Salam Oncology Center	ismailemam@yahoo.com
29. Ms Fetcu	Cristina	Romanian	Baia Mare / General County Hospital	cfetcu@spitaljbm.ro : cfetcu@yahoo.com
30. Mr Favre-Bulle	Parcal	Swiss	Neuchatel / Hopital de la Chaux-de-Fonds	pascal.favre@ne.ch
31. Ms Germano	Sara	Portuguese	Lisboa / Hospital de Santa Maria	galatchi@univ.ovidius.ro

32. Ms Gimenez-Navaro	Eva Nerina	Spain	Valencia / Hospital de Santa Maria	nerina.gimenez@ific.uv.es
33. **Mr Hiltbrand**	Emile	Swiss	Geneva / Hôpitaux Universitaires de Genève	emile.hiltbrand@hcuge.ch
34. Ms Hristova	Diana	Bulgarian	Sofia / German Cancer Research Center	d_t_hristova@yahoo.com
35. Ms Ionescu-Berechet	Ioana	Romanian	Bucharest / National Comission for Nuclear Activities Control	ioanaberechet@yahoo.com
36. Ms Iordache	Andrea-Maria	Romanian	Cluj-Napoca / Physics,University Babes-Bolyai	andres_iro2002@yahoo.com
37. Mr **Jannin**	Pierre	French	Rennes / IDM, Medical School, University of Rennes	pierre.jannin@univ-rennes1.fr
38. **Mr Kaftanov**	Vitaly	Russian	Moscow / ITEP	Vitali.kaftanov@cern.ch
39. **Mr Kamm**	Karl-Friedricht	German	Hamburg / Philips Medizin Système	karl-friedrich.kamm@philips.com
40. Mr Kesner	Adam	USA	Los Angeles / UCLA	alkesner@mednet.ucla.edu
41. Mr Lafon	Cyril	French	Lyon / INSERM U281	lafon@lyon.inserm.fr
41. **Mr Lazeyras**	François	French	Geneva / Hôpitaux Universitaires de Genève	francois.lazeyras@hcuge.ch
42. **Mr Leach**	Martin	British	London / Institute of Cancer Research and Royal Marsden Hospital	martin@icr.ac.uk
43. **Mr Lemoigne**	Yves	French	Archamps / ESI	psf1@compuserve.com
44. Mr Masiuk	Mariusz	Poland	Gdansk/ Instytut of Radiology and Nuclear Medicine	masiuk@amg.gda.pl or mariuszmas@poczta.onet.pl
45. Ms Mastantuoni	Antonella	Italian	Roma / Università "Tor Vergata"	antonella.mastantuoni@virgilio.it
46 .Mr Mathot	Gilles	Belgium	Louvain-la-neuve / IBA	mathot@iba.be

47. Mr Minkin	Victor	Russian	St Petersburg / ELSYS – NRI	minkin@elsys.ru
48. Ms Muntean	Maria-Lucia	Romanian	Cluj-Napoca / "Iuliu Hatieganu" University of Medicine and Pharmacy	luciamuntean@yahoo.com
49. **Mr Noel**	Alain	French	Nancy / Centre Alexis Vautrin	a.noel@nancy.fnclcc.fr
50. Mr Okutan	Murat	Turkish	Istanbu l/ Oncology Institute University of Istanbul	muratokutan@yahoo.com
51. **Mr Ott**	Robert	British	London / Physics Department, Institute of Cancer Research and Royal Marsden Hospital	bob@ICR.ac.uk
52. Ms Ortenzia	Ornella	Italian	Roma / Regina Elena Institute	ortenzia@na.infn.it
53. Ms Pasquali	Gabrielle	Italian	Pavia / Fondazione Salvatore Maugeri	ortenzia@na.infn.it
54. Ms Peycheva Stefanova	Svetlana	Bulgaria	Sofia / Sofia Cancer Center	sstef@abv.bg or svetlana_stefanova@yahoo.co.uk
55. Mr Popa	Marcel Sabin	Romanian	Cluj-Napoca / Technical University	marcel.popa@tcm.utcluj.ro
56. Ms Rahal	Ghita	French	Archamps / France	Ghita.Rahal@cern.ch
57. **Ms Robinson**	Teresa	British	Bristol / Bristol Oncology Centre	teresa.robinson@ubht.swest.nhs.uk
58. Ms Rossi	Raffaella	Italian	Parma / Azienda Ospedaliera di Parma	rrossi@ao.pr.it
59. Mr **Salvadori**	Pietro	Italian	Pisa / CNR Institute of Clinical Physiology Cyclotron and Radiopharmaceutical Chemistry Department	salvador@ifc.cnr.it
60. Mr Sayed-Afifi	Mohamed	Egyptian	Cairo / Minia Oncology Center	mbahaa@hotmail.com
61. Ms Seror	Veronique	Israeli	Beer-Sheva / Ben Gurion University	veronique.seror@gmail.com

62. **Mr Sharp**	Peter	British	Aberdeen / University of Aberdeen, University Hospital NHS Trust	p.sharp@biomed.abdn.ac.uk
63. Ms Shishkova	Natalya	Ukrainian	Donetsk / Donetsk Physics Techn. Institute of National Academy of Ukraine	shishkov@kinetic.ac.donetsk.ua
64. **Mr Souchon**	Remy	French	Lyon / INSERM UMR	souchon@lyon.inserm.fr
65. Mr Tavora	Luis	Portuguese	Leira/ ESTG-Instituto Politecnico de Leiria	imtavora@estg.ipleiria.pt
66. **Mr Thijssen**	Johan	Deutch	Nijmegen / University Medical Centre Nijmegen University Children's Centre Clinical Physics Laboratory	j.thijssen@cukz.umcn.nl
67. **Mr Todd-Pokropek**	Andrew	British	London / University College London Department of Medical Physics and Bioengineering	a.todd@ucl.ac.uk
68. **Mr Tortoli**	Piero	Italian	Firenze / Università degli Studi di Firenze Dipartimento di ingegneria ed elettronica	piero.tortoli@unifi.it
69. Mr Tsekov	Bogomil	Bulgarian	Sofia / University of Sofia	btsekov@fmi.uni-sofia.bg btsekov@gmail.com
70. Mr Turan	Ercan	Turkish	Denizli / Ege University Nuclear Sciences Institute	ercant35@hotmail.com ercant@pamukkale.edu.tr
71. **Mr Turchetta**	Renato	Italian	Oxford / Rutherford Appleton Laboratory Instrument. Dept.	r.turchetta@rl.ac.uk
72. Ms Ungureanu	Felicia	Romanian	Bucharest / National Institute for Laser Plasma and Radiation Physics	feci_comin@yahoo.com
73. Ms Yankova	Yana	Bulgarian	Sofia / Nacional Centre of Radiobiological and Radiation Protection	ian4eto@abv.bg

1. A sunny day at Archamps.

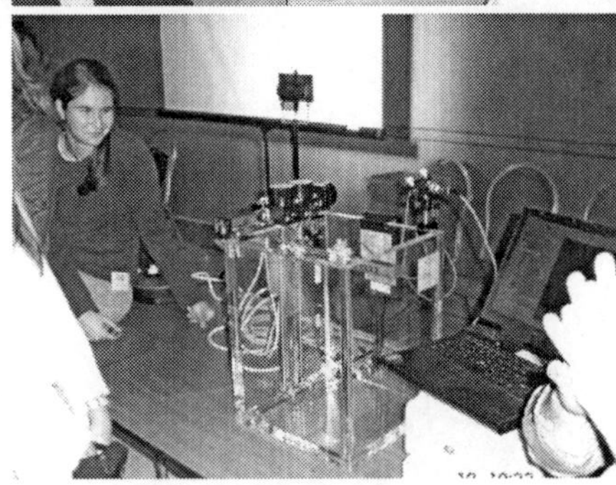

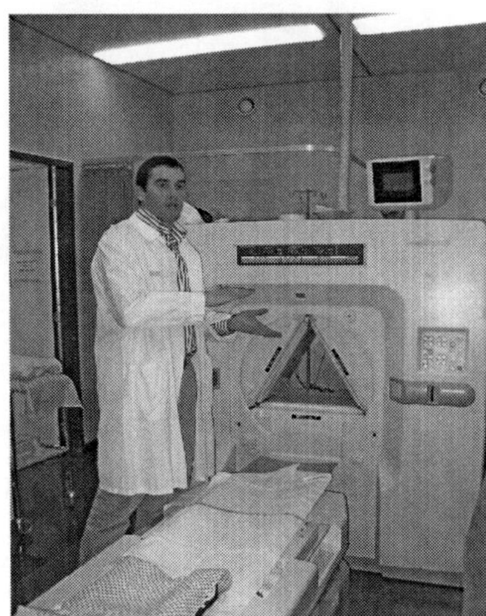

1. School auditorium. 2. Improvised snack. 3. Visit at CERN. 4. Experimental setup at the school. 5. Day off at Chamonix. 6. Visit at Geneva University Hospital.

Part I

General Imaging and Ultrasound Principle

FUNDAMENTAL ASPECTS OF DIGITAL IMAGING

KARL-FRIEDRICH KAMM*
Philips Medizin Systeme, Roetgenstrasse 24, D-22335 Hamburg, Germany

Abstract. Of most applications conventional film based methods may be replaced by digital imaging. This trend is influenced technically by the advent of more efficient detectors, improved image processing methods, faster computers, brighter and sharper displays and larger systems for image storage and archiving. The evolution of digital imaging reflects the fast development of information processing technology. In this context measures to describe the quality of medical images are very important. Intuitively we use the terms sharpness, contrast and the amount of noise in an image dependent on the applied radiation dose and the biophysical properties of the examined objects. Quantitative methods for specifying and evaluating the imaging capacity of digital radiographic systems have become an important tool for users, researchers, engineers and service specialists.

Key words: detector principles.

1. Introduction

During the last two decades Digital Imaging DI became a widely used replacement of conventional imaging techniques using X-rays, amplifying screens and films and gave further insight in the anatomy and function of the human body. In addition DI comprehensively changed the work in an imaging department. To demonstrate that, a few examples should be mentioned.

In Digital Radiography DR digital subtraction of images in real time revolutionized the film based angiographic methods preparing the advancement of interventional methods.

* Karl-Friedrich Kamm, Philips Medizin Systeme, Roetgenstrasse 24, D-22335 Hamburg, Germany; e-mail: karl-friedrich.kamm@philips.com.

Y. Lemoigne et al. (eds.), Physics for Medical Imaging Applications, 3–22.

To summarize, all fields of medical imaging rely on or at least offer the use of digital imaging techniques. In some areas digital imaging methods became the standard procedure.

- What are the major features of digital imaging and what are the advantages compared to conventional screen-film radiography?
- Will it totally replace films in radiography?

Learning from the forces that influenced the fast development of digital imaging methods in recent years this paper will address needs for further improvement, technological innovations, application trends and organizational aspects. It will try to describe the future of X-ray imaging.

2. Factors influencing the progress

2.1. TECHNICAL ASPECTS

2.1.1. *Characteristics of digital imaging*

In conventional radiography the film was serving as image detector, display and storage medium. Digital Imaging systems split the imaging process in several steps: detection, digitization, image processing, storing, display and film recording. By partitioning the imaging process the performance of each stage may be optimized individually.

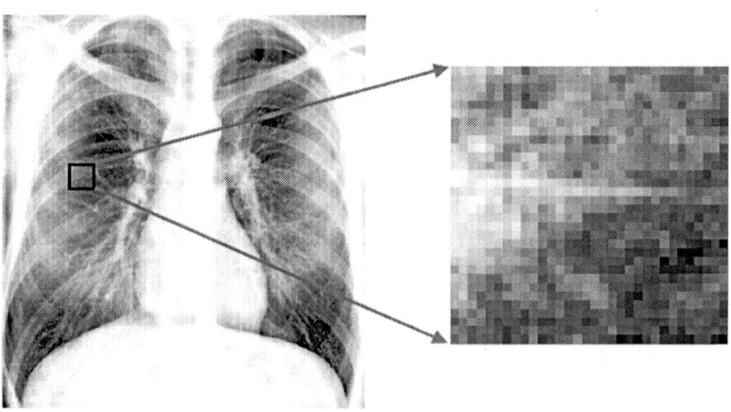

Figure 1. Digitization of a Radiograph.

As an example the detector should detect a high percentage of the incoming quanta, have a large dynamic range and it should show small objects with low contrast without adding artifacts. The image is digitized by dividing it into single pixels and discrete intensity values.

This representation of an image as a standardized, binary data set leads to manifold processing possibilities. As a consequence the appearance of the digital image can be changed with much more flexibility as the image on film. Opposite to screen-film systems digital detectors offer a large dynamic range and the linear relationship between dose and intensity signal. Therefore signal normalization is applied to get an optimal image. As a result there is no relation any more between applied radiation dose and intensity of the image. Well known dependencies are no longer valid, as: low exposure leads to an underexposed image. In addition the contrast of selected objects may be enhanced individually by image processing.

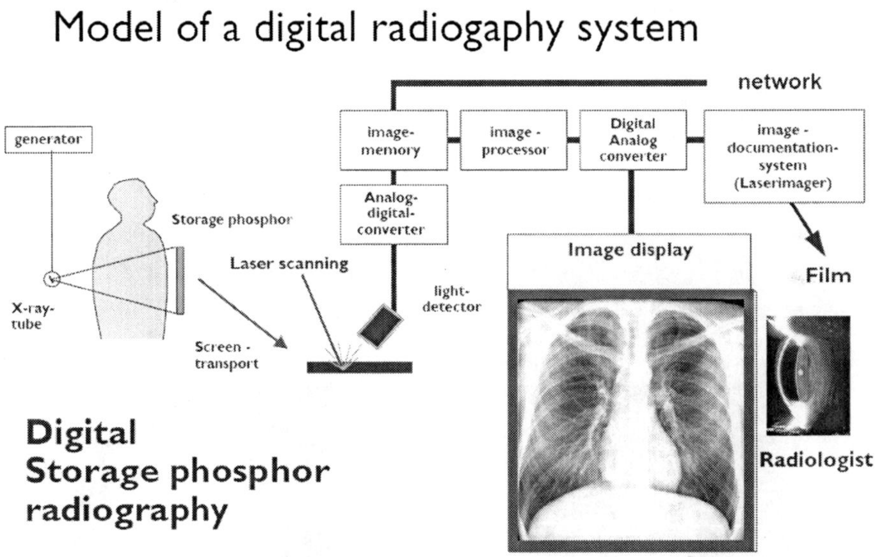

Figure 2. Model of a digital imaging system using storage phosphor plates.

2.1.2. *Aspects of image quality*

The quality of medical images is generally described in terms of sharpness, contrast and amount of noise in an image dependent on the applied X-ray dose and the biophysical properties of the examined object. From the diagnostic point of view the structures such as small fissures and vessels, collaterals, aneurysm, nodules, guide wires and catheters should be clearly visualized. The smallest sizes of these discernible objects have been summarized in medical guidelines.

Image quality of a digital imaging system is primarily determined by the quality of the X-ray equipment and the examination conditions. In addition it is influenced by the technical parameters pixel size, pixel depth, dynamic range and signal-to-noise ratio. If the size of the pixels is much smaller than the smallest detail, which has to be visualized in the image, there is no loss of information (Shannon Theorem).

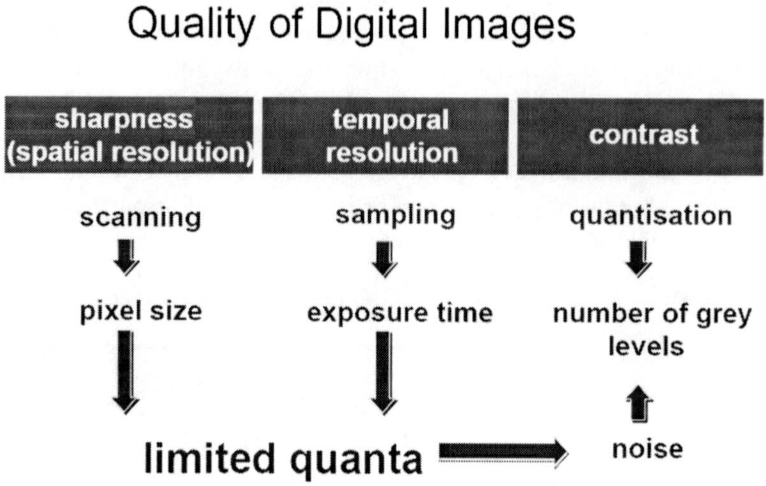

Figure 3. Quality of Digital Images.

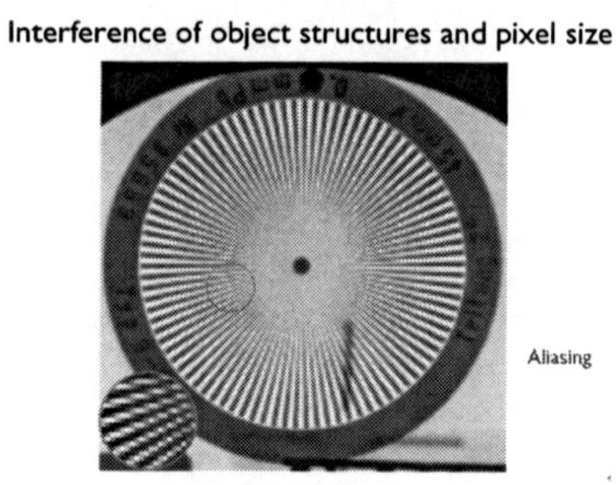

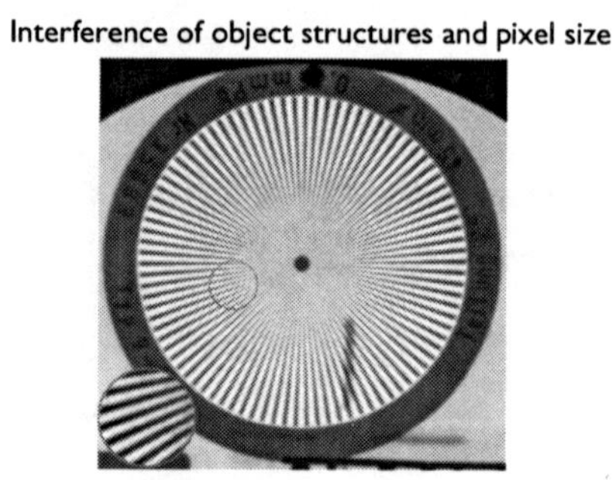

Figures 4 and 5. Interference of object structures and pixel size.

In practice the quality of digital imaging systems is assessed in the same way as with conventional film-screen equipment. High contrast lead bar pairs are used to measure resolution, i.e. the maximally resolvable spatial frequency. Since for a digital imaging system spatial resolution is, apart from the lack of definition due to the projection conditions, only limited by the Nyquist frequency, which is defined by the pixel size of the sampling process, this measurement is of limited value. It does not reflect the main advantage of digital radiography, that the contrast of different spatial frequencies may be selectively enhanced.

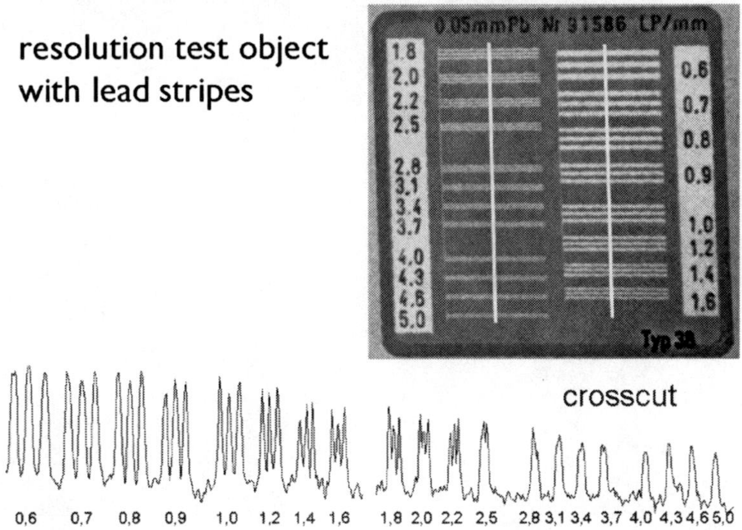

Figure 6. Resolution test object.

The amount of contrast enhancement possible is limited by the noise. The two major noise sources are quantum mottle from radiation and system noise generated by the imaging system. Therefore the amount of noise has to be measured quantitatively.

In image science there is a general agreement that the physical performance of an imaging system is best described by the intensity-transfer-function, the Modulation Transfer Function MTF combining sharpness with contrast and the Noise Power Spectrum NPS. Concepts such as "noise equivalent quanta" NEQ and "detective quantum efficiency" (DQE) have been found useful for normalizing physical noise measurements on an absolute scale and for relating those measurements to the decision performance of a hypothetical "ideal observer". A comprehensive measure of imaging quality is afforded by the NEQ found from a combination of the intensity transfer function, the Modulation Transfer Function and the Noise Power Spectrum. Noise Equivalent Quanta NEQ equals the number of quanta which contribute to the final image. The relation of NEQ to the total number of photons at the detector entrance defines the Detective Quantum Efficiency. The DQE describes the percentage of photons, which are detected by the imaging system and are contributing to the final image. The ideal detection system would give a DQE of 100%. DQE depends on radiation dose and spatial frequency.

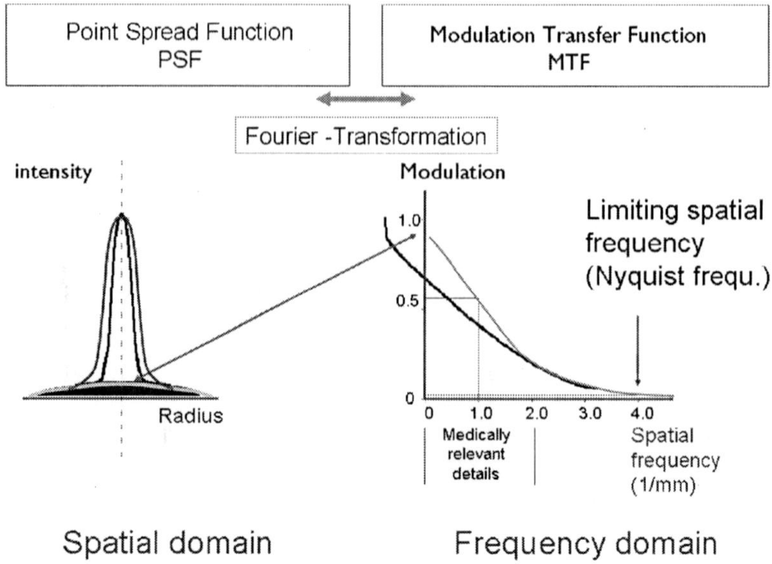

Figure 7. Modulation Transfer Function.

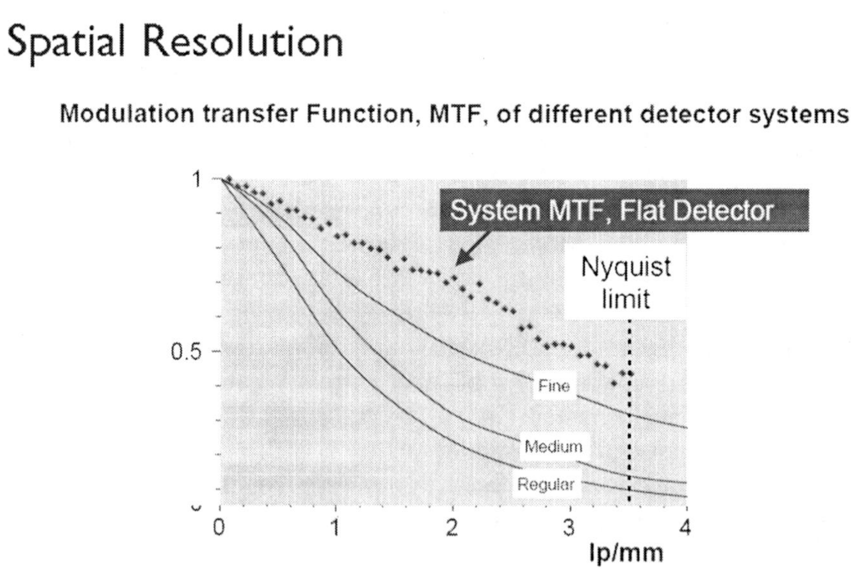

Figure 8. Modulation Transfer Function (practical examples).

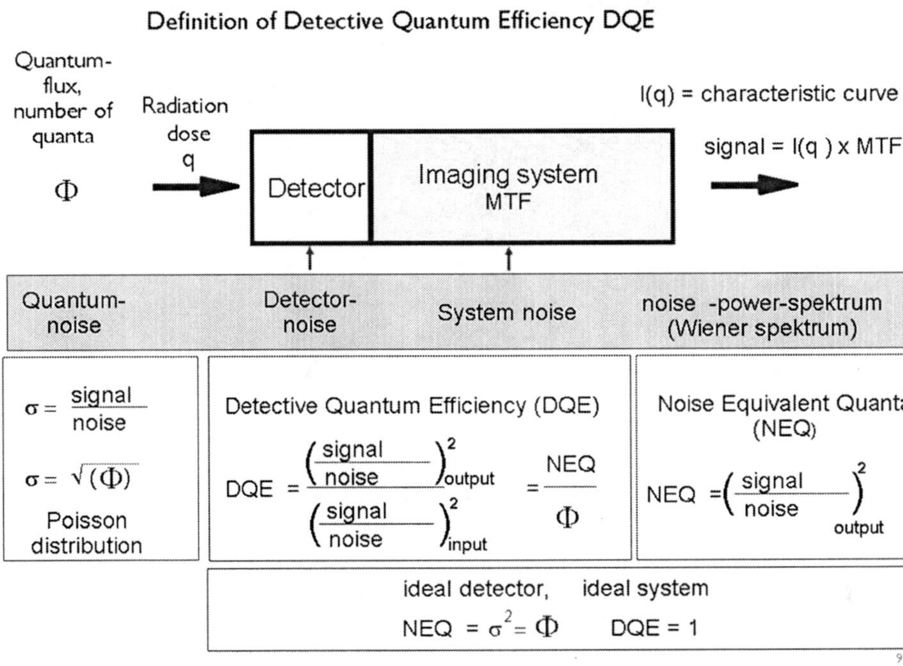

Figure 9. Definition of Detective Quantum Efficiency.

The intensity transfer functions (characteristic curve), describing the transformation of dose values at the detector entrance to intensity values of the resulting image on film or display, differ considerably for conventional and digital systems.

2.1.3. *Improvement of detection*

DR systems can be distinguished by their principle of image detection:

- Image intensifier - video systems
- Storage phosphor screens
- Flat panel detectors (selenium and cesium iodide)

The main objectives of development are high efficiency of photon detection i.e. high DQE and large dynamic range.

Compared with film/screen systems with a dynamic range of less than 1:40 image intensifier systems offer a dynamic range of 1:100 whilst storage phosphor or selenium systems more than 1:1000. This represents an improvement of at least a factor of ten. Practically this means, that over- and underexposures can be compensated after acquisition by signal normalization.

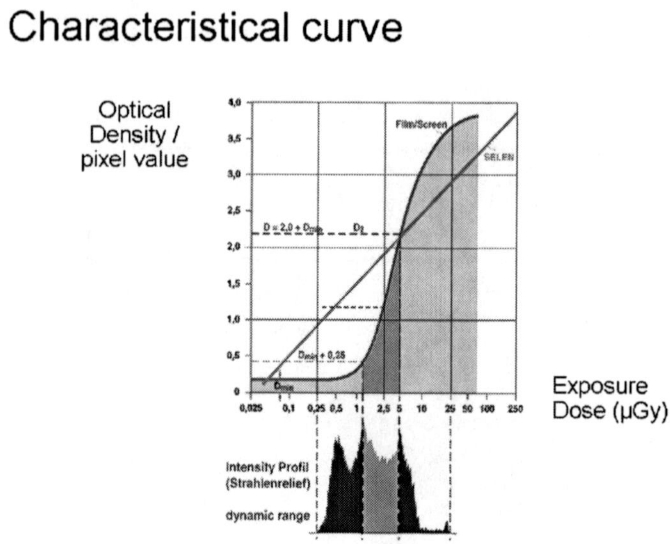

Figure 10. Characteristic curves of screen-film and digital detector.

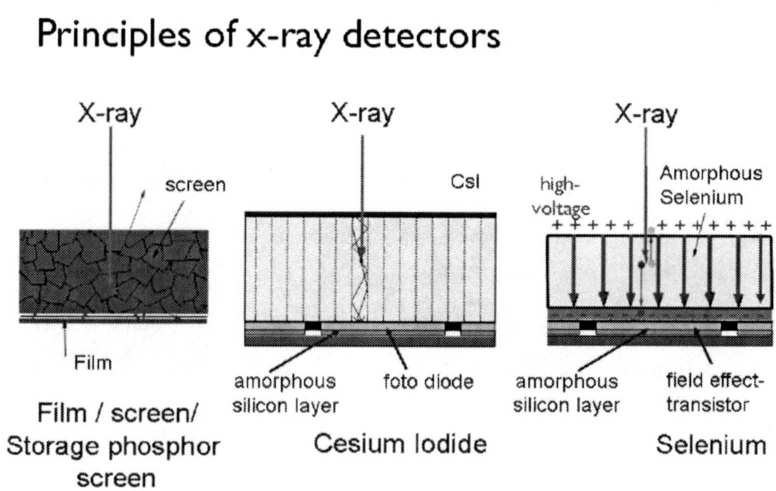

Figure 11. Principles of digital x-ray detectors.

All digital X-ray detectors are combining a layer sensitive to X-rays and means of converting the radiation dependent signal into digital data. At the first stage the image is still a two dimensional continuum of intensities.

During digitization (in the second step) the image area is scanned and divided into pixel elements so called pixels with discrete pixel values. Scintillators such as storage phosphor screens or Cesium Iodide or semiconductors like Selenium are used as X-ray sensitive layer. The accuracy of digitization is determined by the pixel size and the number of bits per pixel. Typical pixel sizes in DR are between 100 μm and 200 μm for storage phosphor plates, 120-200 μm for flat panel detectors and 150-500 μm for image intensifier-video systems. Resulting matrix sizes vary between 512 × 512, 1024 × 1024, 2048 × 2048, 4096 × 4096. Non quadratic matrices are also used. For special exposures like mammography pixel sizes of 25 – 50 μm are needed to resolve fine micro calcifications.

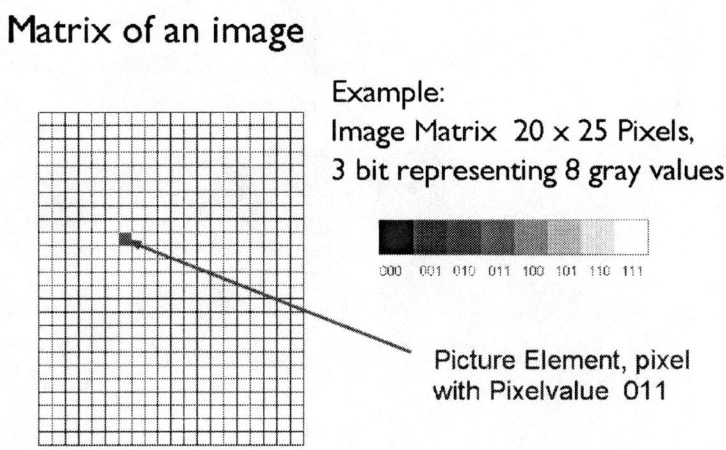

Figure 12. Matrix of an image.

Image noise, in particular quantum mottle, often limits the precision of the gray levels. Therefore, for image intensifier applications images taken at radiation doses below 0.5 μGy 256 – 1024 gray levels (8–10 bits) are sufficient. Exposures made with higher X-ray doses require 10-14 bits for storage phosphor, cesium iodide and selenium systems. Detectors using cesium iodide or amorphous selenium offer the most efficient of quanta detection combined with few inherent artifacts (detector noise). Physically this is measured by the Detective Quantum Efficiency (DQE).

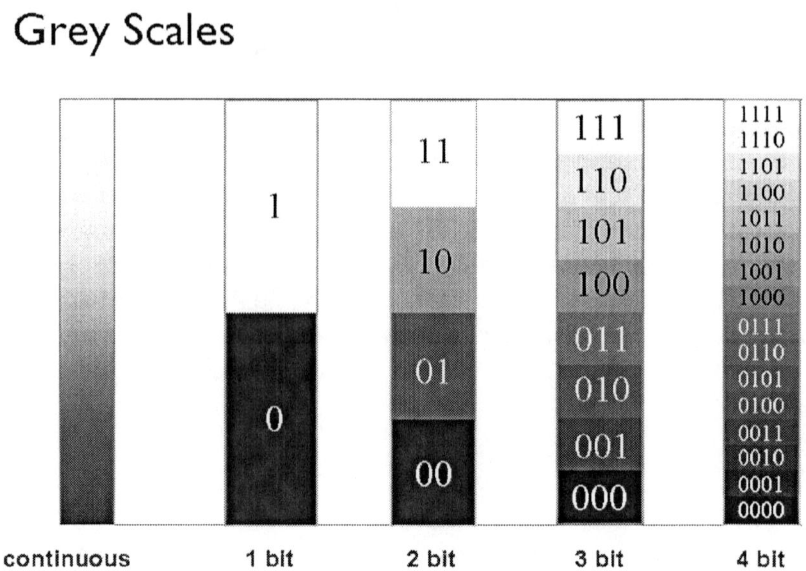

Figure 13. Grey scales.

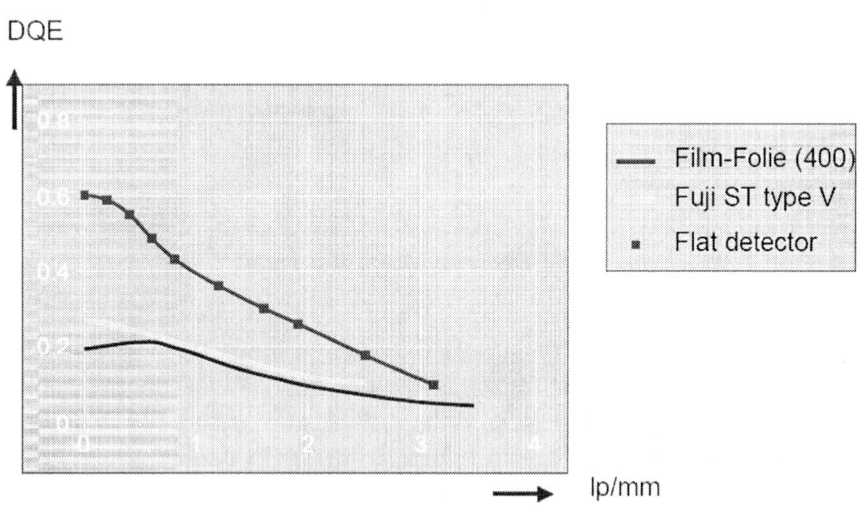

Figure 14. Comparison of quantum efficiency for different detectors.

Ideally we are aiming for a detector with DQE 100%. For low spatial frequencies the DQE of screen-film and storage phosphor systems comes to around 20-40%, for cesium iodide and selenium systems up to 50-60%.

Digital image intensifier and storage phosphor plate systems have been in use since the early 80es. Storage phosphor plates using standard cassette sizes are preferred standard X-ray equipment and for free exposures. The DQE and MTF of the screens has been further improved in recent years. The sensitivity of new screen is equivalent to 400 speed screen-film systems.

Selenium drum systems have been routinely used for chest exposures since 1993. They are the first systems with direct transformation of X-ray energy to a charge distribution, which can be directly scanned and digitized. The intermediate step of light emission can be omitted. This leads to low detector noise and therefore high DQE resulting in high image quality.

The latest development of flat panel detectors will be explained in more detail. The First prototypes are available and tested in clinical applications with small fields. They are built up like a sandwich consisting in a X-ray sensitive layer and an active detector matrix. There are two principles: direct and indirect detection. Like the Selenium drum system direct detection converts X-ray energy to electrical charge patterns. The indirect technique needs light emission as an intermediate step. The X-ray sensitive layer may be a semiconductor (selenium) layer or a scintillator like cesium iodide doped with thallium CsI(TI) used as entrance layer of image intensifiers. The active detector matrix is sensitive for charge in the first case and to light in the case of a scintillator.

This sensor consists of a two-dimensional active matrix, where the sensitive element in each pixel is an amorphous hydrogenated silicon (a-Si:H) photo diode or transistor. The a-Si:H semiconductor material is very suitable for this application because it is a good photo detector in thin-film form (<2μm) and is easily deposited on large glass substrates. Amorphous SiH is very sensitive to visible light, with efficiency close to one and a weak dark current allowing highly sensitive image pick up. The pixel array is a matrix addressed via lines for a row selection and via columns for signal readout. The photo charge level is subsequently transferred to the external drive electronics by a matrix of a-Si:H switching devices. Integrated circuits are connected along the edges of the detector matrix and provide the line scanning sequence and the signal readout amplification and multiplexing. The readout amplifier is optimized to reduce the electronic noise. Specific electronics drive the detector and to provide direct digital data to the processing system. The advantage of flat panel detectors is the direct integration into X-ray imaging systems avoiding cassette transport and allowing direct acquisition and presentation of the images.

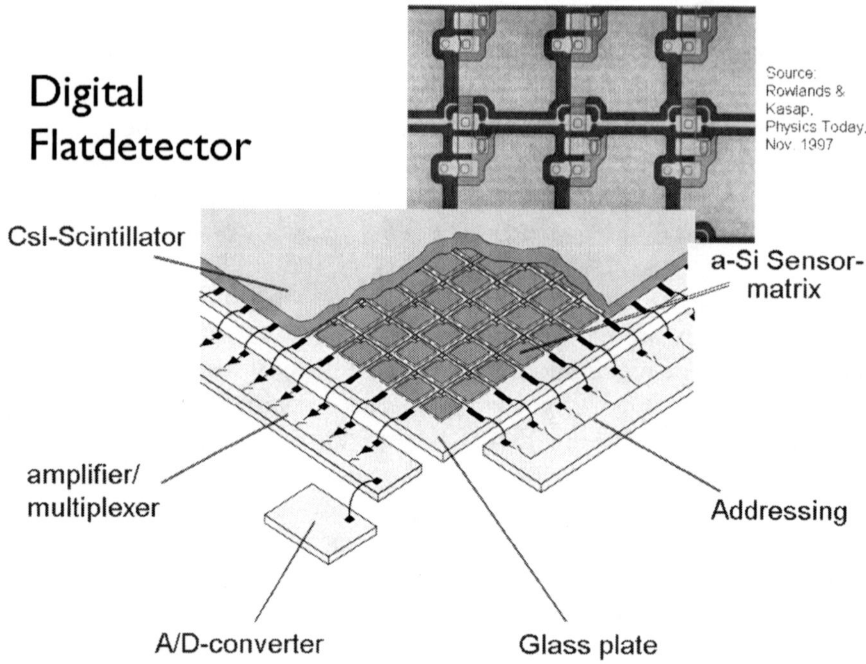

Figure 15. Digital Flat panel Detector.

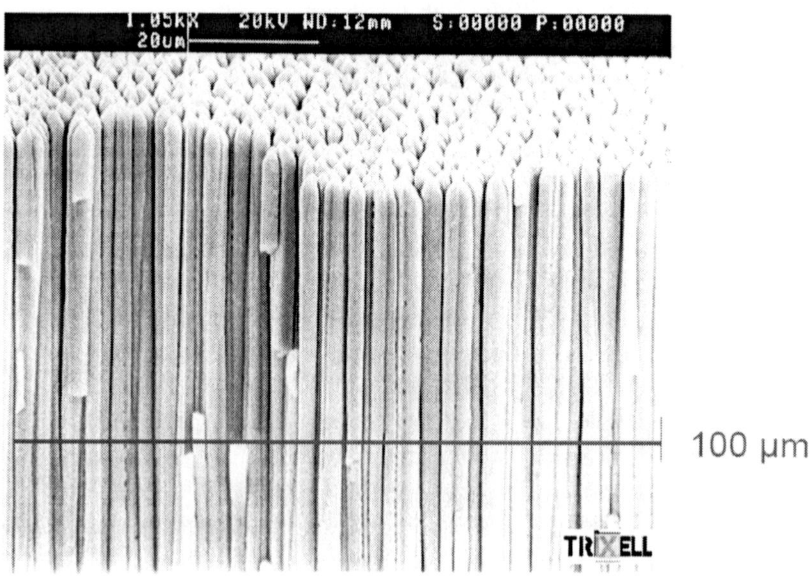

Figure 16. Structure of Cesium Iodide Input Layer.

2.1.4. *Advances of image processing*

The image processor is the core of the digital imaging system. Current computation and transmission speeds and storage capacity allow execution of most algorithms in short time with standard PC based workstations. Workstations with a special hardware base are no longer needed. Large random access and long term storage devices can store basic image information from several days worth of examinations. For high demand of computational power e.g. real time processing and three-dimensional rendering multiple processors are addressed in parallel.

The availability of images in a universal digital format gives enormous opportunities for image processing. From the standpoint of image science film is now widely recognized as a limited image recording and presentation medium.

Image processing occurs at different stages of the image generation process aiming at optimization of detection and display.

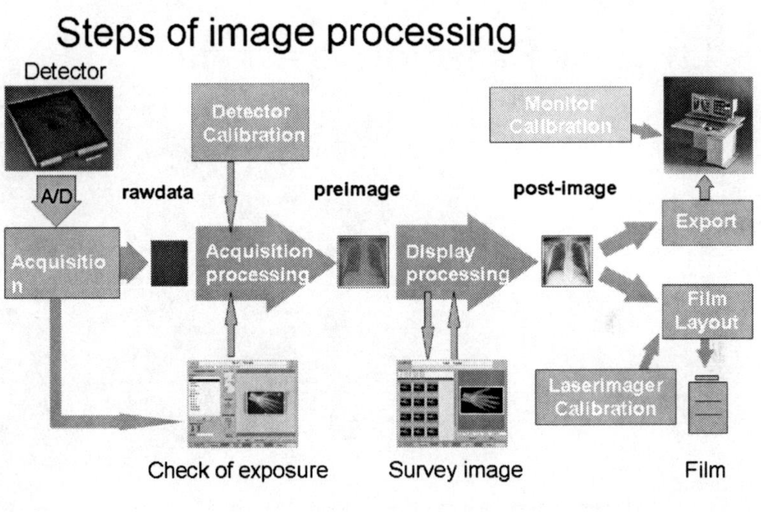

Figure 17. Steps of image processing.

During image acquisition the exposed image area and the characteristic structures are extracted. The result of this first step is a basic image compensated for degradation and varying exposure conditions during the detection process.

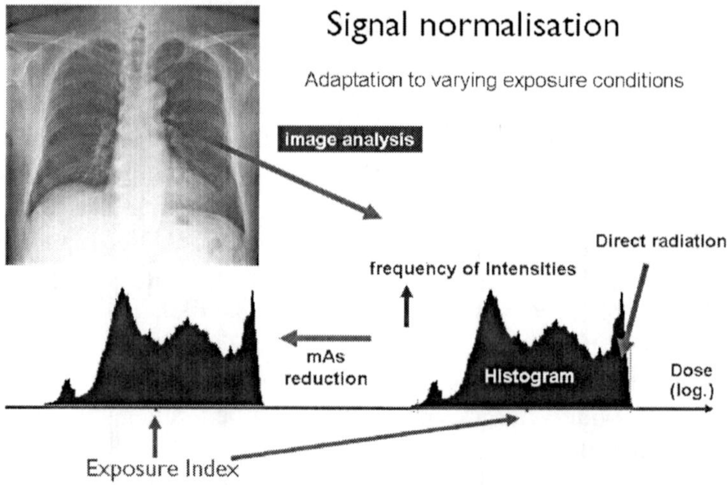

Figure 18. Signal normalization.

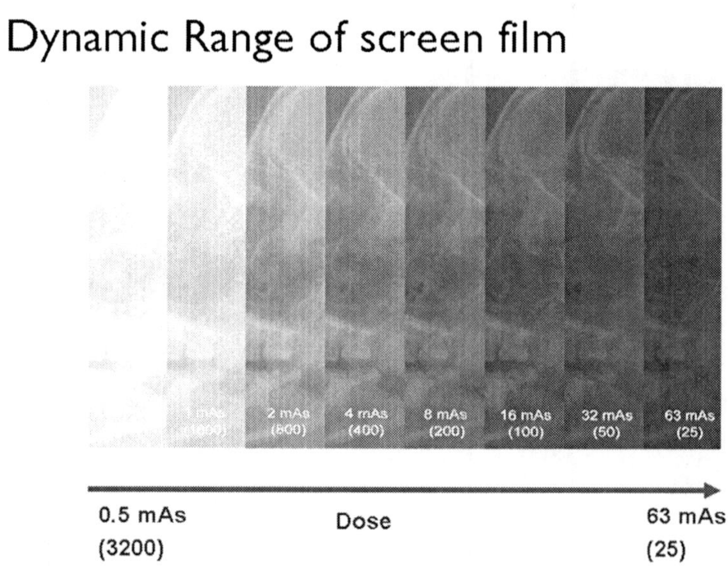

Figure 19. Dynamic range of screen film.

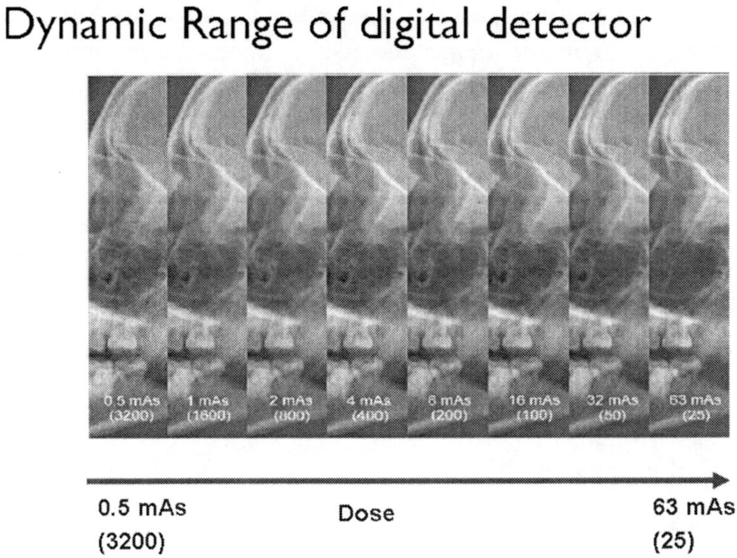

Figure 20. Dynamic range of digital detector.

The information contained in this basic image may exceed the display capacity of either monitor or film displays.

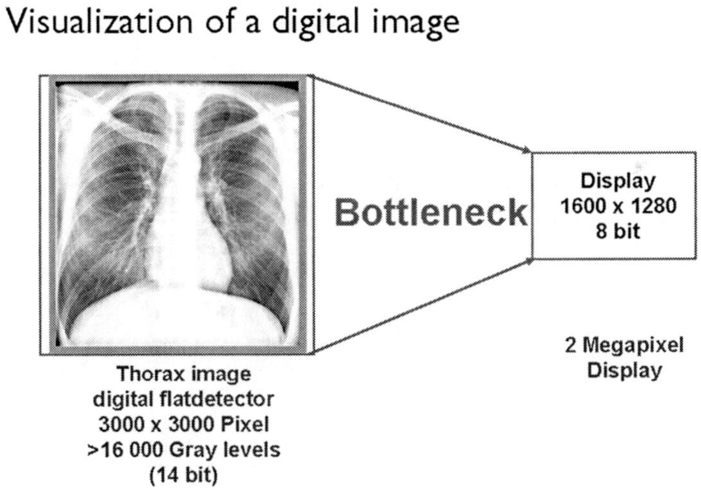

Figure 21. Visualization of a digital image.

To display the information contained in this basic image and to adapt the image presentation to the needs of the observer, digital image processing is applied resulting in an image with modified appearance e.g. enhanced contrast and enhanced sharpness of certain objects. New algorithms allow a reduction of the dynamic range and simultaneous enhancement of local contrast. A critical limitation is the noise inherent to the image. It is the objective to produce images that look similar to conventional film-screen images avoiding over and underexposed areas. Optimal presentation of anatomy should not hide pathology.

The algorithms of image enhancement in digital radiography are:

- Contrast enhancement by specific gradation curves and changing the intensity window and level (Look up table operations)
- Combination of images (Digital Subtraction Angiography)
- spatial frequency processing
- Method of unsharp masking
- Dynamic range reduction
- Multiskalare contrast amplification

For different projections focusing on bones or tissue structures an adaptation of the processing parameters is necessary. Good parameter sets should offer stable image quality even under varying exposure conditions.

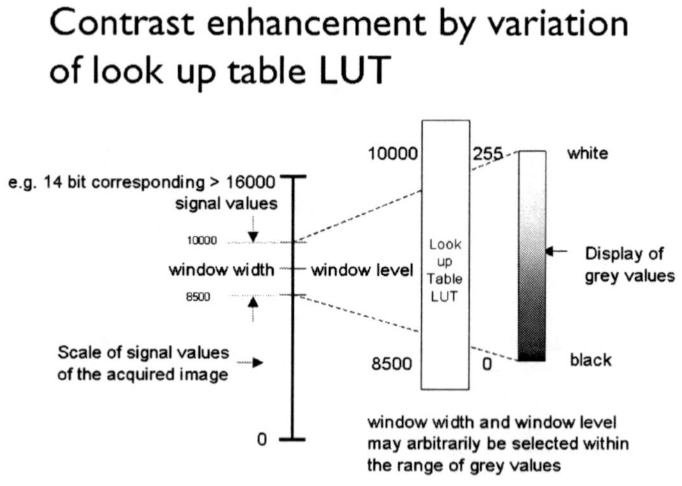

Figure 22. Contrast enhancements by variation of look up table LUT.

Processing artifacts can occur, if inappropriate sizes of the filter kernels and too much enhancement is chosen, leading to overshooting intensities at the edges. These are visible as bright and dark shadows. Small filter kernels (<2 mm) lead to MTF improvement, medium (2-10 mm) to edge enhancement, large kernels (>10 mm) to dynamic range reduction and contrast enhancement and ultra large kernels (>20 mm) to dynamic range compression.

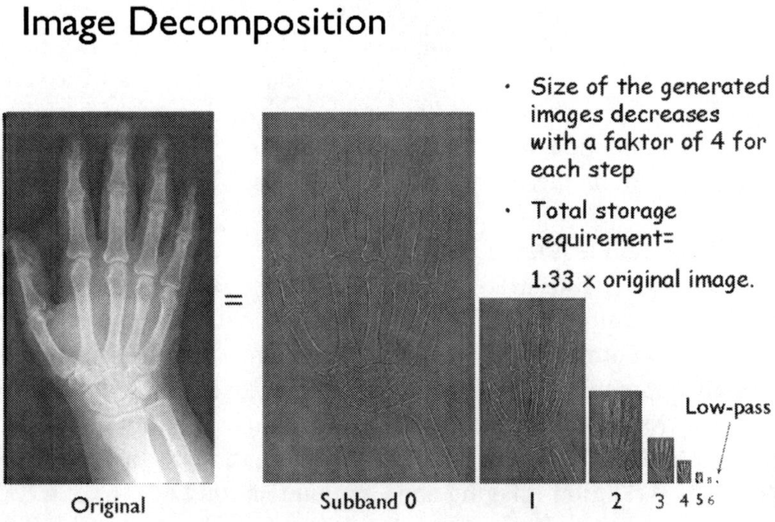

Figure 23. Spatial frequency processing – Decomposition.

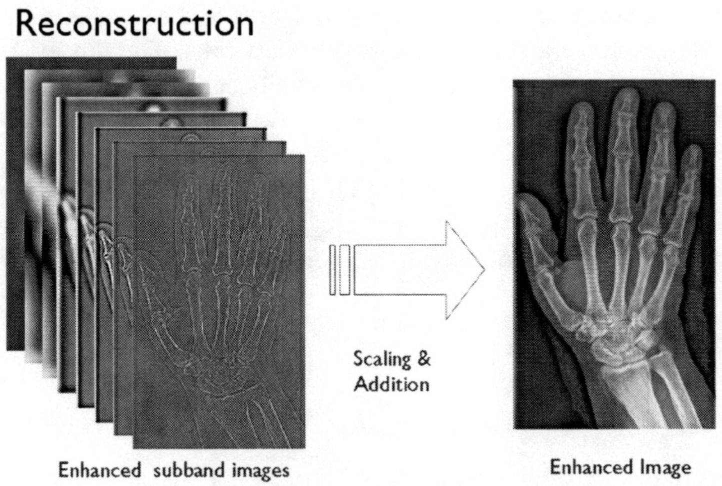

Figure 24. Spatial frequency processing – Decomposition.

If the object is larger than the filter kernel the edges will be enhanced, whereas, contrast enhancement of the whole structure occurs, if the object is smaller than the kernel size. It is a great disadvantage of unsharp masking that large, low contrast structures are suppressed. Combinations of different kernel sizes help to reduce overshooting edges.

2.1.5. *Image storage, display, documentation and archiving*

With the advent of digital radiography images are primarily stored as digital data in the memory of the image processor. After the examination has been completed the images are reviewed on the video monitor of the workstation, their contrast and brightness is adjusted and a set of images is selected for laser printing on film. Normally only the relevant images of a series are selected. Laser imagers produce hard copies of the images with high spatial and contrast resolution in selectable formats. Depending on the national laws and regulations these documents have to be safely archived for 10 to 30 years.

For direct online viewing of images with diagnostic quality a large image memory (>100 MB) and off-line storage capacities of several Gigabytes are required to store all digitally acquired images. Currently There is some controversy as to whether the image should be stored as raw data, as an unprocessed basic image or a processed image. There is also some debate regarding the acceptability of image compression. Standards for storage and exchange of images are internationally agreed upon and been published (Digital Imaging and Communication DICOM).

Several institutions have successfully replaced film reading by soft copy diagnosis at monitors storing all images and reports in a digital form on optical disks or optical tapes. Advantages of a digital archive are the reduction of storage space, the automated access to the images and the possibility of transferring digital images via networks. For the exchange of digital images the data sets may be duplicated without any loss of information. Displaying images of diagnostic quality requires flicker-free monitors with high luminance exceeding 300 cd/m^2, a refresh rate >70 images/s and >1000 video lines. Digital imaging is the first step to digital networking and archiving. Storage devices with several Terabytes of archiving space are state of the art.

2.2. MEDICAL ASPECTS

2.2.1. *Changing working environment*

The advantages of digital radiography are: better visualization of objects with low contrast, easier handling, high consistency of image quality and safe storage. In addition DR offers the potential of multiple copies with original quality, dose reduction in selected cases, fast and loss less transmission of images, direct reporting at an image display and digital archiving. As a result of these advantages a lot of the experience gained from using film as a detector is no longer valid. The direct relation between dose and optical density of the film no longer applies, since under- and overexposure of images can be compensated because of to the large dynamic range and signal normalization. Avoiding under- and overexposed images reduces wrong exposures and leads to a dose reduction by fewer retakes. To better visualize critical structures in the image, digital image processing methods such as contrast enhancement, dynamic range reduction also called harmonization, noise compensation and adaptive edge enhancement are applied. The enhancement is limited by the amount of noise (artifacts) in the image, which depends on both, the quantum mottle of the applied radiation and the noise of the detector system.

A possible advantage of digital imaging systems is the improvement of workflow in the imaging department resulting in higher efficiency and higher throughput. A Prerequisite of this is an operational policy adapted to DI and the realization of a departmental organization system (RIS/HIS). Working with computer based patient work lists reduces administrative work and handling of cassettes and waiting for the films is eliminated. Finally there is direct display of digital images of diagnostic quality. Digital archiving helps to Avoid the loss of images.

With image processing all information should be perceived more readily. This should lead to a higher detection probability and therefore better diagnostic performance of the radiologist. Initially radiologists will use image-processing defaults, which produce images of conventional, i.e. film-like appearance. With time, however, they will become acquainted with stronger filtering algorithms.

Display of digital images is more flexible than that on film. The image can be shown without time delay, it can be zoomed and magnified and the brightness and contrast are set interactively. Various digital filters can be used for contrast and edge enhancement. This implies there may be some image degradation due to operating errors. The ease of operation still varies greatly between different products. Generally spoken digital image processing implies many degrees of freedom. setting up specific guidelines for DI reporting along with training courses should aid standardization.

2.2.2. *Change of paradigms*

In the past, achieving a high spatial resolution was the dominant aim in developing and using an imaging system. Now we have learnt that contrast resolution is of equal importance and that the shape of the modulation transfer function describes best the perceived sharpness of structures. In judging the quality of a medical image spatial and contrast resolution compete with each other. Image sharpness is determined most by those spatial frequencies, for which the modulation (MTF) is still high (>20%). Higher spatial frequencies may also contribute to the image but their influence on image perception is less important. A better measure of resolution would be a characteristic modulation for a typical spatial frequency of an object e.g. at 1Lp/mm. At the advent of digital radiography there were doubts as to whether a system with a limiting resolution of 2.5 Lp/mm would be capable of imaging small structures in a chest radiograph e.g. fine septal lines. A resolution of at least 6 Lp/mm was asked for. Results of practical studies suggest, that medically relevant frequencies are ranging from 0 to 2 line pairs/mm. A vessel with a diameter of 1 mm may be described by a spatial frequency of 0.5 line pairs /mm plus some frequencies of higher order. Studies have shown, that proper processing can compensate for the lower spatial resolution. Recent improvements in detector technology allow pixel sizes of 25 – 100µm and as a consequence very large matrices. These are used for examinations with a need for high spatial resolution like mammography.

New strategies for reading images have to be found and standardized in the medical community. The quality of an image on film cannot be changed afterwards. Digitally stored images virtually include all acquired information of the latent radiation image at the detector entrance. Depending on different diagnostic questions various display processing schemes may be adopted. In the course of a reading session the appearance of an image may be varied considerably. Fixed settings are substituted by processing rules.

Part II

Magnetic Resource Imaging

PRINCIPLE OF MAGNETIC RESONANCE

MARTIN O. LEACH*
Section of Magnetic Resonance, Institute of Cancer Research and Royal Marsden Hospital, Downs Road, Sutton, Surrey, SM2 5PT, UK

Abstract. An overview of five lectures introducing the basic principles of nuclear magnetic resonance, factors affecting relaxation properties and signal, pulse sequences, equipment for magnetic resonance imaging, image acquisition and reconstruction, together with examples of applications, is provided. References to more detailed sources are included.

Keywords: nuclear magnetic resonance; NMR; magnetic resonance imaging; MRI; magnetic resonance spectroscopy; MRS, k-space; relaxation time; gyromagnetic ratio; pulse sequence; free induction decay; saturation recovery; inversion recovery.

1. Introduction

This chapter provides an overview of five lectures presented at the European School of Medical Physics to introduce the basic principles of magnetic resonance imaging (MRI). Sections are arranged in the order of the lectures. Given the extensive scope of the subject, and the limitations on space, reference is provided to more detailed texts that provide a full description of the underlying theory. In particular, much of the theory is covered in Leach (1988). Space limitations preclude illustrations here, which are provided in the lectures and in the books referenced here. Section 2 introduces nuclear magnetic resonance (NMR), and outlines the concepts comprising the development of the field followed by the basic theory of the phenomena, covering classical, quantum mechanical and statistical descriptions. This provides an understanding of the source of the signal. Section 3 addresses how the signal can be manipulated by the use of pulse sequences, exploiting the properties of relaxation processes and discussing how relaxation processes reflect the molecular environment. Section 4

* Martin O. Leach, Section of Magnetic Resonance, Institute of Cancer Research and Royal Marsden Hospital, Downs Road, Sutton, Surrey, SM2 5PT, UK; e-mail: Martin@icr.ac.uk.

Y. Lemoigne et al. (eds.), Physics for Medical Imaging Applications, 25–36.

describes the equipment used in magnetic resonance imaging, relating it to the functionality required to perform measurements. Section 5 addresses image formation, introducing the concept of frequency selective excitation and read out, enabling acquisition of spatially selective projections from which images can be reconstructed. Section 6 extends these concepts to include phase encoding one or more spatial dimensions, describing spin warp imaging, introducing the concept of k-space together with its relationship to data acquisition and image reconstruction. A number of MRI pulse sequences are described. Further information on the principles of MRI are provided in Morris (1986) and details of equipment and imaging sequences in Krestel (1990). Advanced topics are covered in Haacke et al (1999) and Young (2000). Details of applications may be found in Edelman et al (1996) and Stark and Bradley (1999).

2. Nuclear magnetic resonance

2.1. DEVELOPMENT OF FIELD

The effects of atomic magnetism in splitting the optical spectrum of excited atoms in a strong magnetic field was observed by Zeeman in 1891, with Stern and Gerlach demonstrating the quantisation of silver atoms in 1921, measuring their magnetic moment. Rabi determined the nuclear magnetic moment of atoms in a beam in 1935. Nuclear magnetism was first detected in solid hydrogen by Lasarew and Schubnikow (1937) with the independent observations of nuclear magnetic resonance (NMR) in bulk material being reported by Bloch et al (1946) in water and Purcell et al (1946) in paraffin wax. This discovery resulted in the joint award of the Nobel Prize for Physics in 1952 to Felix Bloch and Edward Purcell. The measurement by Bloch of a strong proton signal from a finger was the first *in vivo* observation of nuclear magnetic resonance. NMR played an important role as a tool for exploring the basic properties of nuclei, and was developed as a tool for chemical analysis. Extension of this methodology to imaging the location of materials depended on discovering a means of making NMR signals spatially sensitive. Lauterbur proposed the use of frequency encoding by magnetic field gradients to provide projections of hydrogen atom (proton or spin) density through an object. Convolution and back projection from a set of such projections could then reconstruct a cross sectional image. Mansfield also demonstrated the use of magnetic field gradients to spatially encode signal, and both Lauterbur and Mansfield extended this concept to provide slice selection. These discoveries were recognized by the joint award of the Nobel Prize in Physiology or Medicine in 2003. A range of imaging techniques utilising these approaches were developed, with most imaging now based on spin-warp imaging, introduced by Edelstein and colleagues. This has now led to a very wide range of applications in medicine and medical research, with MRI providing an essential diagnostic technique for most hospitals.

2.2. THEORETICAL BASIS OF NUCLEAR MAGNETIC RESONANCE

A brief overview of relevant theory is given below. For a more detailed description, the reader is referred to Leach (1988). Nuclei with odd numbers of neutrons or protons possess a net nuclear magnetic moment. The hydrogen atom is the most ubiquitous example, but ^{31}P, ^{19}F, ^{13}C and ^{23}Na are further biologically relevant isotopes. MRI predominantly measures ^{1}H present in water of lipids. In the absence of an external magnetic field, these moments have no net alignment, but imposition of an external field B_0 along a direction z results in the nuclear magnetic moment becoming oriented such that there is a component either parallel or anti-parallel to B_0. If a sample is placed in the presence of an external field, the alignment of the magnetic moments will result in a net magnetic moment for the sample. If a radiofrequency (*rf*) electromagnetic field is applied to the sample, a resonant absorption of the rf field occurs when the frequency

$$\omega_0 = \gamma B_0 \qquad (1)$$

where ω_0 is the Larmor precession frequency and γ is the gyromagnetic ratio. The sample moment precesses about the direction of B_0 at the Larmor precession frequency.

An external electromagnetic field acts as a couple on M, the net nuclear magnetic moment of a sample, and will cause the net moment to rotate away from the z axis. The rate of this rotation depends on the strength of the applied rf B_1 field, and the angle through which M rotates. When a component of M is present in the plane orthogonal to B_0, it can induce a detectable signal in a receiver coil. Thus the *NMR* experiment is designed to rotate M through an angle (typically $90°$) away from the z axis, thus allowing a signal to be detected and M measured. This signal following an *rf* pulse is the free induction decay, and the signal generated decays away with a characteristic tissue dependent relaxation rate T2* (discussed further below). As explained earlier, the nuclear magnetic moment of an individual atom (for *spin ½* hydrogen) may have a component m_z aligned with or against the direction of B_0. Under equilibrium conditions slightly more nuclei are aligned with B_0 than against, and this difference is the population detected in MRI. This results in *MR* being a relatively insensitive technique, observing at a B_0 field of $1.0\,T$ some 7 nuclei in a million. The ratio of spin populations is defined by

$$\frac{N(spin-up)}{N(spin-down)}=\frac{\Delta E}{e^{kT_s}} \tag{2}$$

Where ΔE is the nuclear energy difference between the spin-up and spin-down states, and is proportional to B_0, k is Boltzman's constant and T_s is the absolute temperature. Thus the signal available can be increased by increasing B_0.

3. Pulse sequences and relaxation

When equilibrium magnetisation is perturbed by application of an rf pulse, several relaxation processes govern the return of the magnetisation to equilibrium. Following a 90° pulse, M is tipped from alignment with the *z* axis into the *xy* plane. In the laboratory frame of reference, this is seen as the magnetisation rotating in the xy plane at the Larmor precession frequency. This is complicated to present, and more commonly the magnetisation is described in a rotating frame rotating at the Larmor frequency. In this case, if an *rf* pulse with the magnetic component aligned along y is applied, the resulting couple on *M* will result on *M* being rotated through 90° so that it lies along the *x* axis. Initially this will provide a strong in phase signal, but as a result of a number of processes, this signal will lose phase, resulting in the *X* component of the signal being reduced. Factors that cause this include variations in the local magnetic field, causing the Larmor frequency to vary across the sample, and non-reversible effects due to exchange of energy between two nuclei causing a loss of phase information, and the effects of locally varying dipole fields also causing a variation in the net magnetic field experienced. All of these factors together are termed T_2* relaxation, and the non-reversible elements comprise T_2, transverse or spin-spin relaxation. By following a 90° pulse after an appropriate interval with a 180° pulse, the spins will rephase along the $-x$ axis, and a spin echo will be formed. This spin echo will recover much of the signal seen at the beginning of a free induction decay (*FID*), except for that lost due to T_2 relaxation. The spin echo sequence forms the basis for many of the imaging sequences used in MRI.

In addition to the loss of signal experienced in the transverse or xy plane, eventually, the signal rotated out of the *z* direction, the equilibrium magnetisation, will recover. The recovery of this magnetisation results from thermal processes exposing nuclei to energy at the Larmor precession frequency, and thereby causing transitions from spin-down to spin-up states. This energy comes from dipolar magnetic fields in the sample oscillating at the Larmor frequency. The process is characterised by the T_1, longitudinal or spin-lattice relaxation, and can be thought of as a loss of energy from the excited spin system to the lattice. It is characterised by recovery of the *M* component along *z*, which following a 90° pulse will take a period of $3T_1$ to fully recover. For samples containing materials with differing T_1s, varying the repetition time *TR* between repeated 90° pulses can adjust the relative signal received from the different materials, based on their T_1 relaxation times. Materials with a T_1 much shorter than *TR* will have recovered their equilibrium magnetisation completely. Materials with a T_1 similar to or longer than *TR* will only partially recover their equilibrium magnetisation in the *TR* period, resulting in an attenuated signal. This approach is used in saturation recovery sequences, where *TR* is adjusted to maximise the signal difference between different tissues.

A further extension of this process is used in the inversion recovery sequence. In this case instead of an initial 90° pulse to move *M* into the *xy* plane, an initial 180° pulse is employed, orienting the signal along the –*z* axis. The magnetisation will then return to equilibrium along z, taking $5T_1$ to return. This recovery can be monitored by applying a subsequent 90° pulse at a time *TI*, the inversion time, following the initial 180° pulse. During the recovery, the magnetization for a given tissue will at some point be zero, and making a measurement at this value of *TI* will eliminate signal from that tissue, providing a useful way to eliminate signal from particular tissues in images. The approach is commonly used to eliminate signal from fat or from cerebrospinal fluid. Again the overall *TR* of the sequence will also affect the relative signal of different tissues.

Relaxation times of tissues reflect the mobility of hydrogen atoms and of the molecules containing hydrogen. This is governed by the frequency of motion of the atoms, or their correlation times. Materials with short correlation times will cover a wide range of frequencies, resulting in relatively little spectral density in any particular frequency band and will have long T_1 and long T_2 relaxation times. Materials that are relatively immobile also will have low power density at the Larmor frequency and so have long T_1s, but short T_2s. Water in tissues may exhibit a range of mobilities, depending on its environment, as close to surfaces and proteins it may be bound, or have a highly ordered structure with reduced mobility. Further away from potential binding sites it may be much more mobile, providing for the range of relaxation times seen in human tissues. The behaviour of magnetisation under the influence of rf pulses and relaxation processes can be described by the Bloch equations.

Relaxation can be enhanced by using contrast agents. These have a strong local magnetic field, either as a result of unpaired electrons, with ions such as gadolinium being frequently employed to cause predominantly enhanced T_1 relaxation, or from using super paramagnetic or ferromagnetic materials, which can strongly perturb the local field, causing enhanced T_2*

relaxation. The compounds will commonly be attached to appropriate pharmaceuticals to render them safe for administration, and to confer particular biological properties.

4. Magnetic resonance imaging equipment

The magnetic resonance imaging system comprises a number of major components: the magnet and shim coils, gradient coils and drivers, radiofrequency system and coils, patient handling equipment, measurement control and computer system, together with a range of supporting equipment. Here the major components will be described. Further details can be found in Krestel (1990).

The magnet is the major component of the *MR* system. The majority of systems employ a superconducting horizontal magnet, with field strengths ranging from *0.5-3T* for clinical systems, and up to *8T* for clinical research systems. Most superconducting systems (up to *3T*) are self shielded, significantly reducing the magnetic footprint of the system. Recently some vertical split superconducting magnet systems have been introduced at *1T*, providing improved access. Lower field vertical electromagnet systems are available between *0.1* and *0.3T*, of particular use for orthopaedic applications. Permanent magnet systems are available with vertical and horizontal field orientations, the latter allowing imaging of patients standing. These have field strengths up to *0.6T*. In order to provide a very homogeneous field, magnets will be shimmed with steel and by adjustable currents in additional field coils, to optimise the field. Most shim provision is not user accessible, but adjustable shims may be provided and utilised by the system or the user to optimise the *B0* field over a specified region, for example to facilitate fat suppression, or for *MR* spectroscopy, where a particularly uniform field is required.

Orthogonal gradient coils are required for spatial localisation. These have to allow rapid switching, and must be able to handle the large currents produced by powerful gradient amplifiers. They generate significant heat and so require cooling. Many modern gradient sets are capable of peripheral nerve stimulation, which although not hazardous can be disconcerting, or painful, depending on the intensity. Maximum gradient amplitudes are regulated to prevent cardiac fibrillation which is life threatening but occurs at a much greater amplitude than peripheral nerve stimulation, and to minimise the likelihood of peripheral nerve stimulation. A further practical issue is the noise generated by gradient switching, which often requires the use of ear protection, and can also limit gradient amplitudes, and the examination of pregnant women, were the foetus cannot be protected. The gradients are driven by signals derived from a digital gradient waveform.

To deliver the planned waveform, gradients are now generally screened, to minimise the generation of eddy currents in the magnet and shim coil structures. In addition pre-emphasis circuits correct for deviations from the planned waveform due to the electronic characteristics of the circuits, and from any residual eddy current effects.

The radio frequency system provides for transmission of modulated rf signals, derived from a digital source and amplified by a power amplifier that may deliver typically 15 kW of power. This signal is delivered to a body coil, and possibly to some other smaller transmit coils. Some advanced systems are now incorporating array coil transmission, to improve rf homogeneity at the high frequencies used in some clinical research systems (e.g., 7 T). Signal is received from the body by a range of possible coils, ranging from a body coil to specialist multi-element parallel acquisition systems, utilising phased array reconstruction methods. The detected signal is demodulated, amplified and digitised. Transmit coils are generally circularly polarised, and receiver coils may provide a single output or incorporate quadrature elements to increase signal to noise. Specialised coils may be used for particular parts of the body, or for particular applications.

In addition to these major components, the scanner will employ sophisticated distributed intelligence, together with one or more consoles, providing for control, communication, archival and analysis of data. Systems may include peripherals to print images, and interface to picture archival and communications systems (PACS) and Radiology Information Systems (RIS). A patient handling system including a motorised couch, alignment aids, physiological monitoring, accessories for biopsy, interactive monitoring of procedures, and possibly interventional facilities, may be provided.

5. Magnetic Resonance Imaging – slice selection and frequency encoding

5.1. SLICE SELECTION

Equation (1) demonstrated that the Larmor precession frequency was dependent upon the magnetic field B_0 applied to the sample. Most magnets are designed to minimise the variation in B_0, leading to a highly homogeneous field (typically of the order of *0.1 ppm* across *20 cm*). However by imposing an additional magnetic field gradient that increases in field linearly with distance, it can be seen that the Larmor precession frequency will similarly vary with spatial position. This frequency dependence with position provides a means of identifying or selecting spatial position. By irradiating the sample with an *rf* pulse containing a narrow band of frequencies, a slice through the object will be excited. For example a gradient in the *z* direction will be switched on, a *90° rf* pulse exciting a range of frequencies present within the sample will be applied, with the effect of selecting a slice through the sample. If a square pulse (in time) is used to modulate the Larmor frequency, this will comprise a range of frequencies, represented by the Fourier transform of the square pulse, in the frequency domain. This range will be described by a *sinc* function, which is the Fourier transform pair of a square pulse. The result of this would be to have applied a non uniform excitation angle across the slice. This is not satisfactory for slice selection, but if a more complex time modulation of the Larmor frequency is used, such as a *sinc* modulation, the selected frequencies will approximate to a rectangular function, giving relatively uniform excitation across the range of frequencies, or slice. The function is approximate as a *sinc* function should extend to infinity, but in practice it has only a short duration. To improve the slice profile the *sinc* function may be filtered, for example with a Gaussian filter, or a numerically optimised pulse shape may be employed. There is a considerable literature on different types of *rf* pulse. Slice thickness can be controlled by altering the bandwidth of frequencies excited, or by changing the amplitude of the magnetic field gradient.

5.2. GRADIENT PHASE LOSS AND REPHASING

The process of applying an *rf* pulse in the presence of a gradient, as described above, will effectively excite the sample (flip the magnetisation into the *xy* plane). However once a component of the magnetisation is in the *xy* plane it will be subject to the magnetic field gradient, resulting in a dispersion in resonant frequency across the sample, and therefore a dispersion in the phase of the signal that depends on position the slice. If the signal were then subject to a *180°* pulse, producing an echo, the echo would be reduced in amplitude due to this loss of phase, which is not recovered by the spin echo formation. It is necessary to recover this phase loss by applying an additional period of gradient with the reverse polarity (if the re-phasing gradient lobe follows immediately after the slice selection). This acts by changing the magnetic field experienced by the spins, with the result that those spins with a component of magnetisation in the *xy* plane will experience a phase change dependent on the amplitude and duration of the additional (positive or negative) magnetic field. Alternatively an additional lobe could be placed after the *180°* pulse, but in this case (due to the effect of the *180°* pulse) the polarity would be the same as for the slice select gradient. Typically the area of this rephasing gradient lobe will be about half of that of the slice select lobe, although in practice this might need to be experimentally optimised. In an imaging spin echo sequence, the *180°* pulse will also be slice selective, exciting a band of frequencies in the same way as the *90°* pulse, with the gradient encompassing the *rf* pulse. However in this case although there is a gradient induced dispersion during the first half of the *180°* pulse, this is balanced by an equivalent rephasing during the second half.

5.3. THE GRADIENT ECHO

The dephasing and rephasing action of a gradient can also be employed to generate a gradient echo. The spins in the *xy* plane are firstly dephased by one lobe, the polarity of the gradient is reversed, and those spins in the same position will experience a rephasing effect that reverses signal loss and results in formation of an echo. This spin echo has the advantages that it can be applied very quickly, enabling rapid imaging sequences, and that it does not employ an *rf* pulse, reducing power deposition in the patient, a concern in rapid, multislice imaging. Compared with the spin echo, the gradient echo does not refocus signal loss due to the effects of local magnetic field inhomogeneities, and this signal decays by the more rapid T_2* rather than by T_2 relaxation. Whilst leading to less signal at longer echo times, this sensitivity to local magnetic field gradients is taken advantage of in sequences that are dependent on blood oxygenation levels, and in local vascular networks. Gradient echo imaging is now very widely used in clinical imaging.

5.4. FREQUENCY ENCODING

A similar principle to that outlined for slice selection can be used to encode the spatial position of signals. If the spin echo is read out in the presence of a gradient imposed in say the *x* direction, all spins at a particular value of *x* will experience the same net magnetic field, while those at a different position will experience a different field. This will result in spins having different frequencies across the sample, with frequency depending on position. By collecting the signal and then applying a Fourier transform, A frequency spectrum will be obtained which plots the number of spins at a particular *x* displacement. This is in fact a projection of the object onto the *x* axis. The same process can be repeated along the y axis, or by combining gradients applied in the *x* and the *y* directions simultaneously, and varying their respective amplitudes, projections can be obtained at any orientation in the *xy* plane. Reconstructing a two dimensional image from a set of projections, by techniques such as filtered back projection, is well known in computerised tomography, and these techniques can be applied to produce two dimensional *MR* images. Thus the combination of slice selection and frequency encoding allows *2D* images of slices through an object to be obtained.

6. Magnetic resonance imaging – phase encoding and k-space

6.1. PHASE ENCODING

In section 5.2 the effect of a slice select gradient on signal phase in the *xy* plane was described. While the loss of phase and hence signal required correction, the amount of phase change at a particular location is a function of local gradient amplitude – which dictates the local resonant frequency, and of the period for which the gradient is applied. Thus if a magnetic field gradient is switched on for a specific period, the amount of phase change is a function of the local magnetic field change, and thus of position. Once the additional gradient is switched off, the resonant frequency reverts to that dictated by the magnet's B_0 field. However the phase change that has occurred during the gradient pulse remains, and does not change until a further gradient pulse is applied. Thus the sample position is encoded by a phase change through the sample, with for a *y* gradient, each value of the phase gain representing a different *y* position. This is the basis of phase encoding. More detail on the relationship between acquisition parameters and image properties can be found in Morris (1986).

6.2. K-SPACE.

The way signal is acquired in MRI is often described in terms of *k*-space. K-space is the domain in which the demodulated signal acquired during a measurement is collected. The signal acquired during an imaging sequence is typically a set of echoes, or temporal signals changing in time. Considering a gradient echo sequence, a slice is selected with a slice selective *90°* pulse in the presence of a *z* gradient pulse, with the signal being rephased with a *z* rephasing lobe. The coherent signal in the *xy* plane then experiences an *x* gradient pulse, followed immediately by an *x* gradient pulse with opposite polarity. This generates a gradient echo, and the *x* gradient remains on during the echo, frequency encoding it. The acquired signal therefore has no phase encoding, but is frequency encoded in time. In *k*-space, this would provide a line at $k_y=0$, but extending for the duration of the acquired echo in the k_x direction (eg from $k_x.(-128)$ to k_x $(+127)$. In order to fill *k*-space in the *y* direction, it is necessary to acquire further echoes with different values of the phase encoding gradient (the *y* gradient). Typically this will take *256* values, with equal increments. The stepped *y* gradient pulse will be placed after the slice select gradient, and will step through positive values, starting with k_y (*+127*) through to k_y (*–128*), with each echo filling a horizontal line in *k*-space, starting from k_y (*+127*) through to k_y (*–128*). By completely filling the time domain or *k*-space, the image information is fully specified. A two dimensional Fourier transform converts the time domain data into the temporal frequency domain, equivalent in this case to the spatial domain in *x* and *y*. This results from the direct dependence of the *NMR* frequency on position in the linear magnetic field gradients. Thus *k*-space can also be seen to represent the *2D* Fourier transform of the spatial image, and threfore represents spatial frequencies, which are the same as the originally acquired time signals. Other MRI sequences fill *k*-space in an equivalent way, as will be described in the lectures. Krestel (1990) provides examples of pulse sequences and their *k*-space counterparts.

6.3. MRI SEQUENCES

The gradient echo is a basic building block of *MR* sequences. In order to obtain appropriate T_1 weighting, the sequence may require quite a long repetition time (*TR*), providing significant dead time between each line of image acquisition that is used to fill *k*-space. This time can be employed to sample signal from several other slices in sequence, allowing efficient multi-slice imaging. Another approach is to employ a small flip angle to rapidly build an image following preparation of the magnetisation by a preparation sequence, an approach sometimes called Turbo FLASH. Using this technique an image can be obtained in less than a second. Spin echo sequences operate in a similar way to the gradient echo image, but following the phase encoding pulse, a slice selective *180°* produces a spin echo which is read out in the presence of a gradient. A variant of the spin-echo sequence, RARE or Turbo spin echo, collects a series of echoes for each excitation, and uses a different phase encoding for each echo, which consequently occupies a different line in *k*-space. This allows the spin echo image to be acquired more quickly. Echo-planar imaging techniques use similar slice selection, but a rapidly alternating frequency encoding gradient produces a chain of echoes, each of which is arranged to have a different degree of phase encoding. This results in k-space being filled after a single excitation, providing real-time or single shot imaging. Image contrast can be controlled by suitable magnetisation preparation pulses. The method of phase encoding described above can also be applied in the slice select direction as well as in the phase encoding direction, providing a *3D* imaging sequence, and enabling thinner slices with reduced slice profile effects. There are very many types of sequences, for a range of purposes, and the lecture will illustrate a number of these. Further information on pulse sequences can be found in Krestel (1990), Haacke et al (1999) and Young (2000). Applications are illustrated in Edelman et al (1996) and in Stark and Bradley (1999).

References

Edelman R R, Zlatkin M B, Hesselink J R (Eds). Clinical Magnetic Resonance Imaging. WB Saunders, Philadelphia (1996).

Haacke E M, Brown R W, Thompson M R, Venkatesan R. Magnetic Resonance Imaging, physical principles and sequence design. John Wiley, New York (1999).

Krestel E (Ed). Imaging systems for medical diagnostics. Siemens Aktiengesellschaft, Berlin, (1990).

Leach M O, Spatially localised Nuclear Magnetic Resonance. In Webb S (Ed). The Physics of Medical Imaging, Adam Hilger, Bristol (1988).

Morris P G. Nuclear Magnetic Resonance Imaging in medicine and biology. Clarendon, Oxford (1986).

Stark D D, Bradley Jnr W G (Eds). Magnetic Resonance Imaging. Mosby St. Louis (1999).

Young I R (Ed). Methods in biomedical magnetic resonance imaging and spectroscopy. John Wiley, Chichester (2000).

NUCLEAR MAGNETIC RESONANCE (NMR)

EMILE M. HILTBRAND*
Hôpitaux universitaires de Genève, rue Micheli-du-Crest 24, CH- 1211 Genève 14, Switzerland
CERMA, Site d'Archamps, Archamps, F-74160, France

Abstract. In this introductory chapter to MRI the main concepts useful to obtain a diagnostic image are summarized. This short overview does not pretend to be rigorous nor exhaustive. It is primarily intended to be a guideline to understand the physics that is developed in the subsequent chapters (see M.O Leach chapter[1]). Only the essential is presented. For instance, the concept of echo is not a prerequisite to understand the MRI, though essential to make MRI practical. It will be briefly outlined at the end.

Keywords: nuclear magnetic moment; relaxation time; free induction decay; Larmor frequency; gradients of magnetic field.

1. Introduction

NMR was born in1946 with two separate articles (F. Bloch, E.M. Purcell); For the next two decades NMR was developed and used by chemists to characterize the structure of the various molecules. In the early seventies several researchers (P. Lauterbur, P. Mansfield, R. Damadian) thought about producing an anatomical image based on the NMR signal using the water molecule, the nucleus (proton) of the hydrogen atom having a magnetic moment.

Particularly illustrative is the image produced by Kumar et al.[2]. It is a very simple image of two spots representing two vials containing water, these vials separated by some distance. Here is the basic of the MRI since the two informations (the two signals from the two vials) were obtained in "one shot" thanks to the action of a gradient of magnetic field. The two information's having been extracted from a composite signal by Fourier Transform.

* Emile Hiltbrand, hôpitaux universitaires de Genève, rue Micheli-du-Crest 24, CH1211 Genève 14 ; e-mail : emile.hiltbran@hcuge.ch.

Y. Lemoigne et al. (eds.), Physics for Medical Imaging Applications, 37–47.

2. The origin of the signal: nuclear paramagnetism

Biological tissues are constituted of water (*H_2O*) in a proportion, which falls between 75% to 90% for most of them. Some nucleus such as the proton of the hydrogen have a magnetic moment μ (in this paper symbols in bold represent a vector!). One unique magnetic moment has a very tiny magnetic moment and will be unobservable, but if one considers a cubic centimetre, in that volume there is a number of protons (magnetic moments *μ*) of the order of the Avogadro'number (Figure 1a). Placed in a static magnetic field *Bo*, (Figure 1b) these magnetic moments will align along the direction of this magnetic field. Mathematically expressed, the contribution of each magnetic moment ends up (Figure 1c) as a macroscopic magnetization Mo (paramagnetism) whose magnitude is given by[3] :

$$M_o = \frac{N_{Av}\gamma^2\hbar^2 I(I+1)B_o}{3kT} = \chi B_o \qquad (1)$$

The quantum theory says that there are more individual magnetic moments μ pointing "up" than μ pointing "down". χ is the magnetic susceptibility; it account for the propensity of the matter to organize its structure Mo can be considered as equivalent to a small magnet (like the needle of the compass, though the physics is different !).

<u>Summarizing</u>: Acted upon by a magnetic field ***Bo***, a "paramagnetic material" arranges its internal structure to reach an equilibrium.

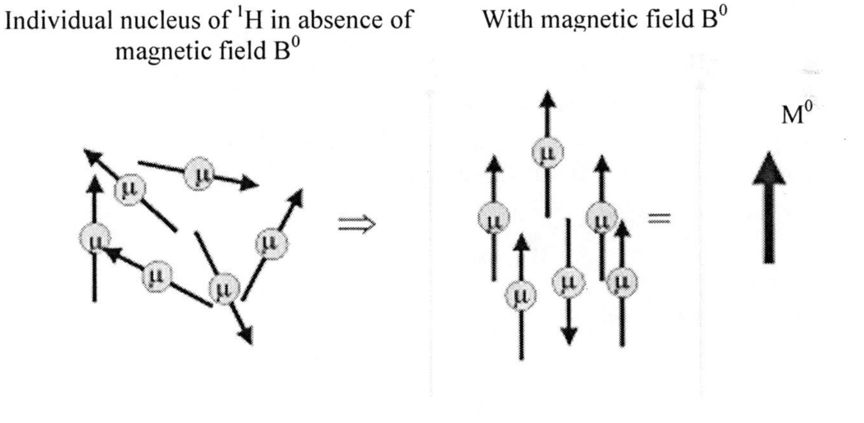

Figure 1. Individual nucleus of *H*: 1a- In absence of magnetic field *Bo*; 1b-With magnetic field *Bo*; 1c- Definition of *Mo*.

This equilibrium needs time to be achieved and the process follows an exponential law with a time constant T_1.

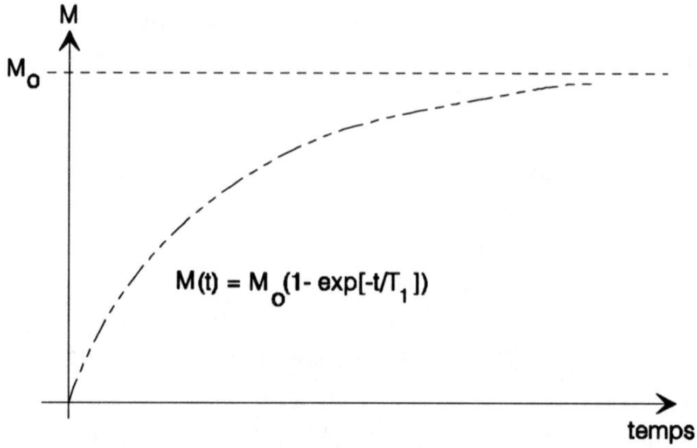

Figure 2. Exponential law with a time constant *T1*.

3. Collecting the NMR signal

A moving magnet close to an electrical conductor (coil of copper's wire for example) induces in this electrical conductor an electromotive force (Faraday's law) (Figure 3).

The movement of $\mathbf{M_0}$ providing the nuclear magnetic resonance signal is a that of a torque and is called PRECESSION (in that respect it is quite different from the one of the needle of the compass ! This precession movement is due to the properties of the proton having: a magnetic moment and an angular moment (SPIN). This implies that its movement is the one of a gyroscope. This movement has a specific frequency that is related to the strength of **Bo** given by the so-called Larmor relation:

$$\nu_0 = \gamma_p \mathbf{B_0}$$

The constant γ_p is called the gyromagnetic ratio, with a specific value for each nucleus considered.

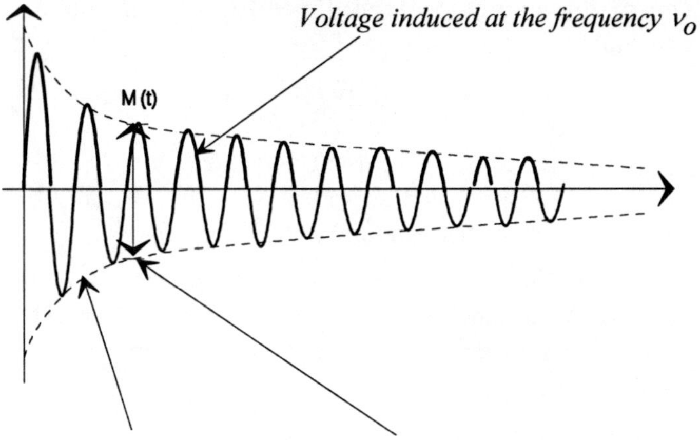

Figure 3. The induced electromotive force decay.

Once the magnetic moments are along **Bo**, (in biological tissues this is a situation that is achieved in less than one minute for 99% of them), nothing will happen since it is an equilibrium situation. Therefore, in order to collect a signal, some mechanism has to perturb this equilibrium. One can imagine changing the direction of the field **Bo** that means tilting the magnet by some angle!! Knowing the size and the weight of those magnets this procedure as to be discarded. But here come the "magic" of the RF pulse, a procedure that will be covered in the next chapters. Let us accept here that by applying briefly an RF pulse, **Mo** will be aligned in a direction perpendicular to **Bo**, a non-equilibrium situation, and once more **Mo** restart its movement toward **Bo**.

Remark: The signal is only observable during the interval between the RF pulse and the time at which **Mo** is rebuilt.

It is not easy to figure out that the T_1 and T_2 processes coexist. The RF pulse disturbs the equilibrium and the FID having the time constant T_2 is collected; but at the same time **Mo** is rebuilding itself with a time constant T_1 ! Even more puzzling is the fact that $T_2 < T_1$.

4. Information contained in the NMR signal

The NMR signal (Figure 3) provides three basic information:

1. A frequency $\nu_o = \gamma_p$ ***Bo*** (64 MHz for ***Bo*** corresponds 1.5 Tesla)
2. An amplitude

3. A decay of this amplitude according to an exponential law having a time constant T_2.

The induced signal (called FID for Free Induction Decay) has the shape of a decaying signal.

Summarizing, an NMR experiment is a three steps procedure that is sketched in figure 4 where one concentrates on the *T1* and *T2* mechanism, these mechanisms that are the fundamental of the MRI:

- A state of equilibrium (reached after the patient is installed in the magnet since a common value for *T1* is of the order 1 to 10 *s*. It is conventional to assume that after 5 *T1* the system has reached almost the equilibrium value ***Mo***.
- An RF pulse.
- A signal induced right after the pulse and collected in an antenna (coil).

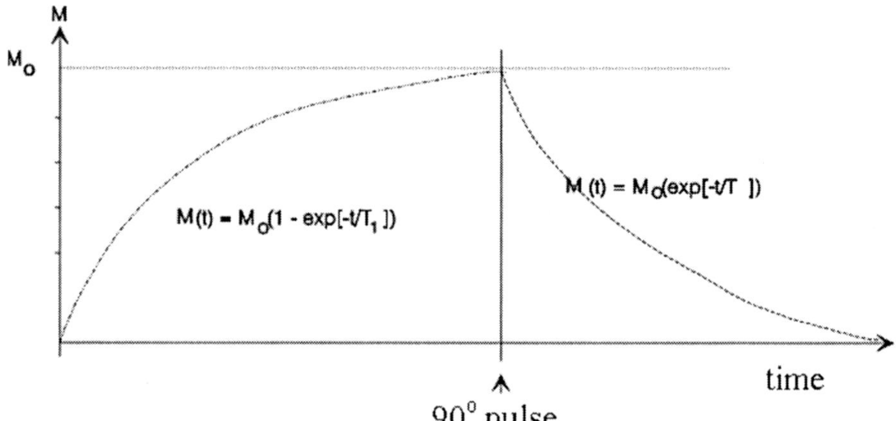

Figure 4. Mo is achieved in T_1 and the FID in T_2.

5. From NMR to Magnetic Resonance Imaging (MRI)

Using gradients of magnetic field MRI can select anatomical slices in any position. Let us think of a particular type of gradient to select a particular slice. Inside the magnet (assuming it is a perfect magnet!) the field is ***Bo*** and *0* (zero) outside. In that crude way a slice having the thickness corresponding to the length of the magnet is selected! Since ***Bo*** is constant over that length, the Larmor relation applies and one can collect the FID all along the axis of the magnet. This kind of perfect magnet do not exist; in the real world, the gradient at the boundaries has the shape sketched on Figure 5a, and the slice (large slice!) is restricted between the dotted lines. Since the Larmor relation provide the frequency the receptor can be tuned at that frequency (resonance), it is possible to restrict even more the slice thickness, playing with gradients and having the shape shown on Figure 5b.

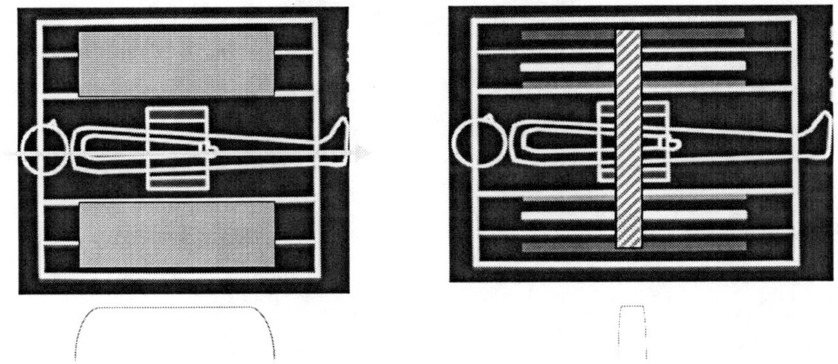

Figure 5a *Figure 5b*

Figure 5. Schematics of the magnet and the gradient coils and antenna.

The selected slice might be represented like the one in figure 6. To create an image, one has to "colour" each voxel (volume element like pixel is a picture element) with a colour corresponding to the information provided by the FID belonging to that particular voxel. Let us see how to distinguish the dark grey voxel (picture element) from the light gray one. This distinction is once again based on gradients.
Here, one assume that a measure of the amplitude of the signal issued by one of the voxel (*i* voxel) 20 ms after the beginning of the signal gives the value *M'* and a similar measure on the next voxel (voxel *i+1*) gives *M"*. To each of these values one attributes a specific colour.

An image in MRIis usually a matrix of 256 by 256 voxels and fortunately it is not necessary to collect one signal for each voxel, a rather time consuming procedure that will have certainly precluded any clinical application. Fortunately, collecting all the signals for all the voxels in line *i* is possible. The signal collected is a composite one made of all the voxels in line *i*. Then, thanks to the Fourier Transform every individual signal is retrieved and attributed to the proper voxel according to the signature provided by its individual frequency (each one having its own Larmor frequency ν. To get the next line (line *j*), a new set of data has to be collected, each line being identified by the "signature" provided by the third gradient in the orthogonal direction (the phase gradient). *These gradients are obtained by additional coils in the three direction of the space wounded around the main magnet.*

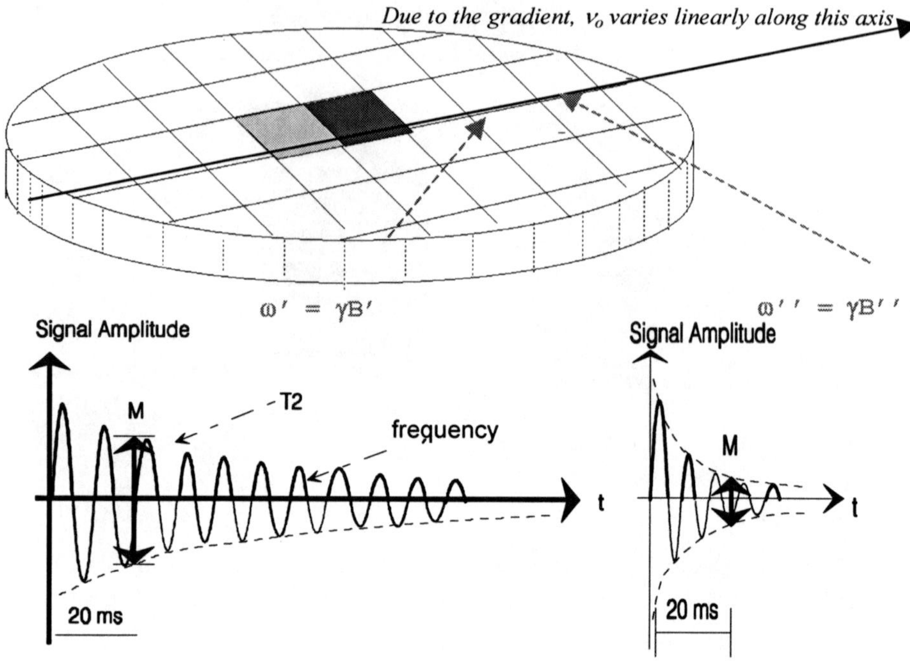

Figure 6. An idealized image and the FID providing the information.

Remarks:

a. The difference in amplitude after 20 ms has observed on the figure above can be attributed to the difference in the number of protons contained in each voxel. One can think that one voxel belongs to the white matter in the brain and the other one to the grey matter. White matter contains more than 80% of water (consequently protons) but the grey matter is around 70% ($Mo = \chi Bo$). If only this difference, the MRI had remained in the laboratory but never as a clinical tool. The environment of the water molecules has a striking influence on the time constant *T2*. Some molecule of water are tight to molecules of proteins for example having various sizes. Consequently, the movement of to the proton that provide NMR signal is different according to the size of these molecules.

b. Before repeating the experiment to collect the next line (figure 6), (the next phase encoding step), it is necessary that the system has again reached the equilibrium state ***Mo***. It is the general practice to wait a time *TR* of about *5 T1*, *TR* being the **R**epeat **T**ime.

6. Making a diagnosis

To be able to make an accurate diagnosis the radiologist needs a "good image". It is therefore important to define what is the meaning of a "good image". A "good image" is an image that provides:

- No artifacts
- Maximum information's
- Maximum contrast
- Minimum noise

It is the beauty of the MRI that it has many parameters with which one can play around to approach the above requirements. Artifacts are nowadays fairly well mastered in most of the system (if properly tuned and calibrated) and let us assume that the service engineer has done a good job.

Having a look at figure 6, let us suppose that one voxel (the dark gray one A) belongs to the white matter in the brain and the light gray (B) to the gray matter in the brain. For each of these tissues one gets a scheme similar to the one drawn on figure 4. This is drawn on the figure 7 below:

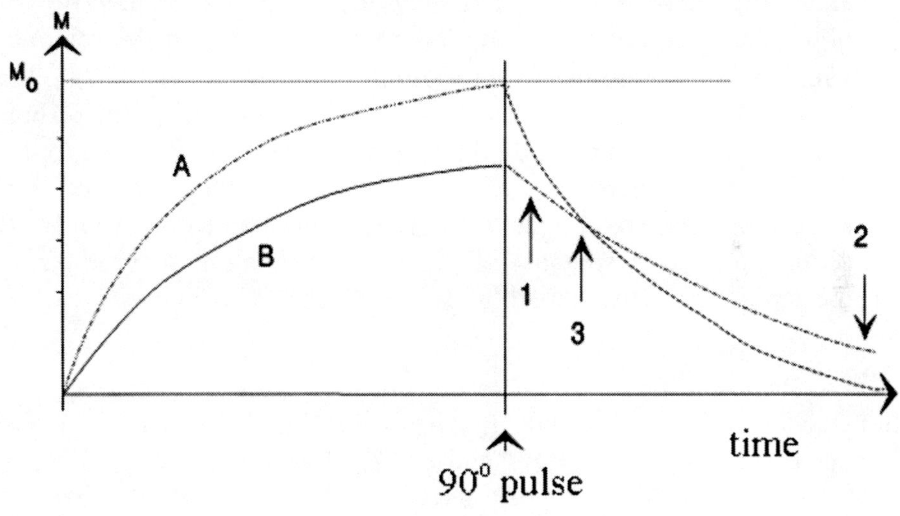

Figure 7. The image quality depends on the right choices !.

7. Comments about Figure 7

One can recognize the three steps of an MRI sequence:

- The T_1 process to create ***Mo***

- The 90° pulse to start collecting the FID
- The FID

The user of the MRI has several choices:

a. He can decide at which time to have the 90° pulse. For instance, this time might occur before both tissues have reached a maximum for ***Mo***. In that case the sequence is called: T_1 weighted since the contrast is strongly dependant on this parameter.

b. He can decide to weight a time long enough that both tissues have reached their ***Mo*** maximum. In that case, the difference in amplitude mostly depends of the difference in the number of protons (the quantity of water in each tissue). This kind of image is called: proton density weighted. Consequently, to obtain the required result one has to have an idea of the relative value of T_1 for each tissue considered.

c. Independently of these to choices, one can start collecting the data (let us say that 512 points are collected) at some time after the beginning of the FID (on Figure 6, 20 ms were chosen for example). This is called the echo time *TE*. Consequently, to obtain the required result one has to have an idea of the relative value of T_2 for each tissue considered.

d. It is desirable to have the best signal to noise ratio (S/N) but residual noise is unavoidable. To maximize the S/N the only chance is to increase the signal. This can be obtained by increasing the voxel size (remember expression 1) or/and by piling up several signals, which means collecting several times the same "line" (the same composite signal).

Appendix A: The concept of echo:

In figure 8 one can observe in the first part the FID as shown before, but the time constant is now called T_2* and not T_2. The true T_2 belongs to the intimacy of the matter but the world is not perfect and the physical mechanism by which T_2 is defined is tempered by technical limitations and in practice the decay is shorter than the ideal T_2. Sometimes the decay might be so short that not sufficient points can be collected to accurately represent the FID and this preclude any image reconstruction. To get around that limitation Hahn invented the spin echo[4]. Thanks to an additional 180° pulse the faded FID is reconstructed (with an inherent decrease in amplitude due to the true T_2 process).

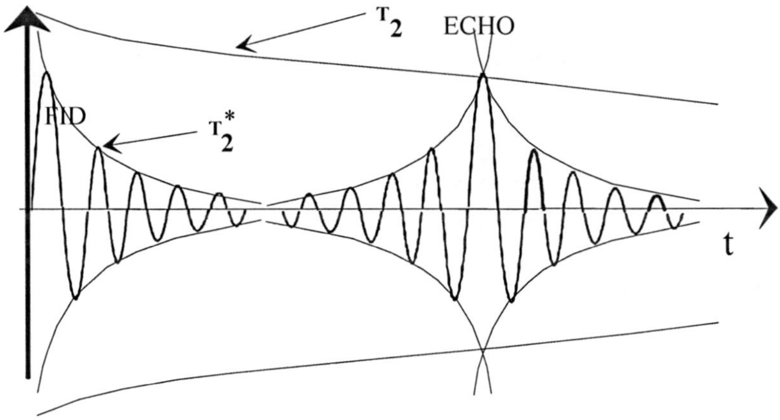

Figure 8. The spin echo phenomenon and the T_2*.

Appendix B: The MRI sequence

The sequence in MRI is the timetable organizing the different "pieces of physics" entering in the game. These several pieces are occurring at some precise time and the quality of the final image strongly depends of this timetable. It looks like the one shown on figure 9:

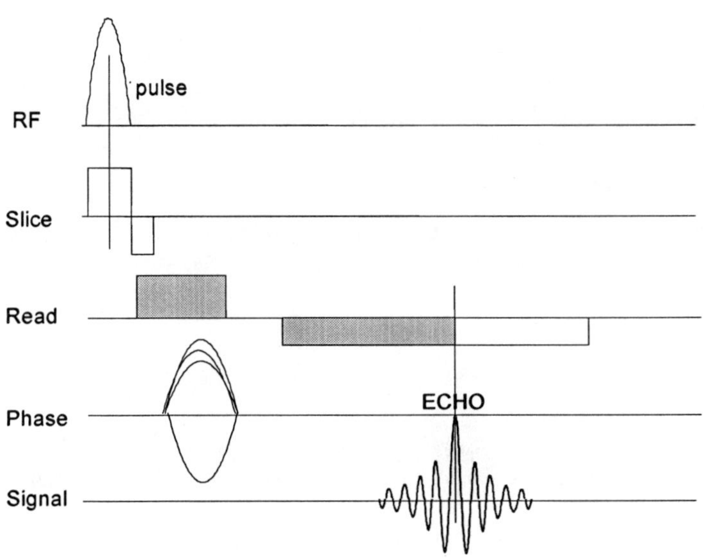

Figure 9. The basic MRI sequence.

- - The RF pulse
- - The 3 gradients
- - The collection of the signal (sampling window)

Tissue	T_1 [0.15]	T_1 [1.5]	T_2
Fat	170	250	80
Liver	250	720	60
White	350	500	75
Gray	500	760	90
Spleen	450	620	75
Kidney-cortex	340		70
Kidney-medulla	650	780	130
Muscle	580	800	40
Blood	720	1000	175
CSF	1500	2650	260
Water	2500		2500

Appendix C: Table I of Relaxation times

Remarks:

1. Some variability for these values of T_1 and T_2 can be found in the literature. Their relative values is what is useful to make the right choice (the right sequence).
2. T_1 depends on the field strength but T_2 not appreciably. It is important to keep in mind this point according to the field of the system one is using.

References

1. M.O. Leach. Magnetic Resonance Imaging, in this book.
2. A. Kumar, D. Welti, R.R. Ernst. NMR Fourier Zeugmatography. J. Magn. Reson., 54: 69, 1975.
3. A. ABRAGAM. The Principles of Nuclear Magnetism . Oxford Clarendon Press (1978).
4. E. HAHN. Phys. Rev. 80, p. 580, 1950.

MRI – QUALITY ASSURANCE

FRANÇOIS LAZEYRAS*
Università de Genève, 24 rue Micheli-du-Crest, CH-1211 Genève 4, Switzerland

Abstract. This paper will address two issues in quality assurance: i) Biological effects and safety in MR imaging; ii) MRI artifacts and quality control.

Keywords: cells blood alteration; vertigo; artifacts; RF power; geometric distorsion.

1. Biological effects in MR imaging and safety

Hundreds of articles have addressed the issues of biological effects and safety in static, and time varying magnetic and RF fields and the potential impact upon magnetic resonance imaging[1-3]. The purpose of this abstract is to briefly summarize potential biologic effects associated with MR imaging and more pragmatically to point out the more likely encountered dangers in day-to-day imaging.

1.1. POTENTIAL BIOLOGIC EFFECTS:

Three interactions have been identified as concerns for MR safety: Static magnetic field strength, rapidly switched magnetic field gradients (dB/dt) and radiofrequency (RF) magnetic fields.

1.1.1. *Static magnetic field strength:*

Numerous studies failed to detect an association between cellular level abnormalities to exposure to static magnetic field, however, a number of other studies have reported conflicting results including increased chromosomal aberrations in human lymphocyte cultures and increased incidence of malformations in chick and frog embryo development. Additionally, alterations of single red blood cells have been demonstrated

* F. Lazeyras, Université de Genève, 24 rue Micheli-du-Crest, CH-1211 Genève; email:françois.lazeyras@hcuge.ch.

Y. Lemoigne et al. (eds.), Physics for Medical Imaging Applications, 49–53.

in vitro (although *in vivo* effects have not been demonstrated). It has been concluded that "although no deleterious biologic effects from the static magnetic fields used in MR imaging have been definitely associated with this modality, all the facts are by no means in yet". The "Food and Drug Administration" guideline limits at 4 Tesla the magnetic field for clinical investigations[4]. No current induced effects have been reported at field strength less than 2 Tesla; however, magnetophosphenes, metallic taste, and vertigo have been reported at 4 Tesla.

1.1.2. *Rapidly switched magnetic fields:*

Thermal effects are considered negligible. The most important non thermal effects include current induction resulting in stimulation of nerve and muscle cells (including cardiac). The necessary current known to produce such stimulation is higher than those produced during a routine clinical MR study, nevertheless the introduction of echo planar MR systems needs to be specifically addressed. While conventional systems typically have one milli-Tesla per meter (1 mT/m) maximum gradient amplitude with 500 to 700 microseconds rise time, echo planar systems may produce 25 mT/m or more (even up to 40 mT/m in prototypes) with a rise time of 150 to 300 microseconds. Reports suggest that direct peripheral nerve stimulation has occurred using such scanners. The FDA defines guidance using dB/dt criteria (dB/dt in T/sec for gradient system of 25 mT/m with a rise time of 300 μsec is 41 T/sec at 0.5 meter). The original recommendation relating to dB/dt was 20 T/sec. With the development of echo planar and similar fast imaging techniques, the recommendation was reviewed such that no painful stimulation is induced with the gradients. Warning of the operator should be made for gradients exceeding 20 T/sec in clinical imaging. A secondary effect due to switching magnetic gradients is the acoustic noise. The FDA guidance limits the peak acoustic noise to 140 dB. The MR fast imaging may produce acoustic noise as high as 80 to 90 dB, and hear protection is recommended when using such imaging sequences.

1.1.3. *Radiofrequency:*

Much controversy surrounds the possibility of radiofrequency (RF) induced magnetic fields causing developmental abnormalities and possibly increased risk of cancer[5, 6]. A far more immediate concern is power deposition caused by RF pulses. Such power deposition is quantifed clinically using specific absorption rate (SAR). The criteria used to define the maximum SAR is that the core temperature should not rise more than 1 degree C. The FDA sets the maximum SAR at 4 W/kg on the whole body for 15 minutes, 3 W/kg averaged over the head for 10 minutes, 8 W/kg in any gram of tissue in the head or torso for 15 minutes, or 12 W/kg in any gram of tissue in the extremities for 15 minutes. In addition, in patients with heart conditions, altered thermal regulation or in pediatric patients, this limit is lowered to 1.5 W/kg. For comparison, human basal metabolic rate is approximately 1-2 W/kg. It is generally agreed that while there is a statistically significant increase in temperature during clinical scanning, there are no physiological consequences. Several studies have also showed that hemodynamic parameters remained within acceptable limits[7].

1.2. POTENTIAL DAY-TO-DAY HAZARDS

Patients have experienced first, second and third degree during routine MR examination. In all cases, it was due to RF interaction with conducting materials in contact with the patients (ECG leads). To avoid such problems, it is recommended to

i) avoid disconnected coil inside the magnet,
ii) avoid wire loops on the patient,
iii) insulate if possible conductors from the patient and
iv) all complaints of patient heating, especially local should be taken seriously.

Another important consideration is the attractive force of the magnet and its gradient on paramagnetic objects. Such attractive force includes both translational and torsional. In terms of screening, care has to be taken for objects on or around the patient as well as objects within the patients. It is mandatory to screen patients for ferrous objects and to constantly re-educate MRI personal about potential dangers associated to ferrous objects. It should be noted that the attractive force is proportional to the spatial magnetic field gradient (always on) and even a fairly light object may cause significant injury to patient or MR personnel. For instance, a paper clip in a 1.5 Tesla system reaches a terminal velocity of approximately 70 km/hr.

2. MR artifacts and quality control

The predominant artifact in MR imaging is motion related artifact. Because little can be done in term of quality control to reduce these kinds of artifacts, little will be said about them. It is nevertheless important to remember that motion artifacts physically occur in all three dimensions but propagate in the image primarily in the so-called phase encoding dimension (ghosting). Errors in the other dimensions also occur but are less visible. Some artifacts due to periodic motion, such as cardiac or respiratory can be reduced by triggering.

Unfortunately, motion is not the only source of artifacts in MR imaging. In some cases artifacts are due to intrinsic factors such as magnetic susceptibility and chemical shift which may be partially "corrected" to some degree by changing the acquisition technique. Others are environmental such as metal clips and implants, which distort the B0, B1 and gradient fields. Another class of artifacts has to due with inadequate sampling of a finite object (spatial aliasing, Gibbs ringing), errors in pulse sequences or applied physics insufficiency. These problems often have no easy cure and a service call does not help either, but they make a job market for imaging physicists. Finally, artifacts may be related to hardware failure and imperfection, which can plague the most careful operator and most of the time they are fixed by a good quality control program:

2.1. RF_POWER SETTING

The determination of the flip angle (the power button of the MRI) checks failure of RF amplifier, coil connection (TR box) and tuning (diodes). Scanning is not allowed with an improper flip angle setting, and it is checked for every new study or anytime a coil is changed. This measurement is used to compute the specific absorption rate for each sequence.

2.2. DETERMINATION OF THE FIELD HOMOGENEITY

The presence of a ferromagnetic object is one of the most pernicious artifacts in MR imaging because it affects the image differently depending of the sequence used. It has long been considered the sine qua non for spectroscopy but not a significant problem in imaging. With the advent of echo planar imaging, field homogeneity became important and is recognized as a key factor nowadays, even in spin echo. Field homogeneity is adjusted on almost every study.

2.3. DETERMINATION OF THE SIGNAL-TO-NOISE RATIO (SNR)

Measurement made every day insuring reproducible acquisition (no gain degradation in the coil, preamplifier and array processor).

2.4. DETERMINATION OF IMAGE UNIFORMITY

Check the RF homogeneity.

2.5. DETERMINATION OF SLICE PROFILE

Check RF amplifier linearity, RF pulse profile, calibrate gradient strength. This test is made at each software upgrade.

2.6. DETERMINATION OF TWO-DIMENSIONAL GEOMETRIC DISTORTION

Check gradient inhomogeneity. The spatial variations of the gradients cause spatial distortions. The deviation from linearity of the gradient fields is corrected by software. Another inhomogeneity is time dependent and is due to Eddy currents. When a time-dependent pulse is "played out" it can induce a current in conducting objects (e.g., shields, cryostat, other metal) which decay with different time constants during excitation or data collection. Such time dependent Eddy current generate a dispersal of phase and frequency to cause structured noise in the image as well as loss of SNR. Pre-emphasis of the gradient waveform may be used to minimize such problem but a more robust means is to shield the gradients.

2.7. DETERMINATION OF IMAGE STABILITY

Important in dynamic experiment such as perfusion MRI or functional MRI. Check the phase, amplitude stability of the RF and gradients as well as the sensitivity of the coil.

References

1. Adey, W. R. Biological effects of electromagnetic fields. J Cell Biochem;51:410-6, 1993.
2. Kanal, E. An overview of electromagnetic safety considerations associated with magnetic resonance imaging. Ann N Y Acad Sci 649:204-24, 1992.
3. Schaefer, D. J. Safety aspects of radiofrequency power deposition in magnetic resonance. Magn Reson Imaging Clin N Am 6:775-89, 1998.
4. FDA. A primer on medical device interactions with magnetic resonance imaging systems. 1997.
5. Adey, W. Tissue interactions with nonionizing electromagnetic fields. Physiol Rev 61:435-514, 1981.
6. Beers, G. Biological effects of weak electromagnetic fields from 0 Hz to 200 MHz: a survey of the literature with special emphasis on possible magnetic resonance effects. Magn Reson Imaging 7:309-331, 1989.
7. Hand, J. W., Lau, R. W., Lagendijk, J. J., Ling, J., Burl, M. and Young, I. R. Electromagnetic and thermal modeling of SAR and temperature fields in tissue due to an RF decoupling coil. Magn Reson Med 42:183-92, 1999.

ADVANCED MRI APPLICATIONS

JEFFRY R. ALGER* AND ANDREW J. FREW
Department of Neurology, David Geffen School of Medicine, University of California, 660 Charles E. Young Drive South, Los Angeles, CA 90095, USA

Abstract. Despite that Magnetic Resonance Imaging (MRI) was introduced as a clinical imaging tool more than 25 years ago, it continues to be characterized by fast-paced technological evolution that has a profound influence on its clinical applicability. This chapter presents a few selected recent technological MRI innovations and new clinical applications.

Keywords: Magnetic Resonance Imaging, Magnetic Resonance Angiography, contrast agent, brain morphometry, high field, gradient echo.

1. Introduction

This chapter will summarize a few of the more advanced clinical MRI applications. The specific topics are 1) Faster and more efficient imaging, 2) MRI contrast agents and 3) Imaging with magnetic field strengths of 3.0 Tesla (T) and higher.

2. Faster and more efficient imaging

The development of such enhanced MRI techniques offer reduced economic costs, lessened sensitivity to physiological motion, a reduced requirement for patients to hold still for extended periods of time and the opportunity to measure more in the fixed amount of time that a patient can hold still.

---

* Jeffry R. Alger, Department of Neurology, David Geffen School of Medicine, University of California, 660 Charles E. Young Drive South, Los Angeles, CA 90095, USA; e-mail: jralger@ucla.com.

Y. Lemoigne et al. (eds.), Physics for Medical Imaging Applications, 55–68.

In the early 1990's, Fast Spin Echo (FSE) imaging[1] (also known as Turbo Spin Echo (TSE) imaging), relaxed the one k-space line per TR requirement and thereby significantly enhanced the time efficiency MRI. FSE techniques collect four or more phase encoded spin echoes during a single TR, making the imaging process considerably faster. The acquisition of multiple echoes tends to restrict FSE to the acquisition of T2-weighted images. Figure 1 provides an example of a typical modern day FSE brain image illustrating that FSE techniques can acquire high quality T2-weighted images.

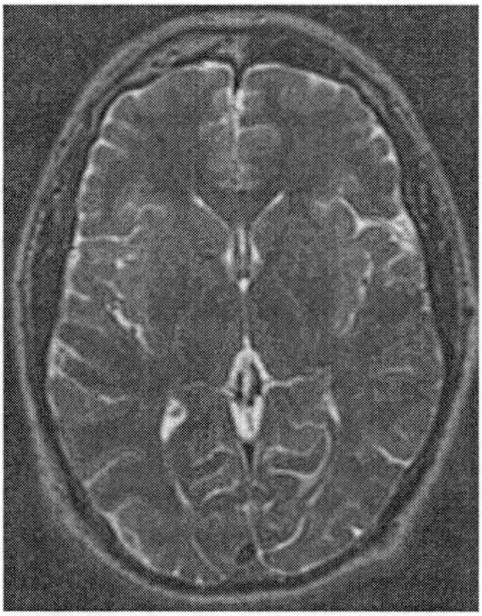

Figure 1. Typical T2-weighted FSE image of normal brain. FSE techniques significantly reduce the imaging time in comparison to conventional spin echo imaging yet provide nearly the same contrast and image clarity.

Gradient echo MRI is alternate approach that can be used to acquire image data more quickly than is possible with conventional spin echo imaging. In gradient echo imaging, the echo signal is formed by gradient polarity reversal at relatively short time-to-echo (TE) rather than by a spin echo formed by the action of a refocusing pulse.

Figure 2 shows the generic pulse sequence design used in two-dimensional gradient echo imaging. The concept of gradient echo imaging was introduced in the 1980's[2]. However, initial applications were relatively limited because gradient hardware systems of the time were not able to reverse gradient polarity without the induction of undesirable spatial and temporal magnetic field variations. Gradient echo imaging became used more extensively in the 1990's as gradient hardware systems improved.

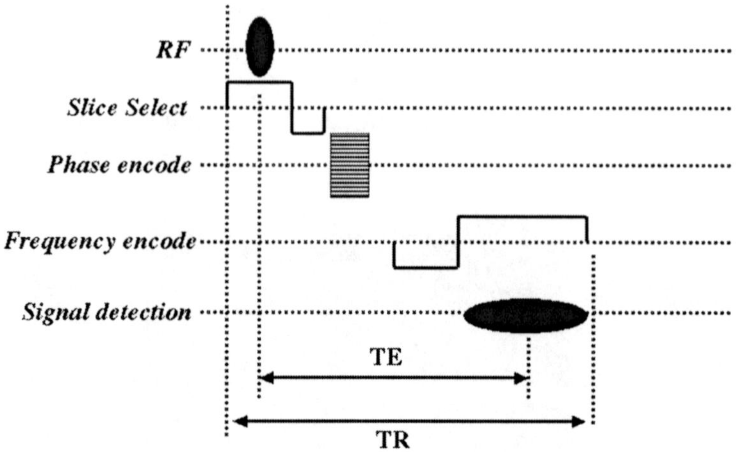

Figure 2. Pulse sequence design for two-dimensional gradient echo imaging.

The use of rather short TE in gradient echo MRI permits TR to be very short. Typical gradient echo imaging pulse sequences have TE of approximately 10-20 msec and this allows TR to be approximately 25 msec. One phase encoded gradient echo is acquired in each TR, so gradient echo imaging is capable of acquiring on the order of 40 lines of k-space per second. Therefore, it is possible to acquire a 2D gradient echo image having a phase encode resolution of 128 in approximately 3 seconds. The acquisition of multiple slice gradient echo image data requires additional time. For example if 10 slices are desired, it is necessary to repeat the 2D acquisition 10 times. Nevertheless for a limited number of slices, gradient echo imaging can still be done rather quickly.

The ability to acquire gradient echo images in a few seconds leads to the use of gradient echo technology for "snapshot" imaging. Here the goal is to complete image acquisition sufficiently quickly to freeze physiological motion. Gradient echo techniques are thereby used in cardiac imaging, vascular flow imaging, kinematic muscle movement imaging and breathhold imaging in the abdomen. Snapshot gradient echo imaging of the liver provides an example[3]. Breathing motion produces artifacts that obscure the liver in images acquired using conventional spin echo or FSE techniques because there is repetitive chest and diaphragm motion during the minutes that are needed for image acquisition. In breathhold imaging, gradient echo techniques are used to acquire an image of the abdomen in 30 sec or less while the patient voluntarily holds their breath, providing an image free from breathing artifacts.

There are many variants of the basic gradient echo imaging scheme. One of the more advanced of these is the "trueFISP" technique. In this technique, all gradient polarity reversals are balanced in time and this allows significantly faster image acquisitions (about 64 msec for a 2-dimensional image) that are relatively free of artifact. The trueFISP technique has been developed as a method for imaging the heart[4]. The image acquisition is completed in less than one heartbeat. Cardiac motion is thereby frozen allowing the heart muscle to be visualized without motion artifact. Furthermore, a time series of images can be collected as the heart beats. This permits the acquisition of a series of image frames during one cardiac cycle. The time series of images can then be formatted into a cine stream that allows the viewer to see the actual motion that occurs during the cardiac cycle.

The short TRs used in gradient echo imaging tends to result in images that have T1-weighting. Excitation pulses of much less than 90^0 are used because the TR is so short that almost complete signal saturation would occur unless rather small pulses are used.

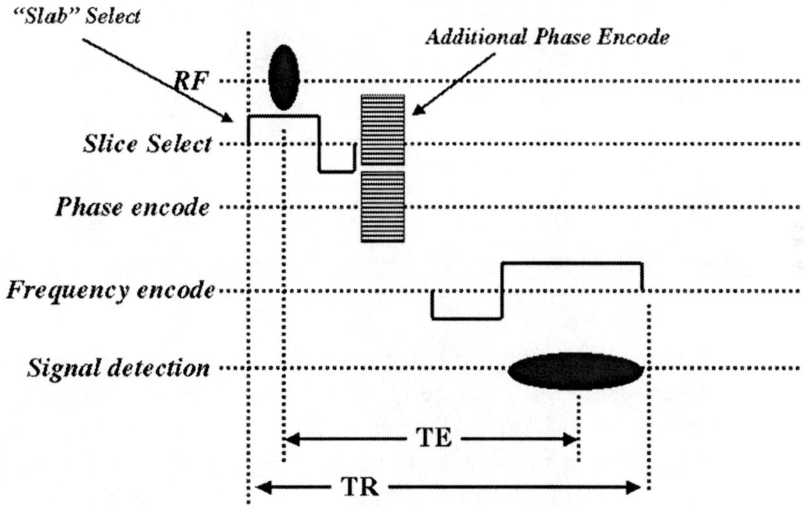

Figure 3. Pulse sequence design for three-dimensional gradient echo imaging.

Typically the excitation pulse is set at an "Ernst Angle" of 20-30 degrees to optimize the signal-to-noise ratio (SNR) for tissue being imaged. In some circumstances, the excitation pulse is adjusted to achieve a specific degree of T1-weighting.

Three-dimensional imaging with gradient echo techniques[5] is possible provided that an addition phase encoding procedure is added (Fig. 3). The typical approach to 3D gradient echo imaging of the head involves using the excitation pulse to select a thick slab that encompasses the entire head with two orthogonal phase encoding processes and frequency encoding of the third dimension (Fig. 4).

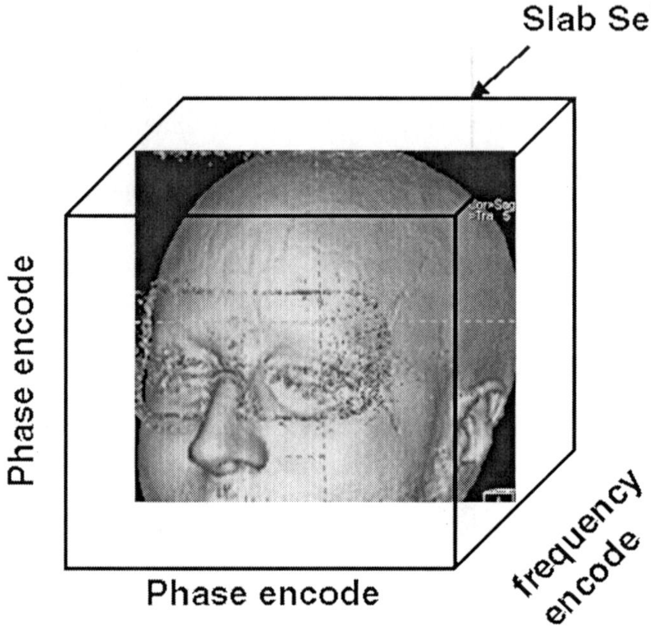

Figure 4. Image acquisition configuration that is typically used in 3D gradient echo MRI of the brain.

The additional phase encoding process lengthens the data acquisition time but, because very short TRs can be used, the total data acquisition time is still within reasonable limits. Here the concept is to use the inherent efficiency of gradient echo imaging to acquire high-resolution T1-weighted 3D image data. The SNR in brain imaging is sufficiently high to permit a spatial resolution of approximately $1.0 \times 1.0 \times 1.0$ mm^3. Three dimensional gradient echo techniques can achieve this resolution in image acquisition times of approximately 10 min by the appropriate adjustment of the frequency and phase encoding resolution and fields-of-view. The value of 3D gradient echo imaging is that resulting image data can be subjected to 3D reformatting and analysis because the resolution is very fine and isotropic. This permits the construction of a variety of pseudo-3D displays that provide a unique visual perspective on brain anatomy (Fig. 5).

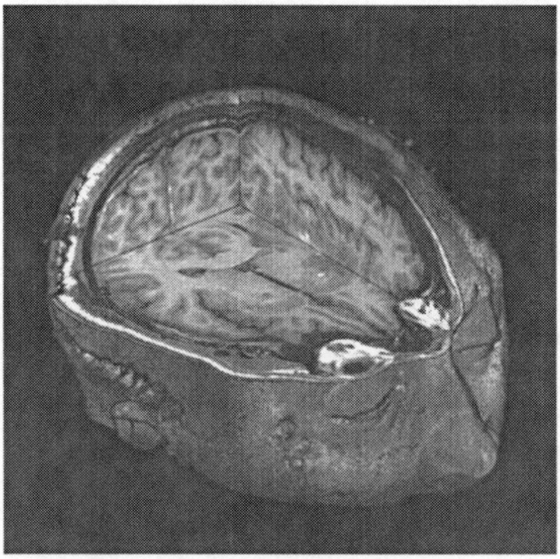

Figure 5. Looking inside of the brain of a patient that has suffered a traumatic brain injury using 3D rendering software to display 3D T1-weighted gradient echo MRI data.

The inherent T1-weighting of 3D gradient echo brain images is also the basis of many of the "morphometric" brain studies that are underway in many laboratories around the world. The T1-weighting provides excellent signal contrast between gray matter (GM), white matter (WM) and cerebrospinal fluid (CSF), which are the predominant tissue types found in the brain. A variety of software tools are available that identify individual image voxels from the 3D gradient echo image data as being most probably GM, WM or CSF. These segmentation tools thereby support automated measurements of tissue content in specific brain regions. Three-dimensional brain imaging with gradient echo technology also supports the development of an atlas of brain anatomy[6]. Software tools that use morphometric distortion operators to match the structures of different brains to a common template brain are available. These tools have been applied to create a volumetric representation of the "average" of many brain 3D gradient echo MRI data sets. Once a representation of the average brain is available, a number of additional further morphometric studies become possible. For instance it is possible to measure how the sizes and shapes of specific brain structures or their tissue contents differ between normal and disease afflicted populations. Moreover, it is possible to perform measurements of how brain structures change size, shape and composition as diseases progress or as people age[7].

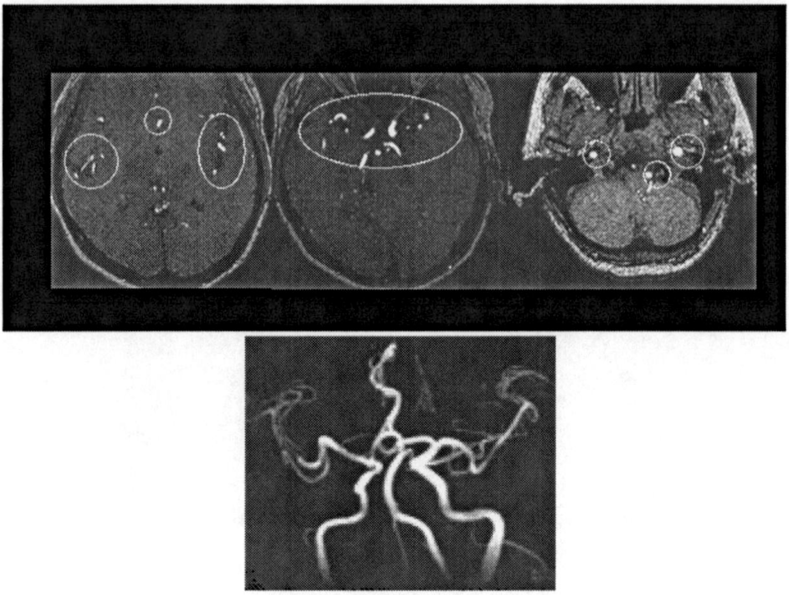

Figure 6. Time-of-flight MRA. The top row shows three brain images selected from a 3D gradient echo image volume collected with a large flip angle. The principal arterial vessels (circled) appear as bright signals over the gray background produced by static tissues. The bottom image is a Maximum Intensity Projection created from the entire 3D image volume.

Three dimensional gradient echo imaging also provides the basic data that is used in image-guided neurosurgery. Here a 3D brain image is acquired prior to a planned neurosurgical procedure. The neurosurgeon then uses 3D image rendering software to visualize and plan the surgical approach. Some relatively sophisticated software systems have evolved for doing this. Of particular note are systems that display the MRI data for the neurosurgeon relative to the patient's orientation in the space during the operation. This effectively allows the neurosurgeon to see "inside" the brain via the previously acquired MRI data as the surgical procedure is performed.

Three dimensional gradient echo imaging also serves as the basis for the "time-of-flight" magnetic resonance angiography (MRA) procedures that are used to non-invasively image vascular anatomy[8]. Here it is the flow that provides contrast that allows the arterial vessels to be visualized. Stationary water proton signal becomes heavily saturated when 3D gradient echo imaging is performed with excitation pulses that are large. This saturation occurs because there is repetitive excitation without sufficient time for T1 relaxation with TR is much shorter than T1. However, the situation is different for flowing water protons in the arterial blood. Such protons do not experience the repetitious excitation because they move into the image volume on a time scale that is much shorter than TR. Accordingly, their full signal intensity is excited and measured and the flowing blood in the arterial vessels thereby appears as very intense signal over a background of saturated signal that is generated by the surrounding stationary tissues (Fig. 6).

The vascular anatomy can be readily visualized in the individual slices of the 3D gradient echo images. However it is tedious for the observer to mentally construct a 3D representation of the vascular anatomy by viewing individual slices of image data. Furthermore this approach does not make the best use of the 3D nature of the data. Hence it is common practice to use Maximum Intensity Projection (MIP) methodology to build a projection of the vascular anatomy onto a viewing plane than is chosen by the user. Such MIP procedures provide a post hoc "virtual reality" approach for cinegraphically viewing vascular anatomy.

The most recent advancement in the area of fast imaging is Echo Planar Imaging (EPI), which provides image data even more rapidly than conventional gradient echo techniques can[9]. EPI can be used to acquire T2-weighted image data at a rate of 10-20 slices per second with a resolution than can be as fine as $2 \times 2 \times 2$ mm^3 (Fig. 7). It is useful for a number of applications for which it is necessary to collect a large amount of image data in a tolerable period of time. Such applications include diffusion weighted imaging, perfusion imaging and functional imaging of blood oxygenation. The foregoing discussion has emphasized that fast efficient imaging methods have been developed and have lead to unique clinical applications. However, there is a cost associated with using these rapid MRI techniques.

There tends to be an inverse relationship between image quality and imaging speed. The faster imaging techniques (gradient echo imaging and EPI) tend to produce a greater degree of artifacts that are associated with inhomogeneities of the static magnetic field. Furthermore, there is a general inverse relationship between the SNR and the speed of image acquisition. Faster image acquisitions produce images that have lower SNR and therefore less clarity. Faster acquisitions also tend to have lower spatial resolution.

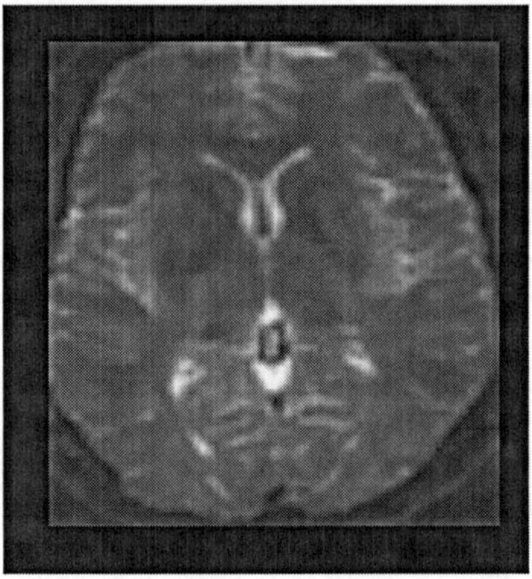

Figure 7. A typical T2-weighted echo planar image of a brain slice.

3. Contrast agents

The use of contrast agents in MRI has been an area of active research during recent years. The most commonly used MRI contrast agents are chelates of metal ions that have unpaired electron spins. These agents are capable of enhancing T1 relaxation in water proton spins because the contrast agent's unpaired electron spin produces a magnetic field that is approximately 2000 times stronger than that produced by the proton spin. The oscillation of the relatively large electronic magnetic field that results from molecular movement or electron spin relaxation is capable of inducing state changes in the proton spins of water molecules that are situated very close (on a molecular dimensions) to the electron spin. This emphasizes the very important requirement that the water proton spin must come into direct molecular contact with the contrast agent molecule in order for the T1 enhancement to occur.

The most commonly used contrast agents make use of the unpaired electrons associated with the gadolinium ion. It is important to remember however that the gadolinium in MRI contrast agents is chelated. Free unchelated gadolinium is actually quite poisonous and could not be used without chelation.

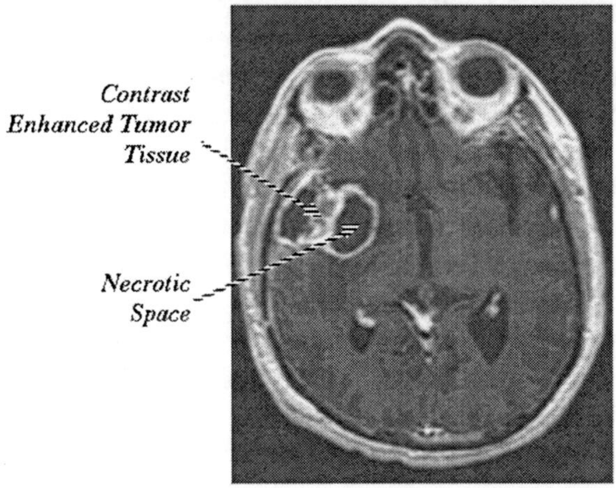

Figure 8. T1-weighted spin echo image of brain in a patient who has a brain tumor.

The most common procedure for enhancing MRI contrast involves the intravenous administration of the agent. T1-weighted imaging using either spin echo or gradient echo techniques can then be used to detect where in the body the T1 of the water molecules has been reduced as a result of the presence of the contrast agent. Regions where contrast agent is present will show enhanced MRI signal because of the inverse relationship between MRI signal intensity and T1 in T1-weighted images. Many tumors, particularly more aggressive tumors, have blood vessels that allow the contrast agent to leak from the blood stream into the tumor. Accordingly, many tumors show MRI signal enhancement after contrast agent has been intravenously administered. A typical example is provided in Figure 8. The actual tumor tissue appears with strong signal because of the relaxation effect that results when the contrast agent leaks from the blood stream into the tumor. Note further that the tumors often have necrotic spaces associated with them. Necrotic spaces have no blood vessels. Contrast agent cannot penetrate into these necrotic spaces and the signal is correspondingly low.

One of the advanced techniques that use contrast agents is contrast enhanced magnetic resonance angiography (CE-MRA) [10]. In this technique, the presence of contrast agent in the vascular system enhances the signal of blood. This relaxation enhancement effect can be much stronger than the flow-related enhancement that is discussed above. Hence, angiograms can be made by performing strongly T1-weighted image acquisitions a short time after the intravenous administration of the agent (Fig 9). Here the problem is that contrast agent labels both the arterial and venous space and unless a more sophisticated procedure is used, the resulting image shows enhancement of both artery and vein MRI signals. In order to acquire images that show only arterial signal enhancement, a rapid image acquisition procedure (usually involving gradient echo imaging) is performed in time synchronization with the rapid bolus infusion of the contrast agent into vascular system.

Here it is important that the image acquisition be completed within the 10 or so seconds that the bolus of agent is present in the arteries. Because of the need for a rapid image acquisition, CE-MRA techniques usually use only 2D thick slab T1-weighted gradient echo acquisitions. The arterial signals are so intense when the appropriate procedures are used, that they can be readily seen over the much lower signal from the non-enhanced tissue. Furthermore, the acquisition of a venogram is possible by simply delaying the image acquisition until the contrast agent is present in the veins.

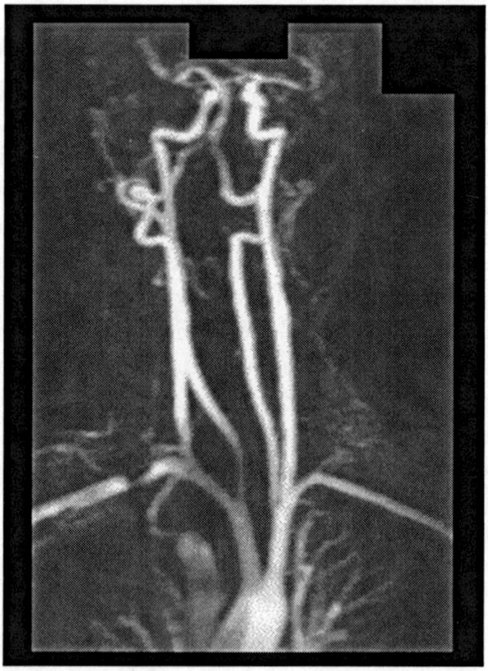

Figure 9. Contrast-enhanced MRA of the arteries of the upper thorax, neck and head.

There is currently a great deal of academic and commercial research directed toward the development of new MRI contrast agents that have specific affinities to gene expression, biological processes or specific biological molecules that might be disease markers[11].

4. Applications with magnetic fields of 3.0 Tesla and higher

During the past 5 years, MRI scanners operating at a field strength of 3.0 T have become available for routine clinical imaging. Prior to this, the strongest field strength that was routinely used for clinical imaging was 1.5 T. Imaging with higher field strength scanners offers advantages. The image SNR is approximately proportional to magnetic field strength. The increased SNR characteristic of 3.0 T in relation to 1.5 T can translate into enhancement of image clarity or reduction of the image acquisition time. The larger field strength can also provide a greater sensitivity in studies that use magnetic field inhomogeneity as a means of developing contrast as is done on blood oxygenation functional imaging and contrast passage perfusion imaging. Furthermore, the T1 values of many tissues are increased at 3.0 T relative to 1.5 T and this leads to improvements for certain studies that utilize T1-weighting for contrast development. Time-of-flight MRA is an example. The increased SNR together with the increased sensitivity to T1 that are characteristic of 3.0 T have resulted in very high quality time-of-flight MRA studies of the brain vasculature (Fig. 10).

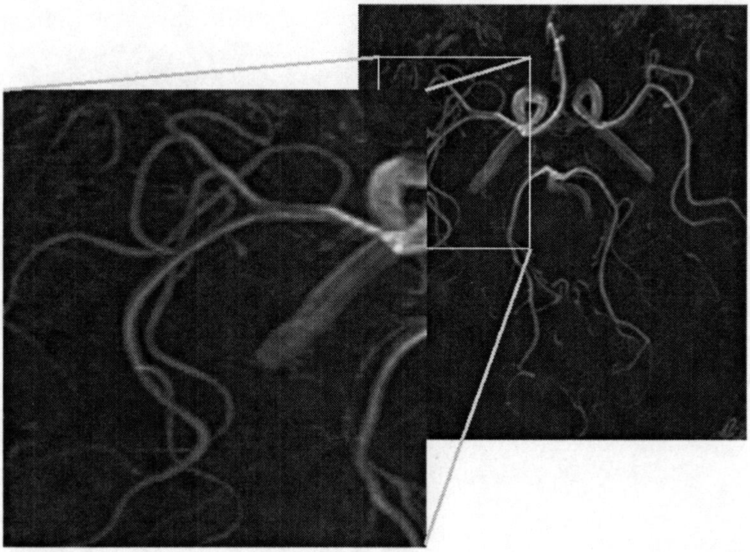

Figure 10. Maximum Intensity Projection MRA of the brain arteries performed at 3.0 T. The magnification shown in the foreground illustrates that submillimeter arteries are well visualized at 3.0 T.

The development of 3.0 T MRI technology has progressed very rapidly and it now appears that 3.0 T will become the accepted standard for clinical neuroimaging. However, it remains somewhat unclear whether 3.0 T will be as valuable for imaging other body parts. The principal technology limitation of higher field strength MRI is related to the higher radio frequencies that are used.

It is technically challenging to design and construct highly sensitive and spatially homogeneous radio frequency coil structures that function effectively at the higher resonance frequencies used at 3.0 T. Higher field images often show undesirable inhomogeneity for this reason and, furthermore, the SNR tends to be somewhat lower than what is expected. There may also be problems associated with excessive tissue heating in higher field MRI when certain pulses sequences are used. These limitations of radio frequency technology become increasingly important as the physical size of the coil structures increase. Reasonable solutions have been found for coil structures that are sufficiently large to accommodate the human head and to resonate at the radio frequency needed for 3.0 T imaging. However it is still unclear whether efficient structures for imaging the thorax at field strengths of 3.0 T and higher can be developed. Nevertheless a number of advanced laboratories are performing research directed at developing MRI technology and applications at field strengths between 4.0 and 8.0 T. [12, 13]

5. Summary

This chapter has summarized what are some of the trends in advanced MRI technology and applications. These trends include faster more efficient imaging which leads to snapshot and cinegraphic imaging in addition to novel types of 3D imaging that support neuromorphological and angiographic studies. Also included are novel uses of contrast agents that rely on faster more efficient imaging and the use of field strengths higher than the present standard of 1.5 T.

References

1. R.T. Constable, R.C. Smith, and J.C. Gore. Signal-to-noise and contrast in fast spin echo (FSE) and inversion recovery FSE imaging. J. Comput. Assist. Tomogr. **16**(1),41–7 (1992).
2. J. Frahm, A. Haase A, and D. Matthaei. Rapid NMR imaging of dynamic processes using the FLASH technique. Magn. Reson. Med. **3**(2):321–7 (1986).

3. R.R. Edelman, B. Wallner, A. Singer, D.J. Atkinson, and S. Saini. Segmented turboFLASH: method for breath-hold MR imaging of the liver with flexible contrast. Radiology. **177**(2):515–21 (1990).
4. J.C. Carr, O. Simonetti, J. Bundy, D. Li, S. Pereles, and J.P. Finn. Cine MR angiography of the heart with segmented true fast imaging with steady-state precession. Radiology **219**(3):828–34 (2001).
5. E.M. Haacke, and J.A. Tkach. Fast MR imaging: techniques and clinical applications. AJR Am J Roentgenol **155**(5):951–64 (1990).
6. J.C. Mazziotta, A.W. Toga, A. Evans, P. Fox, and J. Lancaster. A probabilistic atlas of the human brain: theory and rationale for its development. The International Consortium for Brain Mapping (ICBM). Neuroimage **2**(2):89–101 (1995).
7. A.W. Toga, P.M. Thompson, and E.R. Sowell. Mapping brain maturation. Trends Neurosci 2006 **29**(3):148–59 (2006).
8. E.R. Patel, and R.R. Edelman. MR angiography of the head and neck. Top. Magn. Reson. Imaging.**8**(6):345–65 (1996).
9. F. Schmitt, M.K. Stehling, and R. Turner. Echo-Planar Imaging. Theory, Technique and Application. Springer 1998.
10. D.A. Leung and J.F. Debatin. Three-dimensional contrast-enhanced magnetic resonance angiography of the thoracic vasculature. Eur. Radiol. **7**(7):981–9 (1997).
11. A. Louie. Design and characterization of magnetic resonance imaging gene reporters. Methods Mol. Med. **124**:401–17 (2006).
12. K. Ugurbil, G. Adriany, P. Andersen, W. Chen, M. Garwood, R. Gruetter, P.G. Henry, S.G. Kim, H. Lieu, I. Tkac, T. Vaughan, P.F. Van De Moortele, E. Yacoub, and X.H. Zhu. Ultrahigh field magnetic resonance imaging and spectroscopy. Magn. Reson. Imaging **21**(10):1263–81 (2003).
13. A. Kangarlu, F.G. Shellock, and D.W. Chakeres. 8.0-Tesla human MR system: temperature changes associated with radiofrequency-induced heating of a head phantom. J. Magn. Reson. Imaging **17**(2):220–6 (2003).

PHYSIOLOGICAL AND FUNCTIONAL MRI

JEFFRY R. ALGER* AND ANDREW J. FREW
Department of Neurology, David Geffen School of Medicine, University of California, 660 Charles E. Young Drive South, Los Angeles, CA 90095, USA

Abstract. This chapter summarizes the background for and the applications of several MRI methodologies that image physiological and functional parameters in the brain. These include diffusion MRI, perfusion MRI and Blood Oxygen Level Dependent functional MRI.

Keywords: magnetic resonance imaging, brain function, physiology, perfusion, stroke, blood oxygen.

1. Introduction

Magnetic resonance imaging is traditionally viewed as a method for acquiring images that depict the internal structure of the body for medical purposes. This chapter describes MRI applications that go beyond the mere depiction of anatomy to the imaging of physiological or functional parameters. The most significant applications have been in brain imaging. Although physiological and functional imaging can pertain to other parts of the body, space limitations prevent a broader discussion of application outside of the brain. The chapter illustrates that physiological and functional MRI procedures can be useful for basic neuroscience studies and for practical clinical purposes. Readers interested in more detail on these topics should consult the text by Buxton[1].

2. Diffusion MRI

Diffusion MRI is based on the fact that the signal-generating water molecules within biological tissue exhibit thermally-driven translational

* Jeffry R. Alger, Department of Neurology, David Geffen School of Medicine, University of California, 660 Charles E. Young Drive South, Los Angeles, CA 90095, USA; e-mail: jralger@ucla.edu

Y. Lemoigne et al. (eds.), Physics for Medical Imaging Applications, 69–81.

motion on a timescale similar to that of an MRI pulse sequence. Therefore, MRI pulse sequences can be designed to produce image intensity that depends on the translational diffusion coefficient[2]. The microscopic tissue structure can restrict the speed of diffusive translational motion as well as constrain it in certain physical directions. As a result, diffusion MRI is sensitive to the microscopic structure of biological tissue. The restrictions of diffusion speed are measured in diffusion weighted imaging (DWI) and in apparent diffusion coefficient (ADC) imaging. Directional restrictions are the basis of diffusion tensor imaging (DTI).

MRI signal intensity is usually sensitized to diffusion by use of a spin echo pulse sequence into which additional gradient pulses are incorporated in the periods before and after the refocusing pulse[2]. Diffusive movement of the water molecules causes signal attenuation with this type of pulse sequence. Stronger or longer gradient pulses lead to increased sensitivity to diffusion and increased signal attenuation. The degree to which the pulse sequence is sensitive to diffusion expressed by the "b-factor", which is dependent on the strength and duration of the diffusion-sensitizing gradient pulses and the separation between them.

Diffusion MRI requires the acquisition of a relatively large number of image volumes in a limited amount of time. It is, therefore, desirable to acquire the data using a very fast imaging method. Accordingly, diffusion sensitized spin echo echo planar imaging is commonly used.

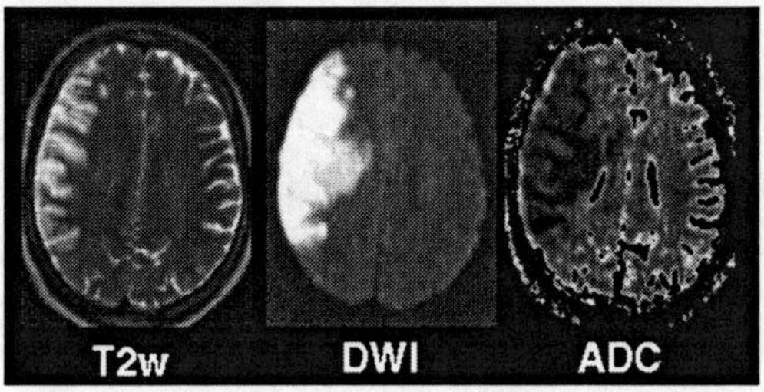

Figure 1. T2-weighted (T2w) , diffusion-weighted (DWI) and apparent diffusion coefficient (ADC) imagse of a patient who recently had an ischemic stroke. The area affected by the stroke is hyperintense on DWI and hypointense in ADC. The T2-weighted signal in the affected area is also elevated so the DWI includes both the diffusion and T2 effects. The ADC image shows only diffusion effects.

Diffusion weighted images are acquired using the diffusion sensitization technique as described above (Figure 1). The more intense signal in such images identifies tissue in which the water diffusive motion is relatively hindered and the less intense signal arises from tissues in which the water molecules can move more freely. A practical problem of DWI is that it is necessary to use relatively long TE values to attain adequate diffusion sensitization and, as a result, DWI signal intensity depends on the transverse relaxation time (T2) in addition to the diffusion coefficient. The combined diffusion and T2 weighting can be confusing in some contexts. ADC imaging is used to obtain image intensity that depends only on diffusion (Figure 1). In ADC imaging a series of DWI acquisitions are performed (hence the need for a fast imaging technique like EPI). In this series, a variety of unique b-factors are used creating a group of images that have variable diffusion weighting but constant T2 weighting. This allows the diffusion weighting to be separated from the T2-weighting. If the diffusion is spatially isotropic (i.e. there is no directional restriction on the diffusive motions), then the following log-linear equation describes the relationship between the measured signal intensity, the ADC and the b-factor.

$$\ln (S_b) = \ln(S_0) - (b)(ADC) \qquad (1)$$

Here S_b is the signal intensity measured at each b-factor and S_0 is the signal intensity measured at $b = 0$. Log-linear fitting of the measured signal versus b-factor is performed for each voxel and an image is formed by using a gray level or color to specify the ADC in each voxel.

The microscopic organization of tissue can be very important in diffusion MRI studies, because the tissue microstructure may impose greater constraints upon diffusive motion in certain physical directions (i.e. the constraints are anisotropic). This is the situation for brain white matter, because it is composed of arrays of parallel microscopic fibers. Water is more likely to diffuse parallel to the fibers as opposed to the orthogonal direction because the myelinated membranes that surround the fibers impose microscopic physical barriers to diffusive motion.

Diffusion Tensor Imaging[3] is an extension of DWI that specifically takes into account the possibility of anisotropic constraints to diffusive motion. DTI amounts to measuring ADC as a function of direction. Spatial sampling can be accomplished because the gradient hardware used in MRI systems is capable of diffusion sensitization along any arbitrary spatial direction by using linear combination of magnetic field gradients along the three principal spatial directions for the diffusion sensitizing gradient pulses. DTI studies are performed by acquiring a series of diffusion weighted images with a variety of b-values and a variety of diffusion sensitization directions. In this situation, a more general form of eqn 1 is used in which $\mathbf{b}$ is a matrix of b-factors (the elements b_{ij} that give the extent of diffusion sensitization along the spatial direction i,j) and $\mathbf{D}$ is the diffusion tensor (the elements D_{ij} give the diffusion coefficient pertaining to the spatial direction i,j). It is widely assumed that molecular diffusion is symmetric (diffusive movement in a positive sense along a direction is equally probable as diffusive movement in the negative sense along the same direction). Under this assumption the minimum number of directional measurements needed to uniquely determine the diffusion tensor is six. However measurements in more than six directions provide additional information that leads to a more precise estimation of the diffusion tensor. Therefore DTI measurements involve collecting at least seven volume images. One of them has no diffusion sensitization (b = 0) and the remaining six are performed with finite diffusion sensitization along six unique spatial directions. The diffusion tensor is estimated from this series of measurements. It is then customary to use numerical methods to diagonalize the diffusion tensor[4]. The resulting eigenvalues (λ_1, λ_2, λ_3) are the diffusion coefficients in the directions specified by the eigenvectors (v_1, v_2, v_3). Two unique rotationally invariant parameters are then derived. The average of the diagonal elements

$$D_{av} = (1/3)(\lambda_1 + \lambda_2 + \lambda_3) \quad (2)$$

gives a "directionally averaged" diffusion coefficient. The fractional anisotropy defined as

$$FA = ((\lambda_1 - \lambda_2)^2 + (\lambda_2 - \lambda_3)^2 + (\lambda_1 - \lambda_3)^2)^{1/2} / (2(\lambda_1^2 + \lambda_2^2 + \lambda_3^2))^{1/2} \quad (3)$$

is an index having values between 0 and 1 that conveys the degree to which diffusion is anisotropic. D_{av} and FA are scalar values that can be represented as images using gray scale or color (figure 2). In addition it is becoming increasingly common to represent the data as red-blue-green "color maps" in which the red intensity conveys the magnitude of the left-right diffusion coefficient, the blue intensity represents magnitude of the superior-inferior diffusion coefficient and the green intensity represents magnitude of the anterior-posterior diffusion coefficient[5]. More sophisticated "fiber tractography" techniques are also used[6]. Fiber tractography techniques attempt to represent "connections" between individual image volume elements by connecting the principal directions (the eigenvector that has the largest eigenvalue) with smooth "stream lines"[4]. Such procedures allow one to represent the most probable orientation of the microscopically-organized white matter fiber structures over large scale dimensions.

DTI has applications in clinical and basic neuroscience. Clinical studies have used DTI fiber mapping procedures to identify and visualize the spatial relationships between white matter structures and brain lesions (such as tumors).

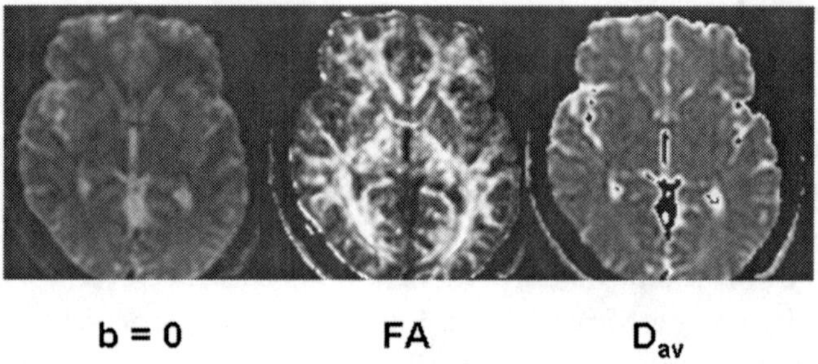

Figure 2. Parametric images derived from DTI. The b=0 image is a T2-weighted spin echo image. The FA image shows high image intensity for white matter. The D_{av} gives the directionally averaged diffusion coefficient.

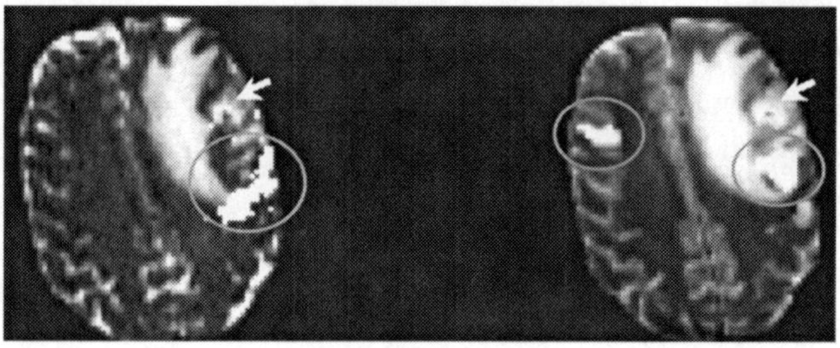

Figure 3. DTI fiber tracing (left) and BOLD fMRI (right) results from a brain tumor patient. The tumor location is identified by the arrows. The left image shows white matter fibers (highlighted in white and circled) that originate near the cortical speech activation areas (right image highlighted in white and circled).

Figure 3 illustrates the use of DTI fiber tractography to identify white matter fibers that lie on the periphery of a brain tumor. These procedures offer the potential for improved lesion resection planning in individual patients. Basic neuroscience studies using DTI include evaluating how white matter measures of D_{av} or FA or white matter connections relate to functional capabilities of individuals. For example a recent paper by Nagy et al.[7] reports a correlation between FA values and mental performance in a few white matter areas.

It may appear from the forgoing discussion that diffusion MRI is just another form of structural imaging because it provides information about tissue microstructure. This is certainly true, however as will be discussed below, the ADC value is sensitive to physiological changes that occur when brain tissues experience interruption of blood flow (ischemia) and therefore diffusion MRI can also be categorized as a physiological imaging procedure.

3. Contrast bolus passage perfusion MRI

Contrast bolus passage perfusion MRI[8] measures several physiological parameters related to perfusion. Paramagnetic MRI contrast agents are used as "tracers" to obtain this information. Paramagnetic contrast agents within the microscopic blood vessels induce a distortion of the static magnetic field homogeneity in the immediate vicinity of the vessels. The presence of such "small scale" magnetic field distortions can be measured with gradient echo MRI techniques because these techniques produce signal level that is dependent on the microscopic field distortion.

The fundamental procedure used in contrast bolus passage perfusion MRI is to collect a time series of gradient echo MRI images during the passage of contrast agent. The contrast agent is injected intravenously as a bolus during the time series acquisition. Echo planar gradient echo imaging is frequently used because there is a need for the time resolution to be on the order of a few seconds per image volume acquisition. The image intensity measured in the time series decreases as the bolus passes and then recovers to almost its original level. The temporal dynamics of the intensity changes can be used to estimate a variety of physiological parameters related to microvascular perfusion. Included are the time required for the contrast agent bolus to pass from a major feeding vessel into the microscopic vasculature of a particular brain area. Also included are quantitative estimates of the amount of blood that is flowing through a particular tissue volume per unit time (the Cerebral Blood Flow) and the steady state amount of blood that is present within a particular tissue volume (the Cerebral Blood Volume).

4. Diffusion and perfusion MRI in stroke

The development of diffusion and perfusion MRI during the past decade has lead to remarkable advances in the imaging of acute ischemic stroke (the type of stroke caused by interruption of the blood supply to part of the brain.)[9]

DWI and ADC imaging provide easily recognized high contrast indications of which specific brain tissues have been affected by stroke. Figure 1 illustrates that brain areas affected by stroke show hyperintense signal on DWI and hypointense ADC. No other MRI procedure provides stroke signatures that are as readily recognizable. The hyperintensity in the ischemic area on DWI reflects that molecular diffusion within the affected territory has slowed. It is still not proven exactly why this happens, although many studies in animal models of ischemic stroke and in human stroke patients have shown that high intensity DWI signal is strongly correlated with the presence of ischemia. Furthermore, the high intensity DWI signal develops within a few minutes of the time the that the blood flow to a particular tissue volume is completely interrupted, allowing DWI to be definitive as soon as the blood flow interruption occurs. ADC imaging is possibly more descriptive of ischemic stroke than is DWI. Brain tissue affected by ischemic stroke eventually show altered T2 that develops over the first day after the interruption of blood flow. DWI of such "older" strokes may be confusing because the altered diffusion coefficient and the elevated T2 both produce hyperintensity on DWI images (see Figure 1). ADC imaging can be used to eliminate the T2 contribution allowing the sole visualization of the effects of ischemia on the diffusion coefficient.

DWI developed as a stroke imaging tool before pulse sequences that support DTI became widely available. As a result, it became customary to make an abbreviated directional diffusion coefficient measurements. Stroke specific ADC imaging protocols typically perform diffusion sensitization only in the three principal gradient directions (left-right (x), anterior-posterior (y) and superior-inferior (z)) to provide a measure of the diffusion coefficient in these three principal directions (D_{xx}, D_{yy}, D_{zz}). It has been found that to a reasonable approximation

$$D_{av} \approx ADC = (1/3)(D_{xx} + D_{yy} + D_{zz}) \tag{4}$$

and therefore for many studies relevant to stroke eqn (4) specifies the definition of ADC that is used.

While DWI and ADC imaging delineate which tissues have been affected by early stroke, it is believed that perfusion MRI provides an indication of which tissues may eventually become involved as the stroke worsens[10, 11]. Here the primary observation is that in early stroke the amount of brain tissue that shows delayed or reduced perfusion is typically much larger than the region that shows a positive DWI or ADC signature of stroke and that over time the volume of the DWI-positive tissue grows to the size of the initial perfusion abnormal volume. This correlates well with the current concept of how strokes engulf more brain tissue as time passes. The concept is shown in Figure 4.

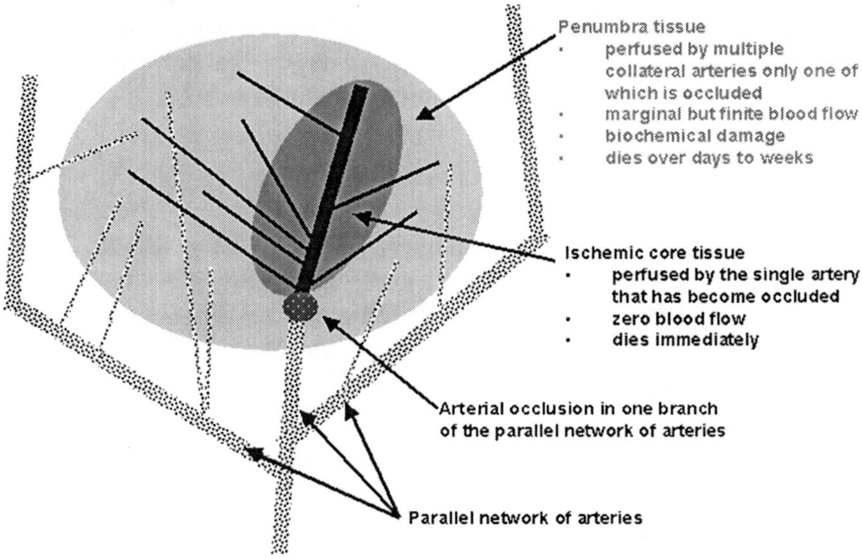

Figure 4. Conceptual description of the core and penumbra of an ischemic stroke.

Here the primary anatomic basis is that much of the brain receives blood supply from more than one primary feeding artery, although there are some brain regions that receive blood from only one supplying vessel. If the flow in this supplying vessel is interrupted, the "core" region is likely to show ischemic signatures on DWI and ADC almost immediately. The "penumbral" regions surrounding the core that receive part of their blood supply from other arteries in addition to the one that has been occluded are expected to show reduced perfusion on perfusion imaging, but little or no changes on DWI or ADC because the blood flow has not been reduced to a sufficient degree. It is widely believed that the core territory dies so quickly that it can not be saved by any available treatment. However the situation is different in the penumbral territory. Here it is hypothesized that the reduced but finite blood flow leads to a slower death and that this downward process could be reversed by prompt reperfusion of the obstructed vessel.

The use of perfusion and diffusion MRI together in stroke is based on the concept that spatially correlated measurements of the perfusion abnormality and the diffusion abnormality can delineate the penumbral tissue that might be saved if reperfusion could be accomplished[9-11].

Figure 5 illustrates the concepts. The area showing reduced perfusion is the entire region that has been affected by blockage in a feeding artery. The smaller region that shows elevated DWI signal (or reduced ADC) is the core region that has already been irreversibly damaged by the interruption of blood flow. The difference between these two volumes is the penumbra that might be saved by reperfusion therapies. This concept is being used in clinical trials of new stroke therapies and in everyday patient care situations to evaluate whether certain types of therapy might be helpful.

5. Blood oxygen level dependent functional MRI

Functional magnetic resonance imaging (fMRI)[12] using Blood Oxygen Level Dependent (BOLD)[13] contrast has developed into a widely accepted procedure for identifying which specific brain regions become active during mental processing. The technique uses the fact the deoxygenated hemoglobin that is present in cerebral capillaries is paramagnetic, while oxygenated hemoglobin is diamagnetic. Therefore deoxyhemoglobin containing capillaries have a distorted magnetic field surrounding them. In this sense, the deoxyhemogobin acts in the same fashion as the paramagnetic contrast agent used in perfusion imaging (see above). The distortions of magnetic field homogeneity can be detected by use of gradient echo MRI procedures (as in bolus contrast passage perfusion MRI).

The physiological basis of the BOLD fMRI technique is as follows. Mental processing is associated with increased electrical activity in the portion of the brain that is responsible for the processing. This increased electrical activity increases the need for the substrates of oxidative metabolism (glucose and oxygen) and this requires there be an increase in the blood flow to the "activated" brain region. For reasons that are not well understood, the flow increase seems to overcompensate for the increased utilization of oxygen. Even though the increased oxygen consumption leads to an increased rate of conversion of hemoglobin to deoxyhemoglobin, the increased blood flow appears to wash out more dexyhemoglobin than is formed by metabolism and the total content of deoxyhemoglobin actually decreases as the electrical activity increases.

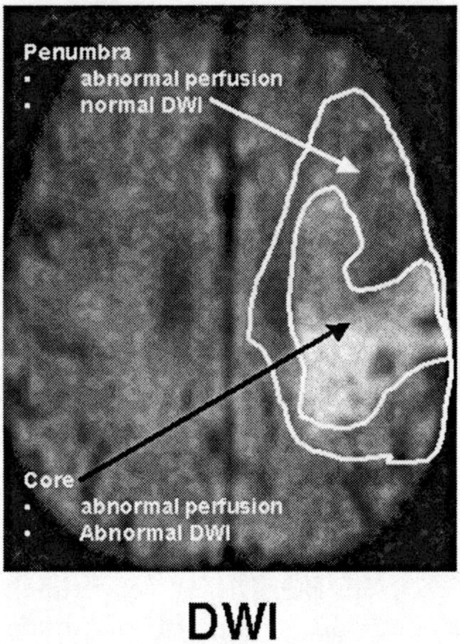

Figure 5. Illustration of how diffusion and perfusion MRI is used to identify the core and penumbral tissues in ischemic stroke.

This activity-induced deoyhemoglobin decrease leads to an increase in gradient echo MRI. It is important to emphasize that the activity-induced signal changes are rather small in comparison to the resting signal level. They tend to be on the order of a few percent and are often of the same order of magnitude as the noise. Therefore BOLD fMRI studies almost always use time averaging, statistical techniques and elaborate post-acquisition signal processing software to identify where in the brain there have been small signal increases that correlate with mental processing. Figure 3 provides illustrative BOLD fMRI results that show how the technique can be used to identify the spatial relationships between a brain tumor and brain areas used in speech processing.

The general procedure used in BOLD fMRI studies is to collect a time series of gradient echo EPI brain images. Typical protocols use multislice EPI to collect a three dimensional image every 2 to 6 sec. At some point during the time series image acquisition, the subject performs some form of mental processing. This could include any kind of mental work such as speaking, moving specific muscles, thinking and so forth. Post-acquisition processing software is then used to identify where within the brain there have been subtle but statistically significant increases in gradient echo MRI signal intensity that temporally correlates with the mental work and this identifies the specific part of the brain that performs the mental work.

There are several important aspects of BOLD fMRI that must be borne in mind. First, the technique does not specifically measure changes in neural electrical activity. Instead it measures a physiological response (the combined alterations of blood flow and oxygen consumption) that accompany neural activity changes. This means that the "BOLD" fMRI signal can not necessarily be used to measure the level of increased neural activity. Although it is tempting to assert a relationship between the amount of gradient echo MRI signal change and the level of electrical activity change, there is not necessarily any direct relationship. Furthermore, there is not a direct temporal relationship between the electrical activity and the blood flow changes. In fact many studies have shown that the gradient echo MRI signal change is time delayed by several seconds relative to the initiation of the mental activity.

Applications of BOLD fMRI include basic neuroscience and practical clinical uses. Basic neuroscience applications usually involve study of how the brain is organized. Here BOLD fMRI experiments are usually focused on learning which individual parts of the brain interact during the performance of particular types of complex mental activities. Recent studies in which BOLD fMRI has been explored as a possible method for detecting deception are of particular relevance to modern day security concerns[14, 15].

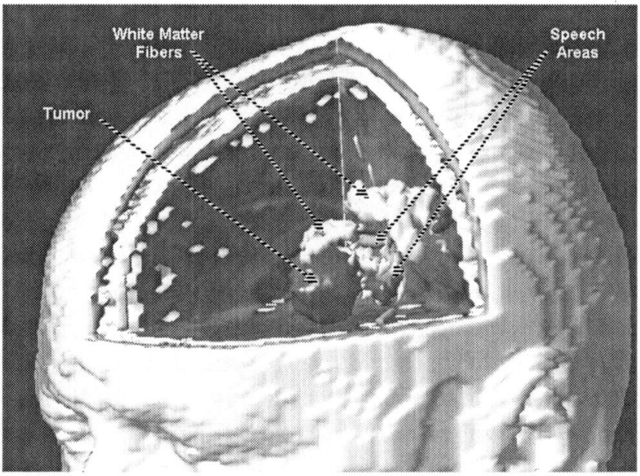

Figure 6. Three dimensional representation of the spatial relationships between brain tumor, speech processing areas (measured with BOLD fMRI) and their associated white matter (measured with DTI).

In the realm of practical clinical uses, BOLD fMRI is being used as a neurosurgical planning tool for brain tumor (Figure 6). Here a BOLD fMRI study is performed before the surgery to identify functioning brain regions situated near the tumor. Knowledge of the location of key functional tissue prior to surgery can lead to more complete resections and, more importantly, less loss of function associated with the procedure itself.

6. SUMMARY

The advanced applications discussed in this chapter illustrate that MRI can produce images that depict aspects of brain physiology or function and that these techniques are valuable to fundamental and clinical neuroscience.

References

1. R.B. Buxton. Introduction to Functional Magnetic Resonance Imaging: Principles and Techniques. Cambridge University Press (2002).
2. D. Le Bihan. Looking into the functional architecture of the brain with diffusion MRI. Nat Rev Neurosci **4**:469-480 (2003).
3. P.J. Basser and C. Pierpaoli. Microstructural and physiological features of tissues elucidated by quantitative-diffusion-tensor MRI. J Magn Reson B. **111**(3):209-19 (1996).
4. R. Bammer, B. Acar, and M.E. Moseley. In vivo MR tractography using diffusion imaging. Eur J Radiol. **45**(3):223-34 (2003).
5. S. Pajevic and C. Pierpaoli. Color schemes to represent the orientation of anisotropic tissues from diffusion tensor data: application to white matter fiber tract mapping in the human brain. Magn Reson Med. **43**(6):921 (2000).
6. P.J. Basser, S. Pajevic, C. Pierpaoli, J. Duda, and A. Aldroubi. In vivo fiber tractography using DT-MRI data. Magn Reson Med. **44**(4):625-32 (2000).
7. Z. Nagy, H. Westerberg and T. Klingberg. Maturation of white matter is associated with the development of cognitive functions during childhood. J Cogn Neurosci. **16**(7):1227-33 (2004).
8. F. Calamante, D.L. Thomas, G.S. Pell, J. Wiersma and R. Turner. Measuring cerebral blood flow using magnetic resonance imaging techniques. J Cereb Blood Flow Metab. **19**(7):701-35 (1999).
9. D. Kim and D.S. Liebeskind. Neuroimaging advances and the transformation of acute stroke care. Semin Neurol. **25**(4):345-61 (2005).
10. C.S. Kidwell, J.R. Alger and J.L. Saver. Evolving paradigms in neuroimaging of the ischemic penumbra. Stroke. **35**(11 Suppl 1):2662-5 (2004).
11. C.S. Kidwell, J.R. Alger and J.L. Saver. Beyond mismatch: evolving paradigms in imaging the ischemic penumbra with multimodal magnetic resonance imaging. Stroke. **34**(11):2729-35 (2003).
12. M.S. Cohen and S.Y. Bookheimer. Localization of brain function using magnetic resonance imaging.Trends Neurosci. **17**(7):268-77 (1994).

13. S. Ogawa, T.M. Lee, A.R. Kay, and D.W. Tank. Brain magnetic resonance imaging with contrast dependent on blood oxygenation. Proc Natl Acad Sci USA. **87**(24):9868-72 (1990).
14. C. Davatzikos, K. Ruparel, Y. Fan, D.G. Shen, M. Acharyya, J.W. Loughead, R.C. Gur, and D.D. Langleben. Classifying spatial patterns of brain activity with machine learning methods: application to lie detection. Neuroimage. **28**(3):663-8 (2005).
15. F.A. Kozel, K.A. Johnson, Q. Mu, E.L. Grenesko, S.J. Laken, and M.S. George. Detecting deception using functional magnetic resonance imaging. Biol Psychiatry. **58**(8):605-13 (2005).

APPLICATIONS OF CLINICAL MAGNETIC RESONANCE SPECTROSCOPY

JEFFRY R. ALGER*
Department of Neurology, Department of Radiological Sciences, Ahmanson-Lovelace Brain Mapping Center, David Geffen School of Medicine at UCLA, University of California, Los Angeles, California, USA

Abstract. Magnetic resonance spectroscopy (MRS) detects electromagnetic signals produced by the atomic nuclei within molecules that are present in living tissues. It can be used to obtain *in situ* concentration measures for certain chemicals in living systems. This chapter will introduce the physics and technology of MRS signal detection. It will also introduce a few basic biochemical concepts that are relevant to MRS.

Keywords: Magnetic resonance; spectroscopy; electromagnetic signals; living tissues; biochemical concepts; concentration of chemicals.

1. Introduction

In magnetic resonance spectroscopy (MRS), radiofrequency magnetic resonance signals are detected from specific nuclei located within specific biochemical species. The MRS detection process is non-invasive and can be applied to living tissues without damaging them. Clinical applications of MRS involve detecting MRS signals from specific molecules that are present within the biological tissues of patients to diagnose or otherwise characterize disease processes that alter the levels of the detectable molecules. Because the MRS-detectable molecules are frequently involved in metabolism, MRS is often described as a "metabolic imaging" technique. Clinical MRS is most frequently used to evaluate brain tissues, although applications to other parts of the human body are growing.

Two broad categories of MRS signal detection procedures are available:

1) Single voxel procedures collect MRS signals from a defined tissue volume and

* Jeffry R. Alger, Department of Radiological Sciences, Ahmanson-Lovelace Brain Mapping Center, David Geffen School of Medicine at UCLA, University of California, Los Angeles, California, USA; email:jralger@ucla.edu.

Y. Lemoigne et al. (eds.), Physics for Medical Imaging Applications, 83–88.

2) Magnetic Resonance Spectroscopic Imaging procedures collect MRS signals from a large number of tissue volume elements simultaneously permitting the creation of images that display the topographic variation in signal intensity of various tissue metabolites.

Each technique has characteristic advantages and disadvantages that are important in specific clinical circumstances. The lecture will develop MRS background physics[1] and illustrate the clinical use of MRS for the evaluation of cancer, epilepsy, inborn errors of metabolism and brain trauma[2].

2. MRS physics and technology

MRS is viewed by the radiology and medical imaging communities as member of a large family of "MRI techniques". MRS detects signals produced by chemical compounds (other than water) with the goal generally being to study the biochemistry of living systems, whereas MRI uses signals produced by the water in living tissue to make images of anatomy. Otherwise MRS and MRI share the same signal detection technology and therefore MRS studies can be performed using the same equipment that is used for MRI. Typical human MRS protocols use a magnetic field of 1.5 Tesla or higher. MRS sensitivity generally improves as the static magnetic field strength is increased, so it is desirable to use the largest magnetic field available.

The primary result of an MRS study is an "MR spectrum", which is a two-dimensional plot of frequency on the horizontal axis and intensity of the resonance signal on the vertical axis.

Many different MRS pulse sequences are available to meet specific experimental needs. These needs typically include defining the anatomic location from which MRS signals are detected (i.e. spatial localization) and simplifying (i.e. editing) the spectrum so that only a few key signals can be detected with clarity and therefore measured with the greatest possible degree of accuracy. Two complementary methodologies are available for attaining volume localization. In localized single volume MRS, a conventional MR image is first obtained and is then used to identify a location of interest, which is typically defined as a rectilinear "voxel". The MRS signal is then acquired from only this location using a pulse sequence that is designed to optimally detect signal produced by the defined region and to suppress signal that arises from other regions. The alternative method is magnetic resonance spectroscopic imaging (MRSI). In MRSI studies, MRS signals are simultaneously acquired from a grid containing a large number of rectilinear voxels that include the tissue of interest.

Sensitivity is an important limiting feature of MRS. MRS signals generated by living tissues are generally similar in magnitude to the noise that is generated by random movements of electric charges within the tissues being studied. Indeed, only the most concentrated materials present in living tissue can be reliably detected above the noise. MRS usually involves time consuming signal averaging to achieve a sufficiently high signal-to-noise ratio (SNR). The need for signal averaging is one of the reasons that MRS studies require more time than many other types of MRI studies. The noise arises from the entire body part that is within the radiofrequency detection coil, while the signal arises from only within the defined volume of interest. Therefore the SNR can also be enhanced by making measurements from the largest feasible tissue volume.

Only certain atomic nuclei (isotopes) of biological significance (e.g. ^{1}H, ^{31}P, ^{13}C, ^{7}Li and ^{19}F) have spin angular momentum and are therefore capable of producing MRS signals. The ability of MRS to detect unique signals from the same type of nucleus (i.e. ^{1}H) on different molecules results from the so called “chemical shift”. Magnetic resonance frequency is directly proportional to the static magnetic field strength at the nucleus. The electrons which surround the atomic nuclei circulate in ways that tend to alter the magnetic field at the nucleus to a small but significant extent. This causes small but detectable alterations in the MRS signal frequency, which are dependent on the chemical structure surrounding an atomic nucleus, allowing the identification of specific resonance signals from individual nuclei within individual molecules. It is a common practice to specify the chemical shift in “parts per million (ppm)” relative to some agreed upon standard signal.

The different nuclei of biological significance have different attributes and practical limitations with respect to MRS detection. The proton (^{1}H) produces the strongest most easily detected MRS signal and is therefore most frequently used for MRS. Moreover, ^{1}H MRS is most convenient in that it can be performed using the same hardware as is used for conventional MRI. All of the above named nuclei have been used in MRS studies of living systems. However for reasons of practicality human MRS examinations are almost always performed with ^{1}H only.

MRS has a strict requirement for a spatially homogeneous static magnetic field. The applied static magnetic field intensity must vary by less than approximately 100 parts per billion over the intended sampling volume. Anatomic features of the anatomy surrounding the tissue of interest can distort the shape and intensity of the applied magnetic field to an unacceptable extent introducing problems with detection of MRS signals from certain body regions. Similarly the presence of certain magnetic materials such as hemoglobin degradation products or paramagnetic contrast agents can also be a problem.

Special problems related to signal overlap apply to ^{1}H-MRS. The tissue water signal is very large relative to the signals produced by tissue chemicals and it overlaps with some chemical signals rendering them hard to identify and measure. Therefore the pulse sequences used in ^{1}H-MRS usually have a water signal suppression component that is designed to selectively saturate the water signal. The fatty acyl components of certain lipid molecules, such as the triglycerides in adipocytes, also produce strong signals that overlap with desired MRS signals and these can be problematic as well. Therefore, ^{1}H-MRS signals can only be detected from tissues that do not contain a substantial amount of adipocytes and triglyceride. Localization procedures that suppress signals from adipose tissue are often the means that are used to suppress lipids signals.

3. Basic biochemistry concepts and applications

MRS can not detect signals generated by molecules having molecular weights greater than a few thousand daltons or by smaller molecules that are bound to macromolecular arrays (e.g. proteins, membranes or nucleic acid polymers). Only the small mobile molecules present within tissue can be detected with MRS. Typically, such tissue-associated small molecules are "metabolites" involved in intermediary metabolism. With appropriate calibration, MRS measurements can be used to obtain the tissue metabolite concentration.

Figure 1 illustrates the 1H-MRS signals that are typically acquired from brain tissue. Included are:

1) N-acetylaspartate, a metabolite found only in neurons,

2) creatines (phosphocreatine and creatine), which are involved in energy metabolism,

3) cholines (phospocholine, choline and gylcerophosphocholine), which are involved in phospholipid metabolism,

4) lactate, which is a product of anaerobic glucose metabolism and

5) certain lipid molecules.

Also shown in Figure 1 are signals (not labeled) produced by certain sugars and amino acids that are used in some clinical and research studies.

Practical applications of MRS usually involve the detection and evaluation of how disease alters metabolite levels, although some investigations have also studied metabolic control and metabolite processing in normal tissue.

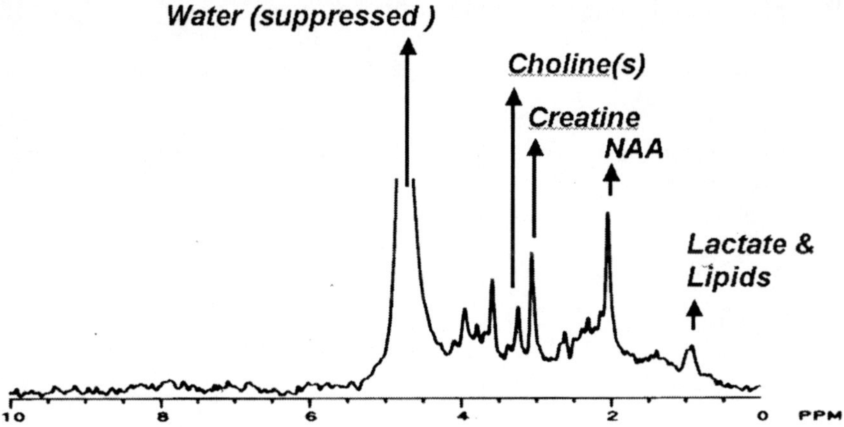

Figure 1. MR spectrum of living brain.

MRS studies can be performed in animal and human subjects, including human patients. In order for MRS to be useful for the clinical evaluation of a particular disease, the disease must alter the concentration of some MRS-detectable tissue metabolite. Frequently a clear understanding of how a particular disease alters the biochemistry of MRS-dectectable metabolites is not available. Accordingly, an empirical approach has been employed. Much of the altered biochemistry that is exploited in clinical MRS studies was in fact discovered by doing exploratory *in vivo* MRS studies. The majority clinical MRS studies have evaluated brain metabolites and the effects various diseases have on these metabolites. The evaluation of cancer has been a particular area of MRS emphasis. Thus far most of this work has been directed at brain cancer. However, there is now growing interest in using MRS to evaluate cancers in the breast and prostate as well.

4. Conclusion

MRS provides a means of measuring the concentrations of the more concentrated tissue metabolites using "MRI technology". For practical reasons, ^{1}H is the nucleus that is most often used for MRS. The inherent sensitivity of MRS limits the number of metabolites that can be measured with MRS and also is responsible for the poor volume resolution of MRS relative to MRI. Detection and evaluation of disease using MRS has been largely based on the study of particular diseases with MRS.

References and further reading

1. Alger JR. MRS of the Brain. in Encyclopedia of Neuroscience (eds. Adelman G, Smith BA). Elsevier Science (2004).
2. Barker PB. Fundamentals of MR spectroscopy. In *Clinical MR Neuroimaging. Diffusion, Perfusion and Spectroscopy.* (eds. Gillard JH, Waldman AD, Barker PB). Cambridge University Press (2004).

CLINICAL MRI

ALESSANDRO ALIMENTI*
Hôpitaux universitaires de Genève, rue Micheli-du-Crest 24, CH-1211 Genève 14, Switzerland

Abstract. The aim of this report is to show the use of MRI in clinical routine, to explain what we can see and what we cannot see with MRI.

Keywords: brain; cardiac; abdominal; uro-genital, musculoskeletal MRI.

1. Introduction

The main goal of this report is to provide some classic examples in daily hospital practice of the use of MRI, to show the advantages, and also the limitations, that this imaging technique adds as a diagnostic tool.

Computed Tomography (CT) Scanner is a powerful imagery technique. It delivers images in very fast time, with high resolution, and allows also for fast 3D reconstruction, However, CT has also important limitations. It's an irradiative technique, limiting the use (e.g. pregnant women). Also, it uses iodinated contrast media that can result in harmful manifestations.

At the contrary, MRI is a safer technique. It's based on magnetic field and radio-frequencies (RF) and not X-Rays emission, and uses lesser toxic media like Gadolinium-based products. It's undoubtedly slower than CT to complete the examination. However, its great advantage compared to CT is the contrast. Contrast in MRI is superior to CT, enhancing better the tissues and the lesions. Another advantage is the possibility to do functional studies of visual, auditory, motor, sensory, speech and other cortical areas of the brain. It provides also the possibility to do spectroscopy, to study the metabolic components of lesions. However, MRI, like CT, has contraindications too. E.g., patients with pace-maker, or with metallic implants not MRI-compatibles, are formally excluded to this imagery technique. Bones and calcifications in general, due to their poor water

* Alessandro Alimenti, Hôpitaux Universitaires de Genève, rue Micheli-du-Crest 24, CH-1211 Genève 14, Switzerland; e-mail : Alessandro.Alimenti@dim.hcuge.ch.

Y. Lemoigne et al. (eds.), Physics for Medical Imaging Applications, 89–97.

contents, are difficult to visualize, and air/bone interfaces give artifacts limiting the usefulness in some areas of the body.

In the next paragraphs, we illustrate some examples of clinical use, subdivided in brain, cardiac, abdominal, uro-genital and musculoskeletal MRI.

2. Brain imagery

2.1. PULMONARY CANCER WITH BRAIN METASTASES

In this case, a man with pulmonary cancer had a brain CT as a complementary examination in search of metastases. At the CT, one metastasis in the brain was discovered. But, as shown in Fig. 1, several others metastases were in fact present, as shown in a subsequent brain MRI, who did reveal many tumoral lesions, invisible in CT. This is an illustration of the greater contrast of MRI compared to CT.

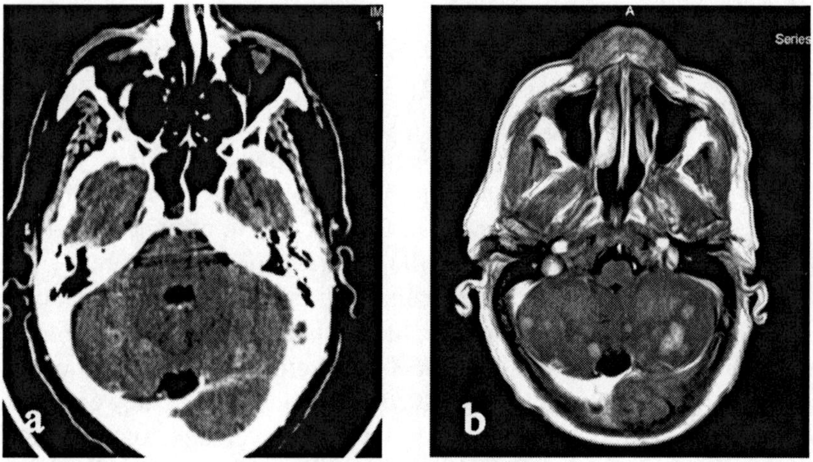

Figure 1. (a) CT scan shows no lesion in the cerebellum; (b) same slice in MRI: many metastases appear (hyperintense nodules).

2.2. STROKE

Another useful use of brain MRI is in the detection of stroke. Usually, with CT, more of 24 hours are necessary before to see an appreciable change in brain tissues. Using diffusion weighted imagery (DWI), it's possible to detect more quickly the stroke, that appears as a hyperintense signal (stroke provokes cell suffering and swelling, restricting the extra-cellular motion of water, and thus provoking a raise in signal in DWI).

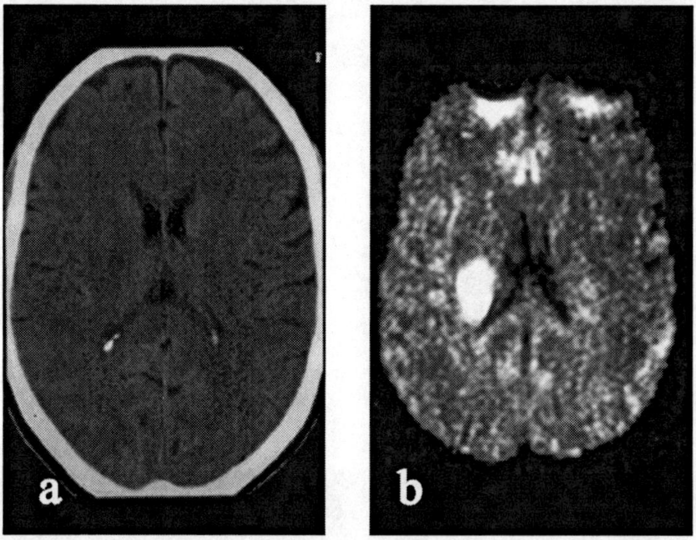

Figure 2. (a) Brain CT scan showing no lesion; (b) brain DWI showing the area of stroke in hyperintense signal.

3. Cardiac MRI

3.1. AORTIC INSUFFICIENCY

In this case, a man suffers of aneurysm of the ascending aorta (aneurysm is a dilatation of a vessel). This dilatation prevents the aortic valve not to close correctly at the diastole (cardiac filling of blood) and blood expulsed at the systole (cardiac contraction) comes back into the heart. Using cine-MRI sequences and phase contrast sequences, it's possible to follow the blood flow (Fig. 3 and 4). Cine-MRI sequence shows the contraction and flow movement like a movie. Phase contrast sequence correlates the speed and direction of the blood with the intensity of signal: the brighter is the signal, the more quickly the flow is approaching; the darker is the signal, the more quickly the flow is going away.

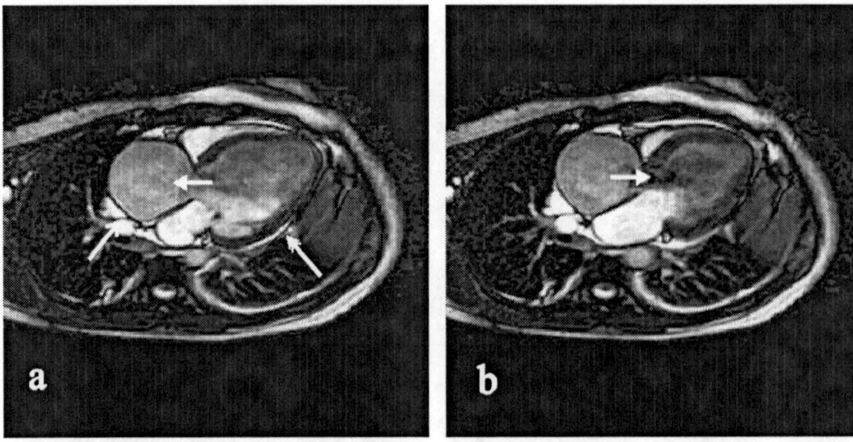

Figure 3. (a) Cine-MRI sequence: at the systole (heart contraction) blood it's expulsed from the cardiac cavity (closed arrow). Open arrow points to the aortic aneurysm, double arrow points to the heart. (b) At the diastole (heart filling) part of the blood.

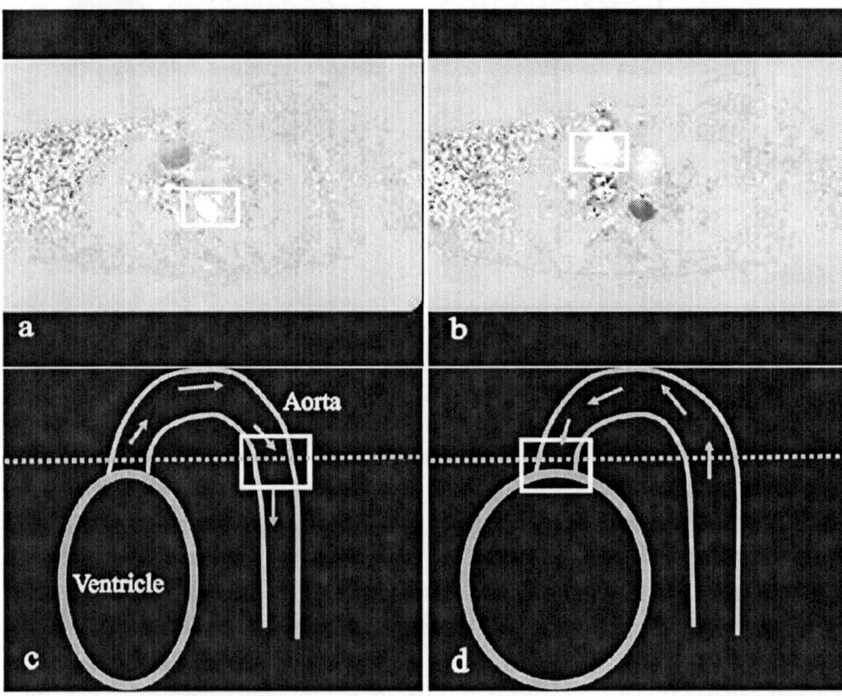

Figure 4. Phase contrast sequence. White square shows the position of approaching flow (hyperintensity) related to the aorta, dotted line in (c, d) shows the level of slice in (a) and (b), grey arrows show blood direction in aorta: (a) and (c): during cardiac contraction blood exits the heart (ventricle). Hyperintense signal is in the descending aorta (a); (b) and (d): during cardiac filling, some of the blood comes back to the heart. Hyperintense signal is in ascending aorta (b).

4. Abdominal MRI

4.1. HEPATIC METASTASIS

In this case, a woman with known breast cancer starts to develop gastric pains. An abdominal CT-Scan shows a suspect lesion in liver. A complemental MRI is made using Endorem, a superparamagnetic constrast medium internalized by macrophages (immunitary cells) of liver. Endorem shortens T2, making appear the normal hepatic tissue in hypointense signal. If there are metastases, because they don't bear macrophages, like normal hepatic tissue, they appear in hyperintense signal (bright spots).

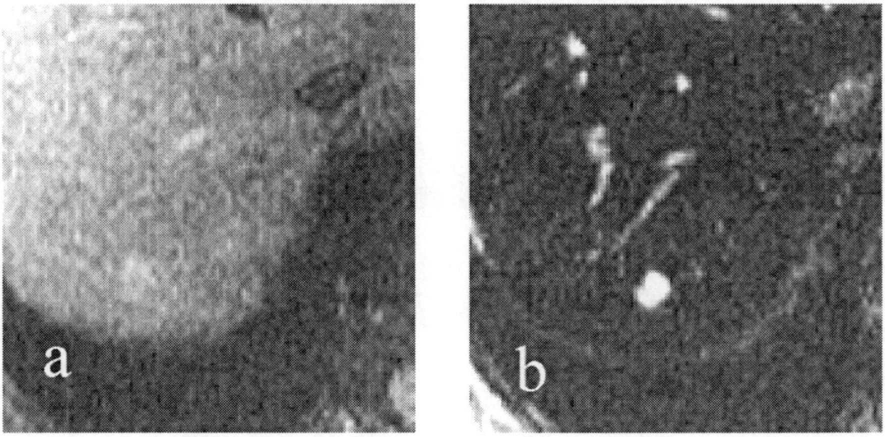

Figure 5. (a) Liver in T2W sequence without Endorem; (b) same liver in T2W after Endorem. The bright spot is a metastasis.

4.2. UNKNOWN HEPATIC LESION

A patient having an abdominal CT-Scan in search of cirrhosis was discovered with a hepatic lesion. To try to characterize the lesion, a lot of MRI sequences were performed. However, in this case, due to the heterogenicity of the lesion even after Gadolinium and Endorem, only biopsy could provide the answer (hepatocarcinoma). This case illustrates well that also even with MRI it's not always possible to characterize lesions.

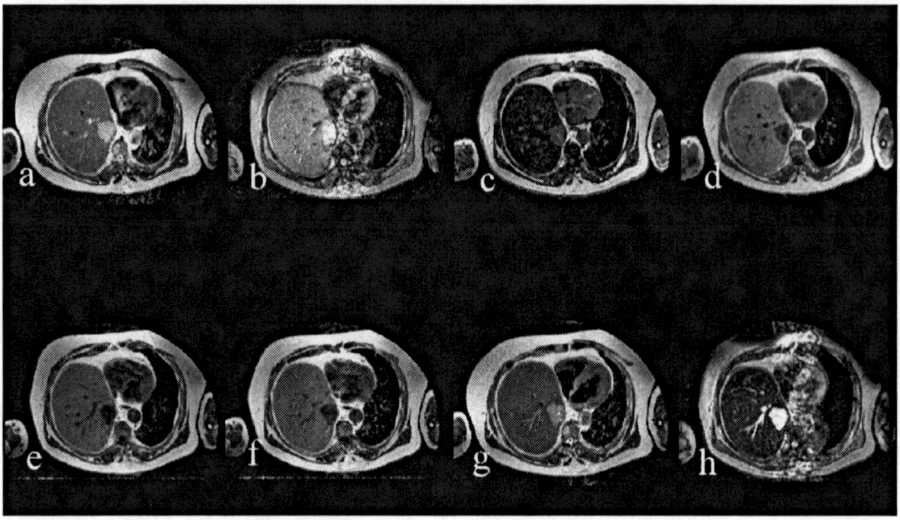

Figure 6. (a) T2W SSh; (b) T2W FFE ; (c) in phase ; (d) out of phase ; (e) T1W; (f) T1W + Gd; (g) T2W SSh + Endorem; (h) T2W FFE + Endorem.

5. Uro-genital MRI

5.1. KIDNEY DONOR

We show here another useful use of MRI. In this case, it is an assessment of good kidneys conditions in an organ donor. One sequence used is angio-MRI. This is because some times anatomical variants of renal arteries exist. In this case, the left kidney has two arteries: the principal renal artery and a polar inferior artery. This is important for the surgeons, because if this kidney is transplanted and only the main artery is connected to the aorta, the inferior part of the organ will not be supplied with blood.

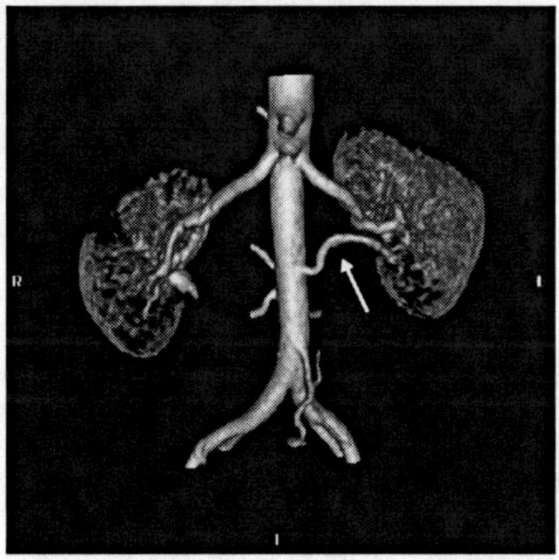

Figure 7. 3D reconstruction of angio-MRI showing the supplementary artery of the left kidney (arrow).

6. Musculo-skelettal MRI

6.1. KNEE CONTUSION

Even with the difficulty of visualize bones and in general calcifications, MRI plays also an important role in the musculo-skelettal system. In this case, after a traumatism of the knee, using T2 FatSat sequences, it was possible to show the bone oedema at the level of the contusion.

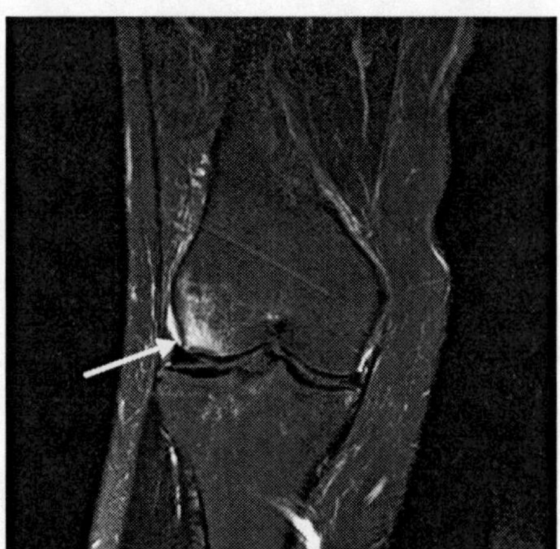

Figure 8. T2W FatSat showing the bone oedema (arrow pointing to the ill-defined hyperintensity).

6.2. OSTEOSARCOMA

In this case, we show the use of MRI in case of a young boy with an osteosarcoma (a malignant bone tumor) of the tibia. T1W and T1W+ Gadolinium (contrast medium) are useful sequences to delineate the location and extension of the tumor. Gadolinium permits to separate also necrotic from tissular parts of tumor.

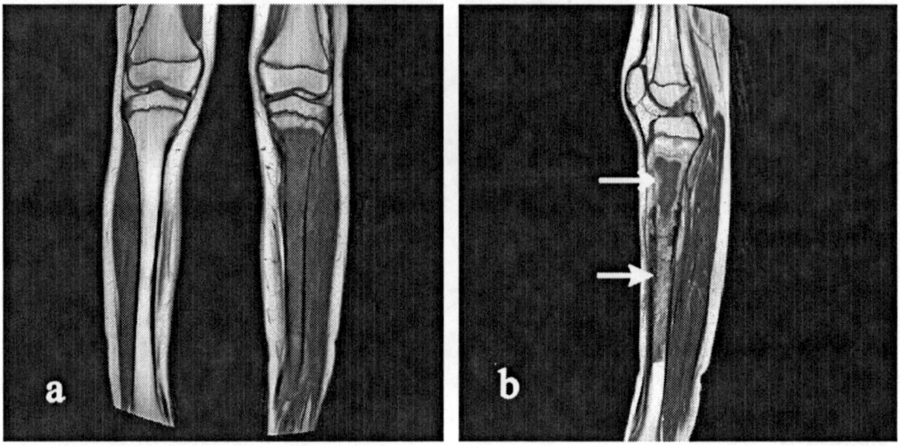

Figure 9. (a) T1W coronal view: the lesion is the dark zone at left leg. (b) T1W+Gd sagittal view: closed arrow points to the lesion enhancing after Gd injection, open arrow points at a zone of necrosis. Note also the extension outside the bone.

7. Conclusion

In this brief review of the clinical use of MRI, we have seen some typical applications in daily practice. As explained, better contrast and safety of this technique compared to CT-Scan are its main advantages. But also MRI is a fast growing technique, and in future new sequences will add new possibilities, as for example:

- Spectroscopy in the study of metabolic components of a lesion
- Diffusion Tensor Imagery (DTI) for the study of the white matter of the brain
- Perfusion imagery to assess the vascularization of a region in brain, heart…
- Cardiac tagging to assess cardiac contraction after an infarct recovery.
- Radiofrequency treatment of liver metastases in interventional radiology

Acknowledgements

I wish to thank all the persons who kindly helped me with the clinical cases: Prof. F. Terrier, Prof. J. Garcia, Prof. C. Becker, Prof. D. Didier, Prof. J.-P. Vallée, Dr. J. Delavelle, Dr. F. Lazeyras, Dr. K.-O. Lövblad, Dr. A. Keller, Dr. J.-P. Imbert, Dr. Duy Nguyen.

Part III

Ultrasound Imaging

BASIC PRINCIPLES OF ULTRASOUND

TERESA M. ROBINSON*
Bristol Oncology Centre, Horfield road,
Bristol BS2 8ED, UK

Abstract. Ultrasound has been used in medicine for at least 50 years. Its current importance can be judged by the fact that, of all the various kinds of diagnostic images produced in the world, 1 in 4 is an ultrasound scan. Ultrasound energy is exactly like sound energy, it is a variation in the pressure within a medium. The only difference is that the rate of variation of pressure, the frequency of the wave, is too rapid for humans to hear. Medical ultrasound lies within a frequency range of 30 kHz to 500 MHz. Generally, the lower frequencies (30 kHz to 3 MHz) are for therapeutic purposes, the higher ones (2 to 40 MHz) are for diagnosis (imaging and Doppler), the very highest (50 to 500 MHz) are for microscopic images. For diagnostic purposes two main techniques are employed; the pulse-echo method is used to create images of tissue distribution; the Doppler effect is used to assess tissue movement and blood flow. The pulse-echo method relies on the partial reflection of a short burst of ultrasound (the pulse) at boundaries between tissues of different acoustic properties. These partial reflections (the echoes) are registered by the equipment on a display in their appropriate anatomical location. Sweeping the direction of the pulse beam through the tissues builds up the scan. The Doppler technique depends on the fact that ultrasound echoes reflected from a moving target have their frequency altered. Targets moving away from the transducer produce echoes with a reduced centre frequency those moving toward it have an increased centre frequency. The change in frequency is the "Doppler shift" and is proportional to velocity. This paper will describe the generation, transmission and interactions of ultrasound which explain its use in medical applications.

Keywords: medical ultrasound; generation & transmission of ultrasound; interactions of ultrasound imaging; Doppler; pulse-echo method; Doppler shift.

---

* Teresa M Robinson Bristol Oncology Centre, Horfield road, Bristol BS2 8ED, UK; e-mail: Teresa.Robinson@ubht.swest.nhs.uk.

Y. Lemoigne et al. (eds.), Physics for Medical Imaging Applications, 101–110.

1. Introduction

Ultrasound imaging is an extremely familiar and widespread diagnostic technique. The basic underlying principle requires sending a pulse of ultrasound into the body and waiting for an echo to return. These returning echoes are then processed to produce an image of internal structures (Fig. 1).

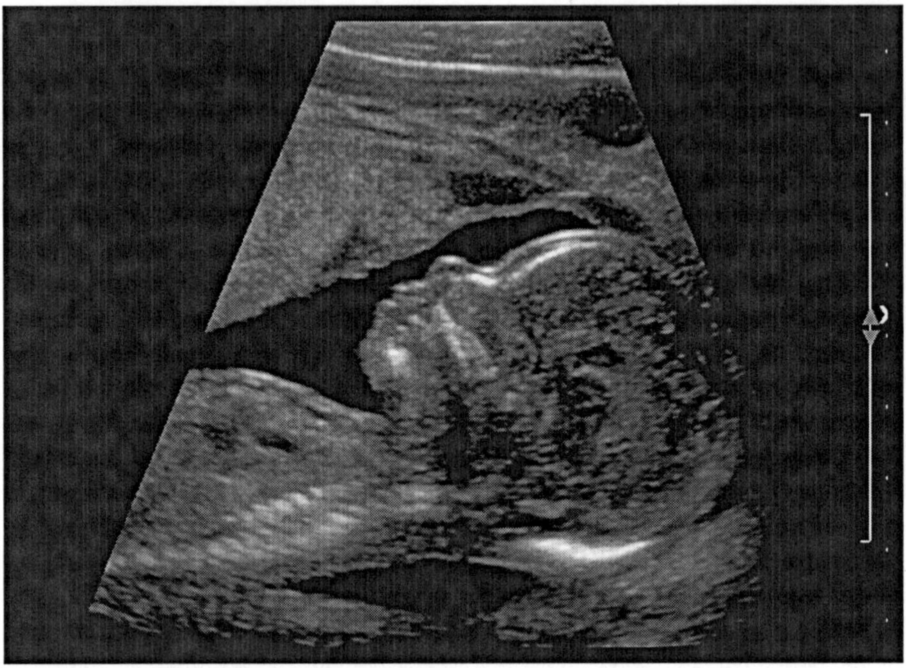

Figure 1. B-Mode image.

2. Sound waves and echo ranging

Sound waves can be described as a mechanical wave created by a vibrating object and propagated by a medium. As the sound wave energy travels through body tissue it is attenuated, scattered and reflected, producing echoes from a range of interfaces. Diagnostic ultrasound waves are longitudinal waves with the motion of particles in a direction parallel to the direction of energy transport. A sound wave consists of a repeating pattern of high and low pressure regions. The wavelength is the distance between two pressure peaks (Fig. 2).

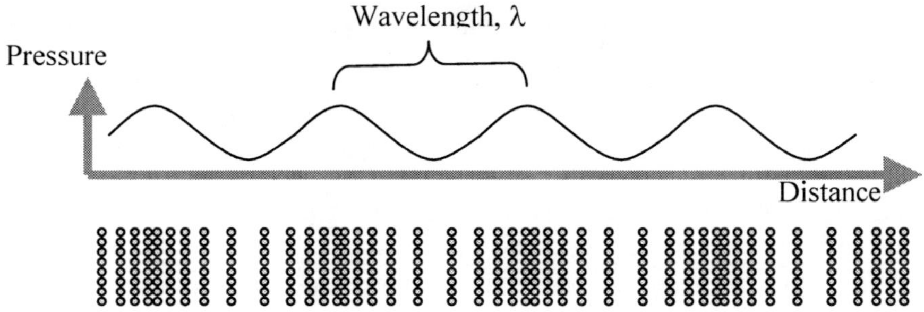

Figure 2. A sound wave consists of pressure fluctuations.

The wavelength λ is the distance over which a property of the wave repeats itself. It is defined by the equation

$$\lambda = \frac{c}{f}$$

where c is the speed of sound and f is the frequency. The speed of sound in soft tissue is taken to be 1540 ms^{-1}. So, for a 5 MHz transducer the wavelength in soft tissue can be calculated to be 0.31 mm.

Ultrasound is defined as a sound wave with a frequency above the human audible range (i.e. >20 kHz). Diagnostic ultrasound applications typically use frequencies in the 1 to 30 MHz frequency range. Whilst using higher frequencies is associated with improved spatial resolution, higher frequencies are attenuated more and that limits the depth of visualisation.

In ultrasound, the vibrating source of the wave is a vibrating piezoelectric material that can both generated and detect ultrasound. The transducer housing the element is designed so that the wave travels in a well-defined beam and direction.

2.1. PULSE-ECHO PRINCIPLE

Ultrasound imaging sends short bursts of sound waves, or pulses, into the body and receives echoes from the tissue interfaces. The time between emitting a pulse and receiving an echo is used to determine the depth of the interface (Fig. 3).

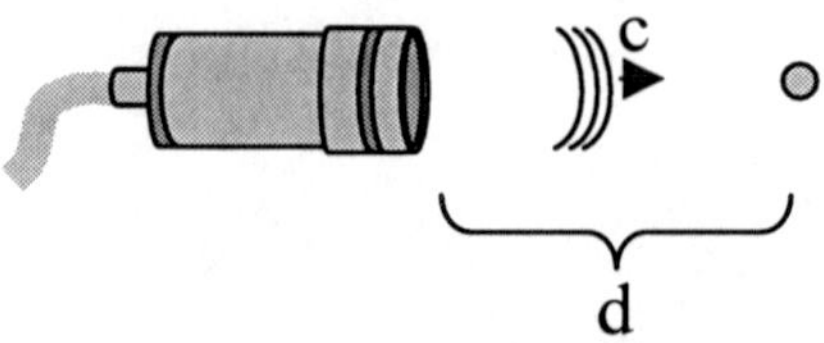

Figure 3. A short pulse of ultrasound can be used to determine the depth of a reflector.

It is the same principle that is used in echo-sounding equipment found on boats. If t is the echo arrival time and c is the speed of sound in the tissue, then d can be calculated from

$$d = \frac{c \cdot t}{2}$$

This is called the range equation. A speed of sound of 1540 m/s is assumed in most scanners when calculating and displaying the reflector depths from echo arrival times.

3. Image generation

To form a B-Mode image a multi-element transducer is placed in contact with the skin and activated to repeatedly emit pulses of sound at a fixed rate, called the pulse repetition frequency. These pulses are directed along narrow beam-shaped paths. As the pulse travel into the tissues of the body, it is reflected and scattered, generating echoes, some of which travel back to the transducer, where they are detected. After sending each pulse the transducer must wait for all the echoes to return before sending out the next pulse. Returning echoes are amplified and processed into a format suitable for display.

The multi-element transducers can be manufactured into different shapes to suit their clinical application (Fig. 4).

In B-Mode scanning elements are activated in turn to produce numerous sound beams that are progressively stepped over the region of interest.

The transmitted field from a single element would spread quickly with depth if it were driven in isolation because the element is so narrow (Fig. 5). However, when a group of elements is excited together a narrower directional beam can be formed. This beam can be focussed by applying time delays to the transmit excitation pulses applied to the individual elements. By exciting the outer elements of a group a little earlier than the neighbouring inner ones the beam shape can be altered and the focus adjusted (Fig. 6).

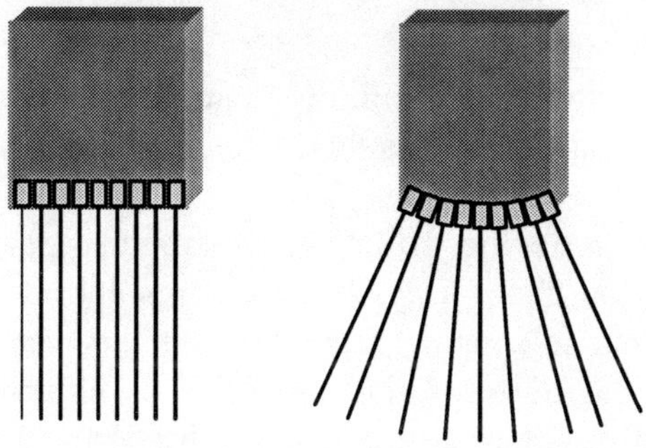

Figure 4. Transducers can be manufactured to suit the clinical requirement.

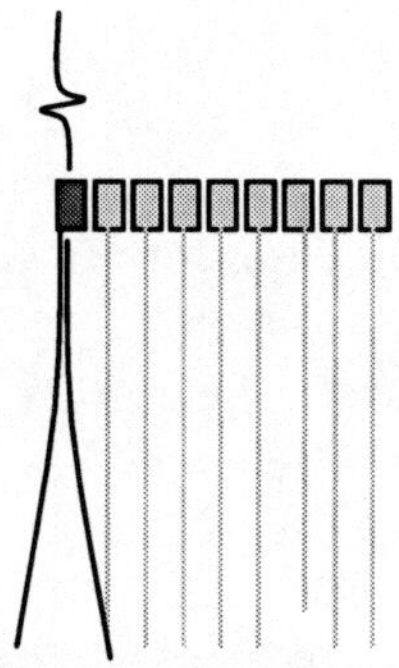

Figure 5. Wide beam produced by a single element.

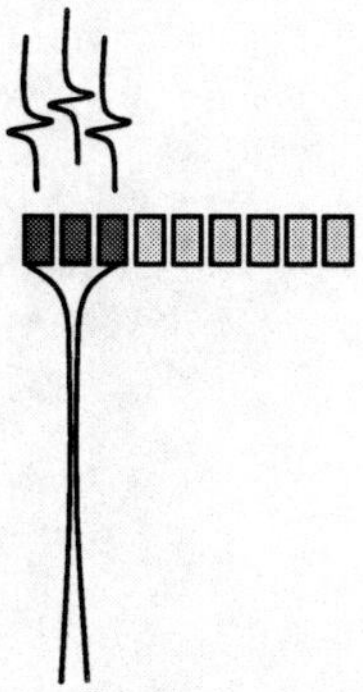

Figure 6. Narrow focused beam produced by a group of elements.

3.1. RESOLUTION

Spatial resolution describes the minimum spacing between two reflectors that can be distinguished on the display. The resolution of the images is different in the 3 dimensions.

The axial resolution is the ability to distinguish reflectors that are closely spaced along the beam axis. The axial resolution is dependent on the pulse length, which is in turn dependent on the wavelength and the number of cycles in the pulse. Axial resolution is improved with higher frequencies because pulses can be made a much shorter due to the shorter wavelength. Figure 7 show the image obtained from am ultrasound QA phantom. The wires marked 'a' and 'b' are 0.25 mm apart vertically and the wires marked 'b' and 'c' are 0.5 mm apart vertically. The echoes from these wires can be seen to overlap vertically and hence they would not be distinguishable as separate wires if it were not for the fact that they are separated horizontally. The wires marked 'c' and 'd' are 1mm apart vertically and the echoes do not overlap vertically. We can therefore say that the resolution of this particular transducer is between 0.5 and 1.0 mm axially.

The lateral resolution refers to the ability to distinguish reflectors spaced perpendicular to the beam. The lateral resolution is dependent on the beam width at the location of the reflectors.

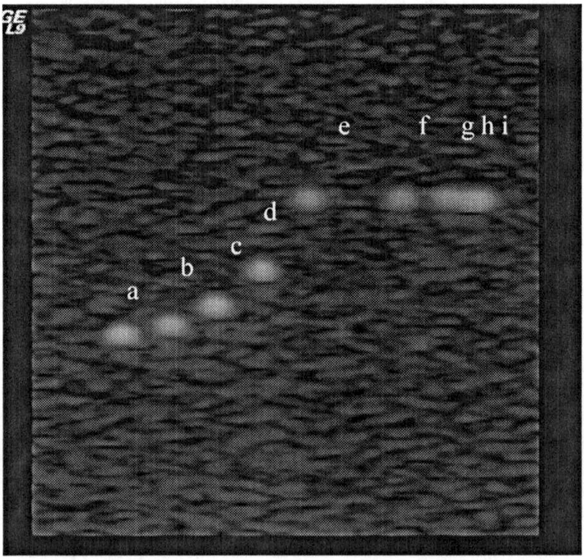

Figure 7. QA Phantom Image.

The wires marked ‘i’ and ‘h’ are 0.25 mm apart horizontally and the wires marked ‘h’ and ‘g’ are 0.5 mm apart horizontally. The echoes from these wires can be seen to overlap horizontally and hence they are not distinguishable as separate. The wires marked ‘g’ and ‘f’ are 1 mm apart horizontally and the echoes do not overlap. We can therefore say that the resolution of this particular transducer is between 0.5 and 1.0 mm laterally.

The slice thickness resolution is the resolution perpendicular to the image plane and dictates the thickness of the section of tissue contributing to the echoes visualised on the image. It depends on the width of the beam perpendicular to the image plane often called the elevation beam width. Many probes still use a lens to provide focussing in this direction. Slice thickness is the worst aspect of the resolution of array transducers. Manufacturers are rapidly developing two dimensional arrays that enable focussing in the slice thickness as well as in the lateral direction.

4. Sound wave properties in tissue

4.1. SPEED OF SOUND

The propagation speed of ultrasound through tissue depends on the properties of the tissue, in particular the stiffness (k) and the density (ρ).

$$c = \sqrt{\frac{k}{\rho}}$$

As a general rule, gases have the slowest propagation speed (e.g. air = 330 ms^{-1}), liquids have an intermediate propagation speed (e.g. water = 1480 ms^{-1}) and firm solids have a high propagation speed e.g. bone = 3400 ms^{-1}).

4.2. REFLECTION

At a boundary between two different types of tissue an ultrasound wave is partially reflected and partially transmitted (Fig 8). The relative proportions of the energy reflected and transmitted depend on the acoustic impedance (z) of the two materials. The acoustic impedance of a material is governed by its density (ρ) and stiffness (k) such that

$$z = \sqrt{\rho \cdot k}$$

The amplitude of the reflected wave is proportional to the difference in the acoustic impedance of the two materials. For perpendicular incidence of the ultrasound beam on a large flat interface the amplitude reflection coefficient (R) is given by

$$R = \frac{z_2 - z_1}{z_2 + z_1}$$

Where z_1 and z_2 are the acoustic impedances of the first and second tissues respectively.

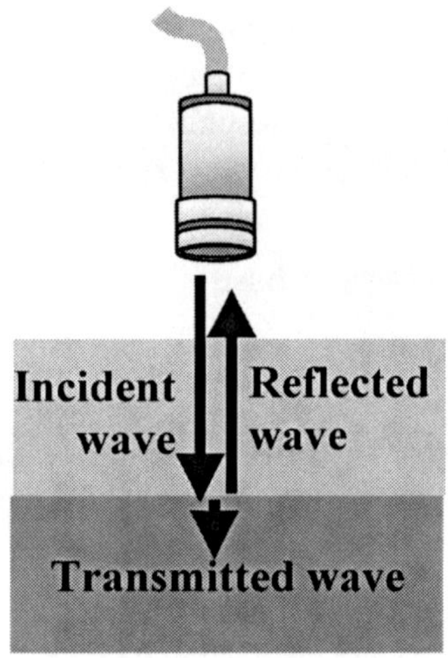

Figure 8. Ultrasound is both reflected and transmitted at tissue boundaries.

The acoustic impedances of various soft tissues are all around 1.3 to 1.7 × 10^6 $kgm^{-2}s^{-1}$. This means the proportion reflected is relatively small. However the acoustic impedance of air is only 430 $kgm^{-2}s^{-1}$ and bone is 6.47 ×10^6 $kgm^{-2}s^{-1}$, which means these soft tissue-air boundaries and soft tissue – bone boundaries reflect a high percentage of the incident ultrasound and hence there is little transmission. Thus imaging beyond these boundaries with ultrasound is usually impossible.

Large smooth interfaces such as the one in the above example (Fig. 8) are called specular reflectors. The direction of the reflected beam is strongly angle dependent. The wave is reflected back to the source only when the incident beam is perpendicular. The amplitude of an echo detected from a specular reflector thus also depends on the orientation of the reflector with respect to the sound beam.

The image of the carotid artery (Fig. 9) shows some areas of strong specular reflection (S) where the wall is perpendicular to the incident beam but weaker specular reflection (W) where the walls are not perpendicular.

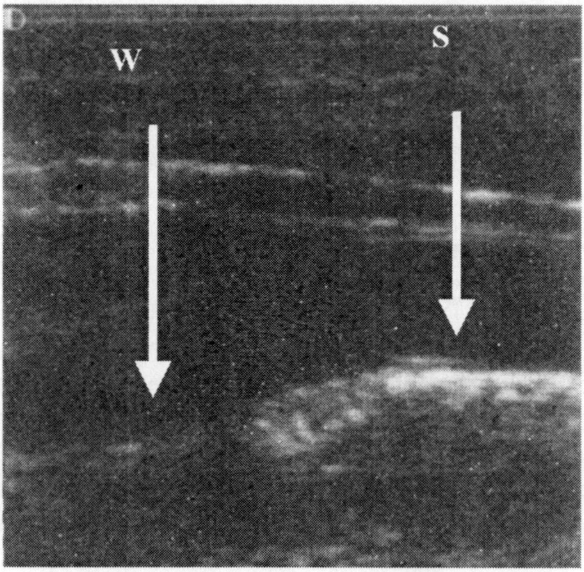

Figure 9. Image of carotid artery.

Some soft tissue interfaces are slightly rough and are better described as diffuse reflectors, with ultrasound reflected over a range of angles.

4.3. SCATTERING

For interfaces whose dimensions are very small (<<λ) reflections are classified as scattering. Much of the background information in the carotid image (Fig. 9) is scatter with no particular interface identifiable. The scattered waves spread in all directions, consequently there is little angle dependence on the strength of the echoes detected. Echoes resulting from scattering within organ parenchyma are clinically important because they provide much of the diagnostic detail seen on the scans.

The total ultrasound power scattered by a very small target is much less than that for a large interface and is determined by the relationship between the size (d) of the target and the wavelength (λ) of the wave. For targets which are much smaller than a wavelength (d<<λ), scattered power (W) is

$$W \propto d^6 f^4$$

This frequency dependence is often referred to as Rayleigh scattering.

4.4. ATTENUATION

The energy of an ultrasound wave is reduced with distance. Energy is lost from the beam by scattering (out of the beam confines) and by absorption (conversion into heat). For each centimetre the ultrasound wave travels the intensity is reduced by the same ratio. This loss of energy can be described by the attenuation coefficient (α) and expressed in decibels. The attenuation coefficient is also dependent on the frequency. The attenuation for soft tissue is usually giving a 0.3-0.6 dB/cm/MHz. Hence for imaging deep organs the returning echoes may be too week to detect, therefore a low frequency (3-5 MHz) must be used to reduce the amount of attenuation.

References

Hoskins PR, Thrush A, Martin K, Whitingham TA. Diagnostic Ultrasound: *Physics and Equipment. Greenwich Medical,* London. 2003.

Szabo TL. Diagnostic Ultrasound Imaging: Inside *Out. Elsevier Inc. 2004.*

Wells PNT. *Biomedical Ultrasonics*. Academic Press, London. 1977.

ULTRASOUND TRANSDUCERS

FRANCO BERTORA[*]
ESAOTE, via Siffredi 58, 16153 Genova, Italy

Abstract. Ultrasound imaging modality has the major advantage, as compared to other medical imaging modes, to provide real-time acquisitions. It is a cheap modality compared to MRI and others, and it is non-invasive. One interest in using ultrasound imaging is to study the dynamic behaviour of various organs such as arteries, liver, heart. Today's scanners allow the visualization of the structures in gray scale images and the visualization of the flow information in color Doppler mode images. Both information can be acquired simultaneously. In this chapter we will present the basic principles leading to the design of probes.

Keywords: ultrasound; transducers; piezoelectricity; beamforming.

1. Probe types and anatomy

The probe is a significant part of an ultrasound scanner, accounting for a good percentage of its cost and performance. It performs the conversion of electrical energy into mechanical energy and, conversely, of mechanical into electrical energy and its properties directly reflect in the quality of the image.

The main purpose of a transducer is the generation of an ultrasound beam that is well defined both in time and in space: to this effect the important factors are the bandwidth and the dimensions of the emitting area, in relation to the focusing distance. This fact also determines the existence of a number of families of probes (see Table 1).

Although there is such a multitude of probes, they all share some common properties: they are made of piezoelectric material, they comprise many active elements and there are various devices for impedance matching and focalization.

[*] Easote, via Siffredi 58, 16153 Genova, Italy, e-mail: bertora@esaote.com.

Y. Lemoigne et al. (eds.), Physics for Medical Imaging Applications, 111–121.

TABLE 1 : Different types and use of probes.

Type of Probe:	Used for:
Linear	Generally employed for superficial parts at high frequency
Convex	Generally employed for abdominal scans (low frequency)
Phased array	Used in cardiac imaging
Specialized	Endocavitary Transrectal, Transvaginal Transesophageal Surgical

2. Piezoelectricity

A piezoelectric material has an asymmetric atomic lattice (typically perovskite): the consequence is that it changes dimensions when subjected to an electric field and generates an electric field when strained. As the figure shows, at temperatures below the Curie point (order of 150 °C), the crystal lattice is elongated and becomes an electric dipole whose potential changes as the crystal is strained

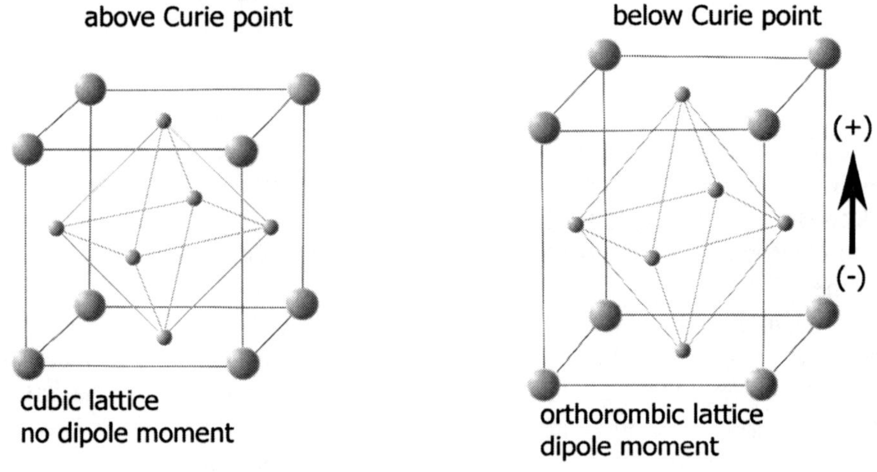

Figure 1. The typical structure of a piezoelectric crystal.

A typical transducer is realized as an array of piezoelectric material elements that are sandwiched between an absorbing material layer (backing) and an impedance matching layer. In fact the acoustic impedance of tissues is close to 1.5 MRayl while that of the piezoelectric material is about 20-30 Mrayl ("composite" materials[1] have a lower impedance that is anyway rather high when compared to that of tissue). If an impedance matching device were not present the efficiency would be extremely low and the transducer would be useless. Moreover by the very nature of the piezoelectric phenomenon the transducer emits energy both from the front and back surfaces of the crystal: it is therefore necessary to suppress the wave emanating from the back side to avoid its reaching the front and cancelling the wave emitted from there. As Figure 2 shows a typical array transducer is obtained lining up a series of elements on an energy absorbing backing layer. Each element has one or more matching layers at its front and is individually driven from the electronics.

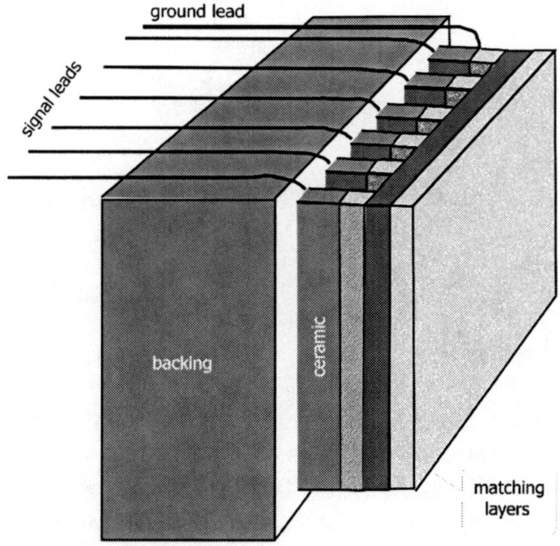

Figure 2. A typical array transducer is obtained lining up a series of elements on an energy absorbing backing layer. Each element has one or more matching layers at its front and is individually driven from the electronics.

3. Array types

Depending on the intended use arrays come in many shapes: the most common are the linear arrays that are most commonly employed for vascular and small parts imaging, the convex arrays generally used for abdominal imaging and the phased array probes generally dedicated to cardiac imaging. Less common formats are the 2-D arrays for real time volume imaging and the so-called 1.5-D arrays for high resolution linear imaging.

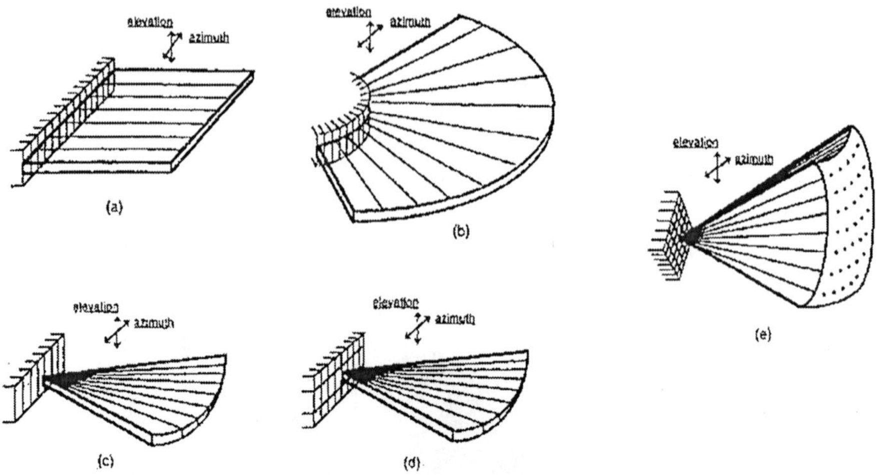

Figure 3. Array formats.

4. Focusing and beamforming

Each individual element of a linear probe, like the one shown in Figure 2, has dimensions that could be from 5 to 10 mm in the vertical (often called the "elevation") dimension and 0.25 mm in the lateral direction. As a consequence the radiation (that has a wavelength less than a millimetre) emitted from a single element propagates as a quasi-cylindrical wave front. If all elements are excited at the same time a quasi plane wave would result, that is not the narrow, well defined radiation pencil that is desired. It is therefore necessary to focalize the beam: this is obtained with a cylindrical acoustic lens that takes care of the focalization in the elevation direction (see Figure 4). In the first days of ultrasound scanning the focalization in the lateral direction was also done physically with the result that it was necessary to mechanically displace the transducer to scan the image area.

Today the focalization in the lateral direction is obtained electronically, with the advantage of being able to combine flexible focusing and beam steering in a completely static assembly that has a much better reliability

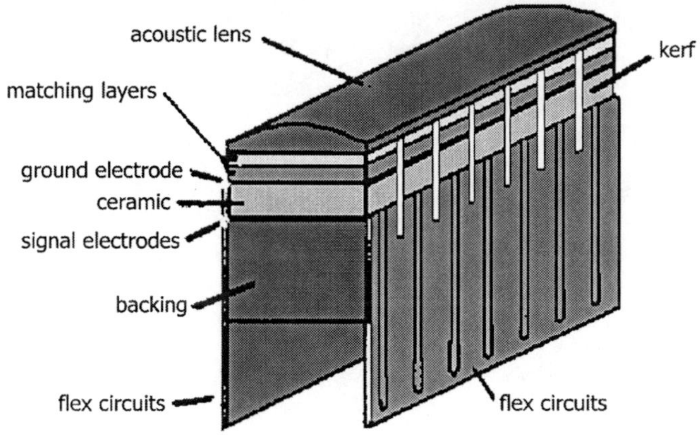

Figure 4. A complete linear array showing the acoustic lens on top of the matching layers.

when compared to the mechanically scanned probe that is now used only in very special applications.

The principle of electronic focusing is shown in Figure 5.

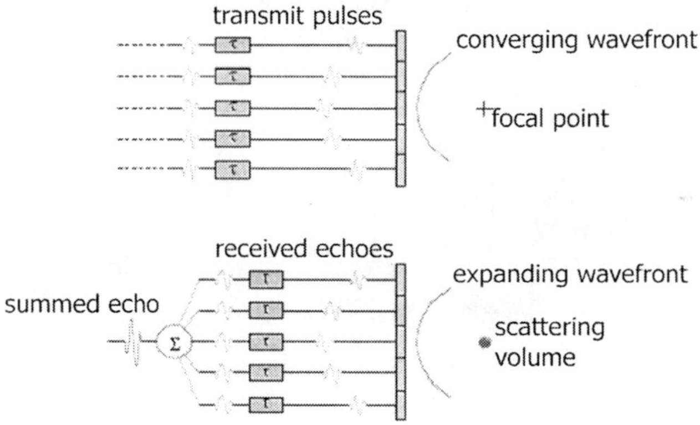

Figure 5. Electronic focusing.

A series of electronic delay elements is inserted in the signal path of every element and the individual delays are computed so that the propagation times from any element to the target are rendered all equal. It is possible in this way to direct all the emitted energy to a single point as well as to sum coherently all the contributions coming from different directions from a given point.

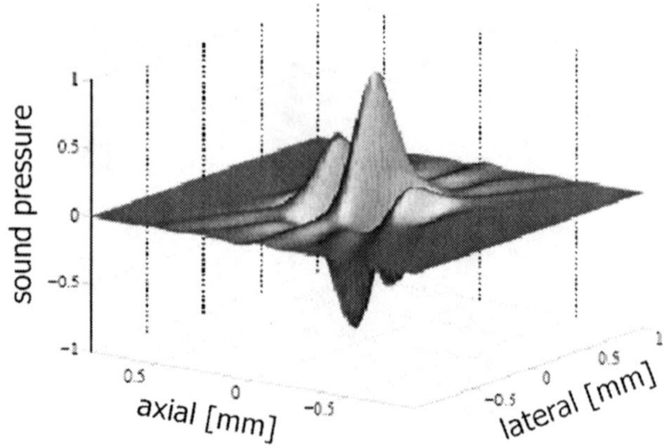

Figure 6. The acoustic pulse of an electronically focused array at the focal point.

Figure 6 shows a representative result of this process. The lateral width of the energy distribution is obtained by means of suitable delays applied to the 128 elements that constitute the array while the axial width of the distribution is closely correlated to the bandwidth (here 60%) of the elements. The overall dimensions of the energy distribution determine the "point spread function" (Figure 7) of the imaging system i.e. the "smearing" factor that alters the ideal shape of a geometrical point and is responsible of the spatial resolution of the image.

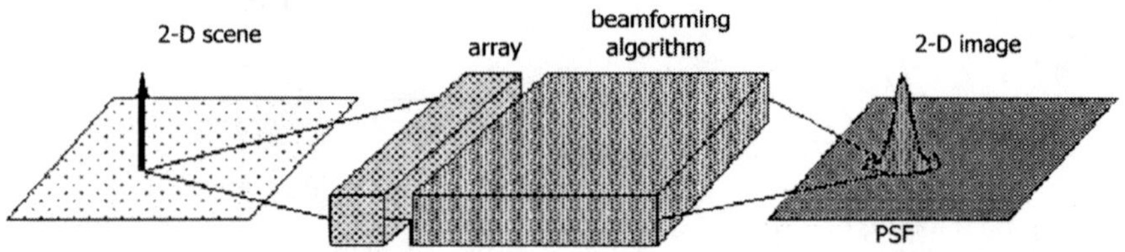

Figure 7. The Point Spread Function (PSF) defines how "smeared" will be the image of a geometric point.

As said before the axial resolution is determined by the pulse duration that is directly correlated to the frequency bandwidth of the probe. The lateral resolution is instead determined by the f-number i.e. the ratio between the lateral extension of the probe and the distance to the focus point. The lower the f-number the wider the angle under which the probe is seen from the focal point and the better the lateral resolution; in fact it can be shown that the lateral resolution is inversely proportional (through a Fourier Transform) to the lateral width of the probe (Figure 8). By the same token the subdivision of the active surface of the probe into individual elements is responsible for the existence of spurious "side lobes" i.e. sensitivity of the array to radiation coming from directions other than "broadside" (the direction perpendicular to the array major dimension).

The interested reader can find a good review of the above topics in references [2, 3, 4].

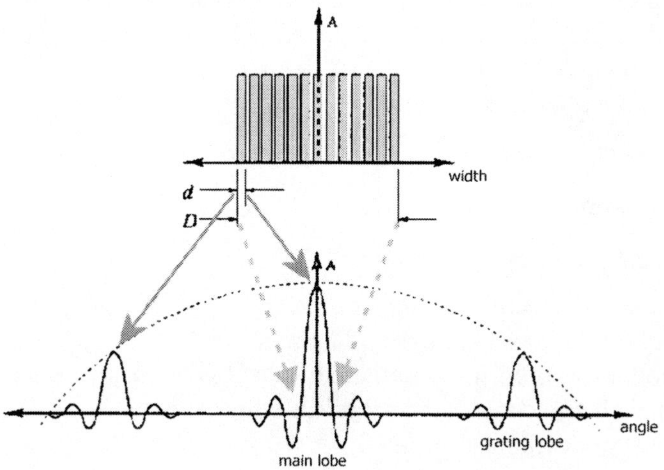

Figure 8. Main sensitivity lobe of a linear array and "grating" lobes arising from the subdivision into elements of the array aperture.

The above considerations shows that a low f-number is a plus in receiving but the same thing cannot be said for transmit. In fact a low f-number also signifies a shallow depth of focus with the result that the transmission would be effective only for a single depth and not for the whole image. The solution generally employed is to use a relatively high f-number (from 5 up) in transmission to obtain a reasonably well define radiation pencil and to employ "dynamic focusing" in receiving. With this technique the delays are continuously modified while the signal is received with the result of being able to focalize using a low f-number at all image depths. If frame-rate is not an issue multi-zone focusing can be used: with this technique a single image is obtained patching together a number of partial images obtained with different transmit focus depths.

5. Technological aspects

Although the basics of transducers are the same all over, each application calls for its own specialized transducer: the main factors that contribute to the final performance of the probe are many times conflicting among them and require some type of compromise. In fact the penetration requirement contrasts the obtainable resolution. In fact, since the maximum beam intensity is limited by safety requirements, tissue attenuation that increases linearly with frequency would suggest to employ the lowest possible frequency; on the other side the achievable lateral resolutions is ultimately determined by the wavelength and is therefore worse at low frequency.

Another conflict arises between resolution and probe footprint. A small footprint probe is many times desirable for ease of handling and is a must in those application, like cardiac imaging, where an obstacle (as the ribcage) limits the "acoustical window" available to access the organ.

Abdominal application requires:

- Deep penetration
- Wide field of view
- Moderate footprint

The solution is to employ a low frequency (typically around 2-2.5 MHz) probe to get the best penetration. The requirement of a wide field of view with a moderate footprint is satisfied with a convex probe geometry that does not penalize too much the achievable f/number.

Cardiac applications require:

- Good penetration
- Wide field of view
- Very small footprint

Here also a low frequency probe (again in the 2 MHz range) gives a solution to the penetration requirement; the very small footprint can only be satisfied by a phased array probe, although the consequence will be a reduction in the lateral resolution, due to the high f/number.

Vascular applications require:

- Good resolution
- Moderate penetration
- Doppler flow imaging

The contrast here is between the resolution requirement, which would call for a high frequency, and the Doppler application that calls for a lower frequency. A medium frequency linear probe is the best choice in this case.

Small parts applications require:

- Very Good resolution
- Moderate penetration

Here the obvious solution is to employ a high frequency, linear probe since, in this case, the requirements are not contrasting.

Beyond the above requisites safety and ergonomic aspects regulate the physical characteristics of the probe: the probe must be waterproof, it must be engineered with the appropriate materials to withstand sterilization, cleaning, etc.

Electrical safety must be guaranteed and requirements of minimal weight, cable compliance and footprint must be observed.

6. Fabrication process

The main steps of the fabrication process are shown in Figure 9. The piezoelectric ceramic is glued to the backing layer and the assembly is cut to separate the individual elements. The cuts (kerf) are filled with an appropriate compound and the matching layers and the elevation-focusing lens are applied on top.

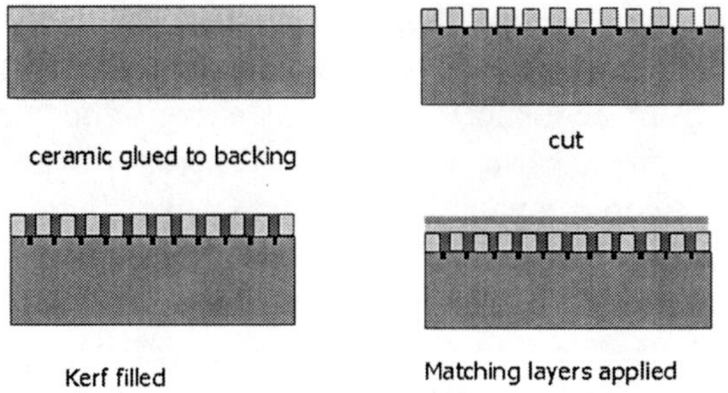

Figure 9. Main steps in the fabrication of an array.

In the next step the signal wires are attached to the individual elements and the assembly is moulded in a plastic block and encased to produce the finished probe.

7. Future developments

What has been described so far reflects the status of the current transducer technology that has, by now, reached its maturity. Other technologies are emerging and only the future will tell us whether their advantages will be enough to displace the current technology.

The most important development areas are two: single crystal transducers and cMUTs.

7.1. SINGLE-CRYSTAL TRANSDUCERS

Single-crystal technology can be considered an extreme development of today's transducers: as seen before the basic material is a ceramic obtained by sintering at high temperatures a finely ground crystalline material, taking care, at the same time, that the prevalent orientation of the crystals is coherent. Obviously this process cannot be 100% effective and the result is a material whose piezoelectric characteristics are inferior to the maximum theoretical limit.

There is a wealth of research going on in producing, sizeable monocrystals, by techniques that are similar to those successfully employed in growing silicon and other materials. The results so far obtained are remarkable in terms of performance[5], but the difficulties still encountered, both in the crystal growth and in its subsequent machining, do not yet allow the use of this technology in the production of competitive probes.

7.2. CMUT TRANSDUCERS

The second technique being actively pursued in many research and, recently, industrial laboratories is based on an altogether different principle. The conversion from acoustic to electrical energy and vice versa is still obtained through an electric field that is here, however, stored in a multitude of microscopic capacitors obtained from a silicon wafer by means of micro machining (hence the acronyms "**c**apacitive **M**icromachined **U**ltrasound **T**ransducers")[6]. The main advantages of this technology lie in its capability of leveraging on the extensive experience gained in silicon manufacturing to obtain low cost devices. Another relevant advantage is represented by the excellent characteristics of the resulting transducers, in terms of bandwidth and impedance matching to the body[7], combined with the possibility of direct integration with the front-end electronics of the scanner.

The development level of cMUT transducers appears today more close to a clinical application, particularly when transducers for real time 3D imaging are considered.

References

1. W.A. Smith – The Role of Piezocomposites in Ultrasound Transducers – Proc 1989 Ultrasonics Symp. 755-766

2. J.W.Hunt, M. Arditi, F. Stuart Foster – Ultrasound Transducers for Pulse-echo Medical Imaging – IEEE Transactions on Medical Imaging, vol. BME-30, n. 8, 453-481, August 1983
3. R.E. McKeighen – Design Guidelines for Medical Ultrasonic Arrays – SPIE International Symposium on Medical Imaging, San Diego, CA Feb 25, 1998
4. Special issue on Ultrasonic Transducers – IEEE Trans. Ultrason., Ferroelec., Freq. Contr., vol. 44 n. 5, Sept. 1997
5. C.G. Oakley, M.J. Zipparo – Single-crystal Piezoelectrics: a revolutionary development for transducers – Proc 2000 Ultrasonics Symp. 1157-1167
6. I. Ladabaum, X. Jin, H.T. Soh, A. Atalar, B.T. Khuri-Yakub – Surface Micromachined Capacitive Ultrasound Transducers – IEEE Trans. Ultrason., Ferroelec., Freq. Contr., vol. 45, 678-690, 1998
7. J.F. Gelly, F. Lanteri – Comparison of Piezoelectric (thickness mode) and MEMS transducers – Proc 2003 IEEE Ultrasonics Symposium 1965-1974

ULTRASONIC DOPPLER MODES

PIERO TORTOLI*, PAOLO FIDANZATI, LUCA BASSI
Dipartimento di ingegneria ed elettronica, Università degli Studi di Firenze, Largo E. Fermi 2, 50125 Firenze, Italy

Abstract. Any US equipment includes Doppler facilities capable of providing information about moving structures inside the human body. In most cases, the primary interest is in the investigation of blood flow dynamics, since this may be helpful for early diagnosis of cardiovascular diseases. However, there is also an increasing interest in tracking the movements of human tissues, since such movements can give an indirect evaluation of their elastic properties, which are valuable indicators of the possible presence of pathologies. This paper aims at presenting an overview of the different ways in which the Doppler technique has been developed and used in medical ultrasound (US), from early continuous wave (CW) systems to advanced pulsed wave (PW) colour-Doppler equipment. In particular, the most important technical features and clinical applications of CW, single-gate PW, multi-gate PW and flow-imaging systems are reviewed. The main signal processing approaches used for detection of Doppler frequencies are described, including time-domain and frequency-domain (spectral) methods, as well as novel strategies like, e.g., harmonic Doppler mode, which have been recently introduced to exploit the benefits of US contrast agents.

Keywords: Doppler ultrasound; non-invasive flowmetry; color-flow-mapping; non-invasive diagnosis.

1. Introduction

The well-known *Doppler effect* consists in a frequency variation (shift) of a wave originated from the relative movement between the source and the receiver [1]. In the biomedical field, such phenomenon is exploited for performing ultrasound (US)-based blood flow velocity measurements, which are of interest for many diagnostic applications. In this case, flowing blood is not a direct source of ultrasonic waves, but, when impinged by a wave originated by a still source, it backscatters US energy, and becomes equivalent to a moving source for any still receiver.

* Piero Tortoli, dipartimento di ingegneria ed elettronica, Università degli Studi di Firenze, Largo E. Fermi 2, 50125 Firenze, Italy; e-mail:ultra@ingfi1.ing.unifi.it.

Y. Lemoigne et al. (eds.), Physics for Medical Imaging Applications, 123–136.

With reference to figure 1, let us suppose to have a single target (e.g., a red blood cell), moving with constant velocity, *V*, impinged by an acoustic plane wave of frequency f_0. Let c be the speed of sound in the surrounding medium and ϑ the angle between the beam propagation direction and the velocity of the target.

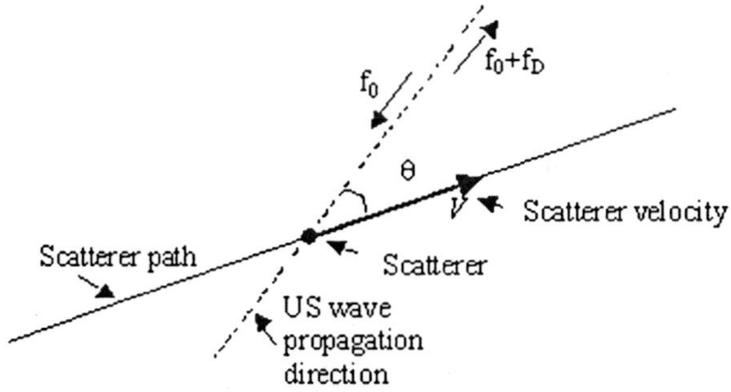

Figure 1. Schematic representation of the Doppler effect.

The frequency shift between the transmitted and backscattered waves is given by the following Doppler equation[2].

$$f_D = 2\frac{f_0}{c}v\cos\vartheta \tag{1}$$

By measuring the frequency shift, it is possible to estimate the axial component, $v\times cos\theta$, of the target velocity. Eq. (1) also indicates that the sensitivity of velocity measurement increases with the transmitted frequency, f_0. The frequency range actually used in Doppler US applications varies from 2 MHz, typical of transcranial (TCD) applications, where the explored region can cover a depth of 10 cm[2], up to 10-20 MHz, used for high resolution non-invasive measurements (1-2 mm) [2, 3]. The expected measured velocities are typically lower than 1 m/s in the peripheral circulation of a healthy subject, and they are never higher than a few m/s [2]. Accordingly, the detected Doppler shift is generally in the range a few kHz.

Doppler US techniques are used in many applications, from blood flow detection in peripheral vascular circulation, in the foetus and in umbilical cords, to cardiac output measurements. The instruments used in such applications range from the portable ones (ideal, e.g., for detection of peripheral arterial occlusive diseases -PAOD- or for bedside echocardiography), up to very complex and expensive systems, capable of accurate stenosis assessment through 2D velocity maps. The most common systems will be discussed and presented in detail in the following sections.

2. Technology

Information that can be retrieved by a Doppler system is strongly dependent upon the transducer characteristics and its electrical excitation.

The first basic flow meter, introduced for the first time in 1959 [2], continuously irradiates the vessel with a continuous ultrasonic wave, and it is thus called continuous wave (CW). In this case the transducer is made of two always active elements. One is the transmitter element, and the other one acts as receiver. Their axes, as shown in Fig. 2, are arranged in order to cover a wide overlapping region. The velocity information that can be obtained is an average over such region, and single Doppler contributions coming from different depths cannot be distinguished. Nevertheless, the low resolution of CW Doppler systems makes easy the research of the vessel of interest. Even if the vessel is not completely irradiated, valid Doppler information can be obtained.

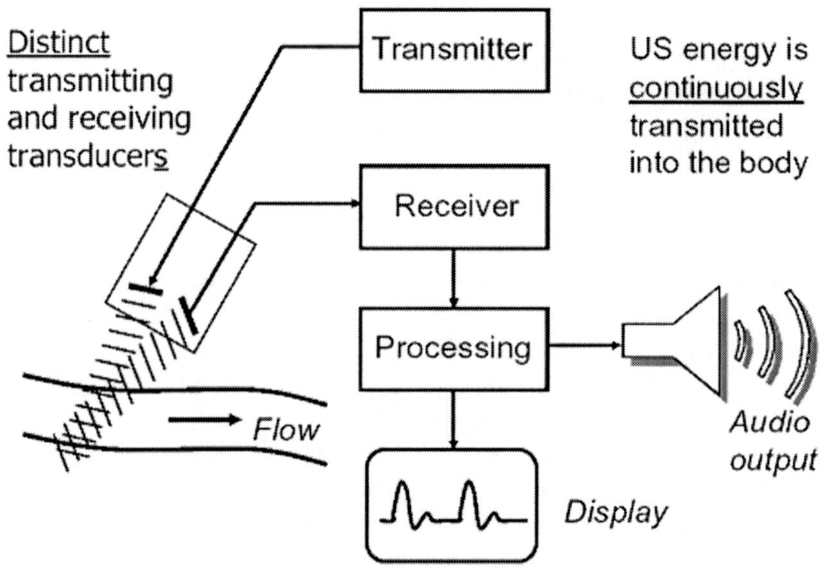

Figure 2. CW system architecture. Doppler information that can be retrieved from the echo-signal is related to the entire overlapping region between the acoustic beams of transducers connected to the TX and RX section, respectively.

The architecture of the receiving module typically includes an In phase-Quadrature (I/Q) demodulator. The direction of the measured velocity component can be retrieved from the in phase/quadrature signal components.

The signal is then both high- and low-pass filtered. High pass filtering allows eliminating the "clutter", i.e., the low frequency Doppler components due to still or slowly moving targets, like tissues or vessel walls. The clutter (the amplitude of which can be 60 dB higher than the useful signal) is generally removed because its presence could saturate the receiver dynamic range. Low pass filtering is useful to limit the wide band noise in the receiver and, when an Analog-to-Digital conversion (ADC) is implied, to avoid possible aliasing effects.

Pulsed Wave (PW) Doppler systems overcome the limit of CW systems given by the incapability of selecting the velocity information from a specific depth. PW systems typically transmit short energy pulses at a rate called "Pulse Repetition Frequency" (PRF). Each pulse is reflected/backscattered from all targets/scatterers intercepted during US propagation, and multiple echoes are thus subsequently generated. A sample volume (SV, i.e. the spatial region which contributes to the Doppler signal received at a given time) is identified by selecting, with an electronic "gate", a specific portion of the received echo-signals coming from different depths. The SV dimensions depend on the temporal length of both the transmitted pulse and the gating signal as well as on the US beam characteristics, while the SV position only depends on the delay between the transmission of the ultrasonic burst and the gating signal.

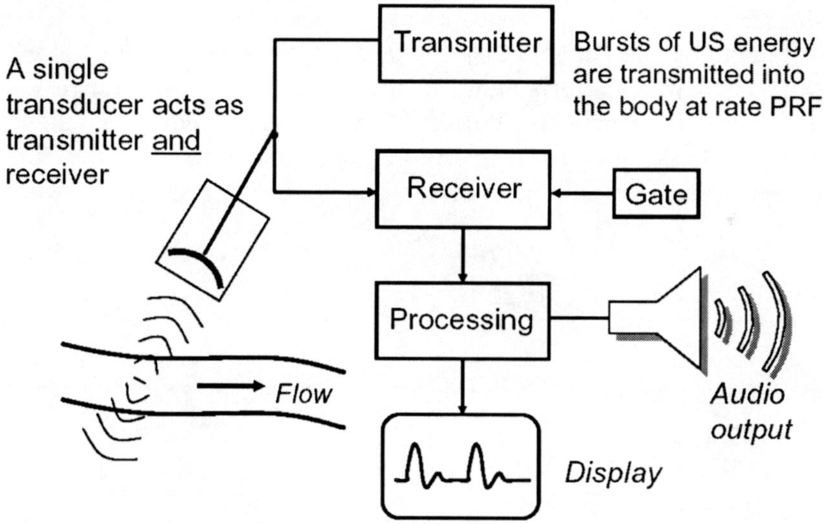

Figure 3. PW system architecture. The Doppler information is here extracted from a single sample volume of position selectable along the acoustic beam using the gating signal.

The high selectivity typical of PW systems involves possible difficulties in positioning the transducer over the region of interest, and mixed PW Doppler and imaging US systems have thus been developed. These systems, called "duplex" scanners, allows to highlight directly on the echographic image the direction of the US beam, and on the latter, the position of the sample volume. An optimal placement can be achieved with small sample volumes (1 mm) too.

The main differences between CW and PW systems can be seen from Figures 2 and 3. The key features of the PW system are the use of a single-element transducer and the presence of a Sample & Hold circuit, which holds the signal up to the next sample. In this way, the Doppler signal is selected and then filtered. As in CW systems, the signal can be sampled before numerical processing.

As shown in Fig 4, the Doppler signals can be taken from a single sample volume or from a group of cells along the US beam. For each transmitted burst we can obtain an entire group of samples, each corresponding to a different depth. If the samples obtained during a same Pulse Repetition Interval are stored in a column of a matrix, after several pulses, each row of the matrix will contain the time evolution of the Doppler signal related to a specific depth (or sample volume). If the line of sight crosses a vessel, the detection of Doppler frequencies from multiple depths allows reconstructing the velocity profile inside the vessel [8, 9, 10].

The same multiple sampling is realized in Colour Flow Mapping (CFM), or Colour Doppler, systems. Based on linear array or mechanical scanning transducers, such systems can sequentially investigate several lines of sights over a suitable 2D-region. Analyzing both traditional echo and Doppler signals corresponding to the same region, a bi-dimensional velocity map is obtained. Velocity values are colour coded and superimposed to the image of the corresponding anatomical structure. In the next paragraph some examples of these types of visualization are shown.

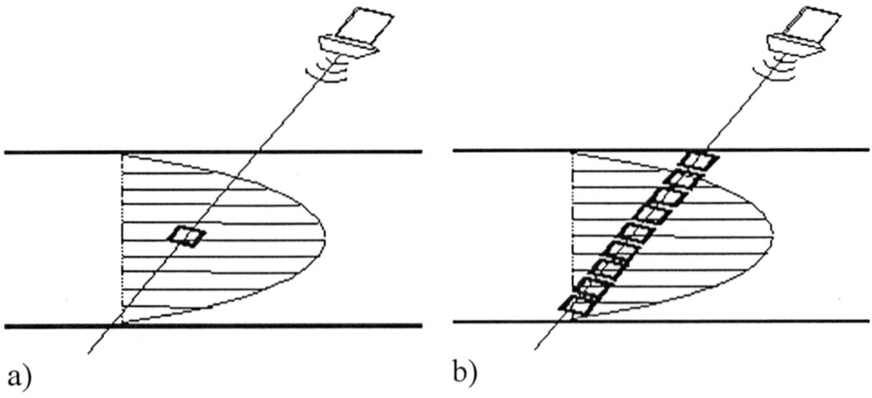

Figure 4. Echo signal selection in PW systems: (a) "single gate", only the signal components synchronous with the electronic gate are extracted; (b) "multi-gate", all the signal components synchronous with the multiple electronic gates are extracted.

3. Processing

As said above, typical values of Doppler shifts lie in the audio band (0-20 kHz). The early systems were only based on the production of sound, as obtained by sending the detected Doppler signal, suitably amplified, to a loudspeaker. By listening to this sound, the operator could "easily" (with CW systems) find the correct transducer position during the exams. Moreover, skilled operators could also distinguish between healthy and diseased arteries just listening at the Doppler sound, a sort of "voice" from arteries.

Even if this "audio" capability is still well appreciated, it is not sufficient for a quantitative Doppler exam. For this, it is at least necessary to estimate the mean frequency of the detected signal. In the first flowmeter, the Doppler frequency was evaluated using an electronic circuit capable of counting every change in sign of the received signal ("zero crossing"). The measure was very raw (thus making the audio output an essential complement) and could only provide a rough estimate of the average frequency.

With the introduction of Digital Signal Processing (DSP) systems, this approach was quickly substituted by more precise and complex methods, based on either frequency domain or time domain processing.

3.1. FREQUENCY DOMAIN PROCESSING

In order to appreciate the complexity of the Doppler signal which has to be analyzed in a flow meter, let us consider the classic situation in which a single sample volume intercepts multiple flow lines with different velocities. This is what typically happens in all cases in which there is a velocity gradient within the investigated vessel.

The signal received by the transducer is the result of the superposition of different contributions: the resulting Doppler spectrum has a width depending on the range of velocities inside the sample volume, while its shape is related to the scatterer density in the different flow lines.

The simplest way to evaluate such a spectrum from the sequence of samples, s_n, received from the sample volume, over subsequent PRIs, is calculating the discrete Fourier transform (DFT) as:

$$S(k) = \frac{1}{N}\left|\sum_{n=0}^{N-1} s_n e^{-j\frac{2\pi n}{N}k}\right|^2 \qquad k = 0,1,\ldots,N-1$$

The latter can be efficiently calculated through the Fast Fourier Transform (FFT) algorithm. Since the blood velocity is typically not stationary, such estimation has to be repeated several times within each cardiac cycle, in order to follow the temporal evolution of the velocity distribution.

The result of such analysis (now present in all commercial instruments) is typically displayed in the form of spectrograms[11], as shown in Fig. 5. Here, each Doppler spectrum estimated from the samples received over the last N PRIs, is reported on a corresponding vertical line, the spectral power detected at each frequency being coded to grey levels. By reporting on adjacent lines the spectra which are subsequently computed, the evolution of blood velocity during the cardiac cycle is clearly recorded (see Fig. 5).

Figure 5 confirms that the signal backscattered from each flow line has an intrinsic strong irregularity, since it comes from a sequence of targets randomly distributed along the same trajectory. A signal like this, called "speckle-like", origins spectra with high variance, as visible in the spectrogram granularities. This phenomenon can be partially limited though alternative frequency estimation methods (see, e.g., parametric methods[2, 12]).

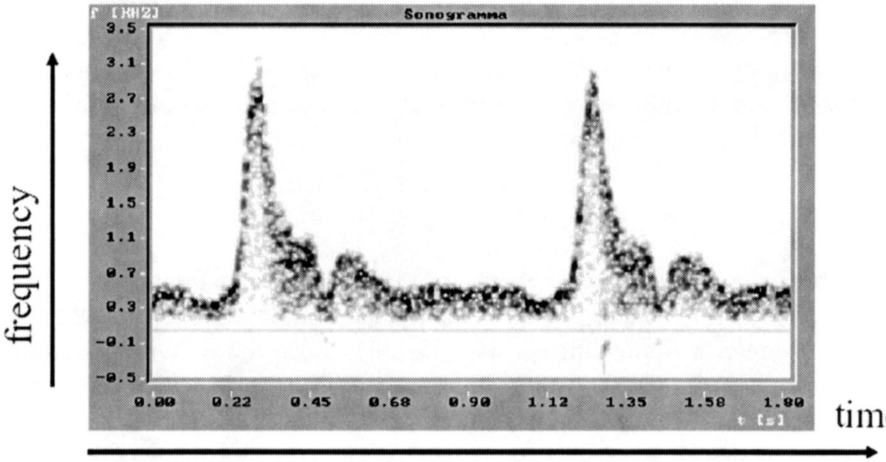

Figure 5. Doppler spectrogram obtained from investigation of a common carotid artery. Different Doppler frequencies (velocities) are detected during the different phases of the cardiac cycle.

When the processing system has the required capabilities, the Doppler analysis can be extended to all the echo-signals produced along the US beam axis. In this "multigate" approach, the flow velocity can be detected for multiple depths and the velocity profile along an entire vessel can be reconstructed. The dynamic evolution of the velocity profiles can be observed in real-time[10], or reported in a display like that shown in Fig. 6, which shows the velocity profiles detected at 4 different phases of the cardiac cycle from the common carotid artery of a healthy subject.

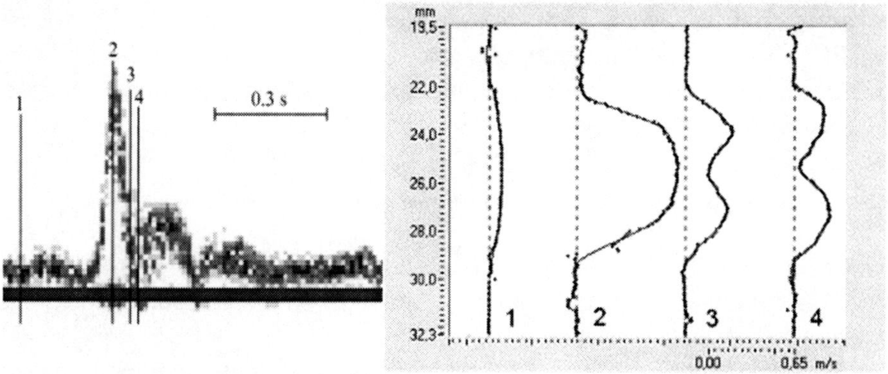

Figure 6. Sequence of CCA velocity profiles detected at the times 1, 2, 3, 4 outlined in the reference spectrogram on the left.

Multigate Doppler analysis is also the key feature of CFM systems, in which multiple lines of sight are scanned in real-time. The most relevant technical problem is here related to the short time available to collect samples from each scan line. This time is influenced by several factors: the PRF (which is, on turn, conditioned by the depth of investigation); the number, M, of lines contributing to each frame; and the frame rate, FR (typically expressed in terms of frame per second - fps). Using typical values (PRF < 20 kHz; M > 50; FR > 10 fps) the time available for each line is reduced to a few US bursts, i.e. to a few samples collection for each gate. For example, when PRF = 10 kHz, M = 100 and FR = 25 fps, only 4 samples are available for each Doppler frequency estimate.

Due to the aforementioned high variance of Doppler spectra, the FFT algorithm is not suitable to analyze a low number of Doppler samples. Other frequency estimation algorithms are used to perform the velocity analysis in two dimensions. Such algorithms can only provide the mean Doppler frequency, and not the whole spectrum, by analyzing the phase difference between subsequent samples.

It is known that the power spectrum of a signal s(t) can be obtained directly from the square of the modulus of its Fourier transform, or from the Fourier transform of the self correlation function[13]:

$$S(f) = \left| \int_{-\infty}^{+\infty} s(t) e^{-j2\pi ft} dt \right|^2 = \int_{-\infty}^{+\infty} r(\tau) e^{-j2\pi f\tau} d\tau$$

where:

$$r(\tau) = \int_{-\infty}^{+\infty} s(t) s(t-\tau) dt$$

It has been proved that when the mean Doppler frequency is directly obtained from the phase of the autocorrelation function (14), and not from the FFT, better results in terms of signal to noise ratio are obtained[14]. This has been, for years, the reference method to estimate the mean Doppler frequency in CFM systems.

3.2. TIME DOMAIN SIGNAL PROCESSING

The flow velocity can also be estimated by evaluating the delay between two subsequent echoes, that is directly related to the space travelled by the scatterer[15, 16, 17].

Let $s_i(t)$ and $s_{i+1}(t)$ be the echo signals relative to two consecutive ultrasonic pulses. $s_{i+1}(t)$ can be seen as a replica of $s_i(t)$, in advance or delayed, depending on the motion of the target structure, approaching or moving away from the transducer. The relative delay, Δt, is estimated by identifying the peak position of the cross correlation function between the two signals $s_i(t)$ ed $s_{i+1}(t)$:

$$c(\tau) = \int_{-\infty}^{+\infty} s_i(t) s_{i+1}(t-\tau) dt$$

The space, Δs, travelled by the moving target during two consecutive PRIs, can be expressed through the "time of flight" of the US pulse during its path from the transducer to the target and back to the transducer:

$$\Delta s = \frac{\Delta t \cdot c}{2}$$

Such value can be directly related to the velocity component along the transducer axis:

$$v \cdot \cos\theta = \frac{\Delta s}{1/PRF} = \frac{\Delta t \cdot c}{2} PRF$$

in which 1/PRF is the interval between two transmitted pulses.

A direct comparison between this approach and the auto correlation method is hard to do, as the results mostly depend on the type of application. Some comparisons have been made on simulated data[18], showing that the cross correlation technique is slightly better in terms of precision and accuracy when the Doppler signals are related to low velocity dispersions. Other comparisons made on real "in vitro" signals[19] have shown that the two methods are almost comparable in realistic situations.

3.3. IMAGING SYSTEMS

In "colour" Doppler systems, a series of colour pixels are superimposed to the black and white (B-mode) image, in all points where a movement is detected. The simultaneous real time presence of anatomic and functional information in the final CFM image lets the operator to easily identify the flow in those regions where colour variations can be seen.

The "colour" functional modalities can be classified into the two categories which are described in the following sections[20] .

4. Colour doppler

In CFM images, the colour of each pixel is decided according to the direction of the velocity: it is red when the flow is moving away from the direction of propagation of the ultrasonic wave, and blue if the flow is moving into the opposite direction.

The colour saturation is associated to the module of the velocity: light colours represent high velocities, while darker colours are used for low velocities.

More complex colour codes are used in some systems, in order to report on the image the velocity variance. This can be useful to accelerate the diagnosis in some pathological conditions in which flow has a high variance, due to turbulent motion[21] .

As an example, Figure 7a shows the flow near the bifurcation of a healthy carotid artery, through a colour Doppler image frozen during the systolic peak of the cardiac cycle. The red region highlights either the direction of the flow, going towards the bifurcation, and the different values of velocity, which is higher in the internal carotid (the light red zone in the centre). In the lower part of the image a small blue zone can be noticed on wall opposite to the bifurcation that divides the flow, generating a flow inversion due to the geometry of the bifurcation. This inversion was predicted before the introduction of colour Doppler techniques[22]. The blue region above the bifurcation is the flow in the jugular vein that moves into the opposite direction.

Figure 7b shows the image of flow in the portal vein. In this case an artefact can be seen, which could be confused with a flow inversion. The high velocity of blood flow, coupled with a low PRF due the need of investigating a deep region, causes an aliasing phenomenon, recognizable from sudden colour changes that do not correspond to any physiological condition. This can be compared to the colour inversions visible in figure 7a, which appear smoothed from light red, dark red, black, dark blue, while moving from the centre to the outer part of the vessel.

4.1. POWER DOPPLER

In Power Doppler systems, the power of the Doppler signal is revealed instead of the mean frequency[23]. In this case, only the integral of the Doppler spectrum is taken in to consideration, independently of its position on the frequency axis.

With respect to the colour mode, in this case the direction and velocity of the blood flow are typically not retrieved. On the other hand, the sensitivity is increased and investigation of slow flows can be made, too. A typical application of Power Doppler analysis is the evaluation of organs perfusion (see, e.g. the power Doppler image of kidney perfusion shown in Figure 7d).

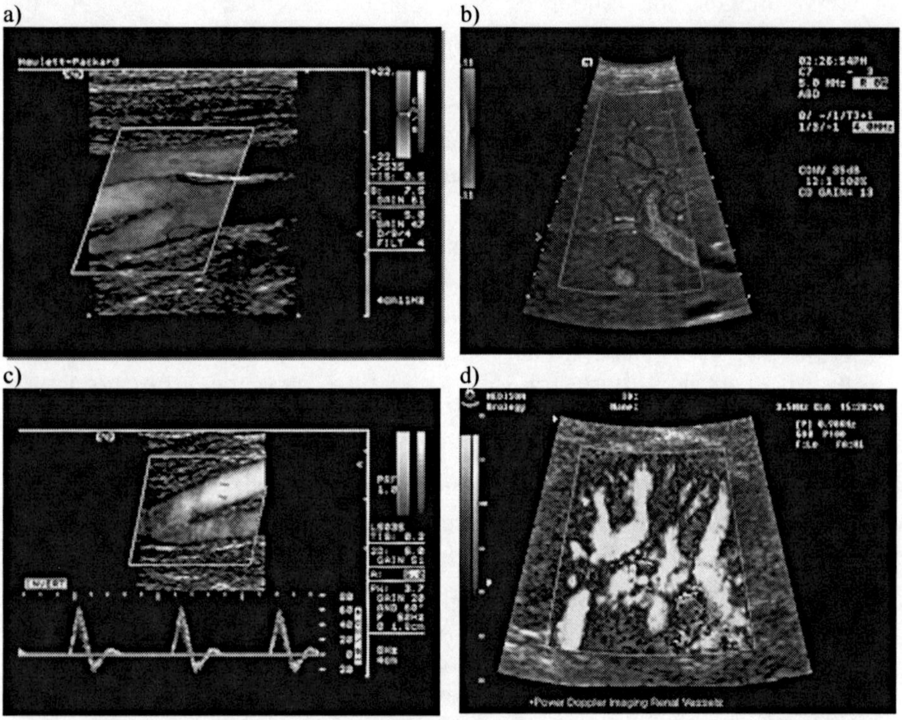

Figure 7. Typical US Doppler images provided by commercial equipment. (a) Carotid artery bifurcation and jugular vein are identified through red and blue colours, respectively, because of the opposite blood flow directions in such vessels; (b) portal vein: sudden colour changes are here due to aliasing; (c) femoral artery bifurcation colour image and spectrogram; (d) renal perfusion estimate through power Doppler. (See color section)

Colour Doppler techniques can also be used to display tissue movements. In this case the amplitude of the echo-signal can be even 60 dB over the signal from blood. Typical problems of some critical situations in flow investigations, related to a poor signal to noise ratio, are not present. The frequency estimate can thus be made on a limited number of samples per scan line (from 3 to 5) and, therefore, a higher frame rate.

One of the most important applications in this field is the investigation, in terms of velocity[24] and velocity gradient[25], of the movements of the myocardium. It is known that ischemic and post infarct regions have anomalous contractions, with respect to a healthy myocardium. Such abnormal contractions are highlighted by colour techniques with the appearance of anomalous colour in the CFM image.

4.2 DISCUSSION AND CONCLUSION

Let us re-write the Doppler equation by expressing the velocity as a function of the other parameters:

$$v = \frac{c f_D}{2 f_0 \cos\theta} \qquad (2)$$

It can be noticed that a small error in the estimation of the Doppler angle, θ, can produce large errors in the velocity estimate, especially when the angle is close to 90°, while for θ=90° the measurement would not be possible.

The latter limitation is derived from the model used for the acoustic beam, that is considered as a plane wave. Actually, the acoustic fields used to form the US images are quite more complex. We can imagine a focused beam as originated by several plane waves propagating to directions slightly different from each other, which intersect in the focal region. This consideration has an important consequence: the Doppler spectrum generated by a single flow line is not a single spectral line, but a full spectrum covering a band of frequencies depending on both the flow velocity and the transducer beam focusing properties[26, 27].

During US flow investigations, poor results are sometimes obtained. This may be due to different causes, like the high attenuation of the US, the acoustic beam distortion introduced by tissue inhomogeneities, the movement of the surrounding tissues capable of masking the blood movement. Such difficulties are more evident in investigations of deep and/or small vessels, intracranial and coronary arteries.

One possible way to improve the signal to noise ratio in extreme situations is injecting US contrast agents[28] in arterial circulation. These are encapsulated gaseous micro bubbles, not dangerous for the human body, with a large difference in acoustical impedance with respect to the plasma, thus acting as efficient scatterers which move exactly like erythrocytes. Micro bubbles are small enough to pass through the natural filter corresponding to the lungs, and stable enough to allow a complete exam for a few minutes. The measured increment of SNR with micro bubbles is typically 10-25 dB. Signals backscattered from small and deep vessels have finally become visible.

The popularity of micro bubbles is not only nor mainly due to this aspect. They in fact resonate, i.e. they can absorb and scatter the US waves more efficiently at certain frequencies, becoming active sources of US. As a noticeable coincidence, their resonation frequency (at the dimensions required) is coincident with the typical frequency range of US equipment (2-10 MHz). Moreover, it has been noticed that with high acoustic pressures, the spatial variations of micro bubbles, induced by oscillations, are asymmetric. Expansions and contractions have different amplitudes. The intrinsic non linearity in this phenomenon leads to echoes extremely rich of harmonics. By “tuning” the receiver on the second harmonic of the echo signal, we can select the information related only to the flow, without any contribution from the tissue. With this so called “harmonic imaging” technique, the ratio between clutter and signal amplitudes is reversed, with a higher signal amplitude for the latter. A typical field of application is heart investigation. The rapid movement of the myocardium originates very strong signals that completely mask the weak ones coming from blood. The use of the second harmonic leads to impressive results, where the colour associated to the velocity of the micro bubbles and, consequently, of the blood, is perfectly separated from the structure of the surrounding tissues.

References

1. C. Doppler, Über das farbige Licht der Doppelsterne und einiger anderer Gestirne des Himmels. Abhandlungen der königlich böhmischen Gesellschaft der Wissenschaften, 5, Folge 2, pp. 465-482, 1843.
2. D.H. Evans, W.N. McDicken, R. Skidmore, J.P. Woodcock, Doppler Ultrasound - Physics, Instrumentation and Clinical Applications, p. 8, John Wiley & Sons, 1989.
3. R. Aaslid, "Transcranial Doppler Sonography", Springer-Verlag, Wien, 1986.
4. A.C. Cristopher, P.N. Burns, J. Armstrong, F. Foster, "A high frequency continuous-wave Doppler ultrasound system for the detection of blood flow in the microcirculation", Ultrasound in Med. & Biol., Vol. 22, pp. 1191-1203, 1996.
5. W. Li, C.T. Lancee, A.F.W. van der Steen, E.J. Gussenhoven, N. Bom, "Blood velocity estimation with high frequency intravascular ultrasound", IEEE Ultrasonic Symposium proceedings, pp. 1485-1488, 1996.
6. H.F. Routh, Doppler ultrasound, the ability to measure and image blood flow, IEEE Engineering in Medicine Biology, pp. 31-40, 1996.
7. S. Satomura, "Study of the flow patterns in peripheral arteries by ultrasonics", J. Acoustic Society Japan n. 15, pp. 151-158, 1959.
8. A.P.G. Hoeks, R.S. Reneman, P.A. Peronneau, "A multi-gated Pulsed Doppler System with Serial Data Processing", IEEE Trans. Sonics Ultrason 28: 242-247 1982.

9. P. Tortoli, F. Guidi, G. Guidi, C. Atzeni, "Spectral velocity profiles for detailed ultrasound flow analysis", IEEE Transactions on UFFC, Vol. 43, N.4, pp. 654-659, 1996.
10. P.Tortoli, G.Guidi, P.Berti, F.Guidi, D.Righi An FFT-based flow profiler for high resolution in-vivo investigations, Ultrasound in Medicine and Biology, Vol. 23, No. 6, pp. 899-910, 1997.
11. P.R. Hoskins, "The sonogram in Doppler ultrasound", Ultrasound International Vol. 3, pp. 134-144, 1996.
12. K. Kremkau, "Doppler ultrasound: principles and instruments", pp. 180-181, WB Saunders Company, 1995.
13. J.G. Proakis, D.G. Manolakis, "Digital signal processing: principles algorithms, and applications", Macmillian, 1992.
14. C. Kasai, K. Namekawa, A. Koyano, R. Omoto, "Real-time two dimensional blood flow imaging using an autocorrelation technique", IEEE Trans. SU, Vol. 32, pp. 458-463, 1985.
15. O. Bonnefous, P. Pesque, "Time domain formulation of pulse-Doppler ultrasound and blood velocity estimation by cross-correlation", Ultrason. Imaging, Vol. 8, pp. 73-85, 1986.
16. J.A. Jensen, "Implementation of ultrasound time-domain cross-correlation blood velocity estimators", IEEE Trans. BME, Vol. 40, pp. 468-474, 1993.
17. P.J. Brands, A.G.P. Hoeks, R.S. Reneman, "The effect of echo suppression on the mean velocity estimation rangeof the RF cross-correlation model estimator", Ultrasound in Med. & Biol., Vol. 21, pp. 945-959, 1995.
18. A.G.P. Hoeks, T.G.J. Arts, P.J. Brands, R.S. Renemann, "Comparison of the performance of the RF cross-correlation and Doppler autocorrelation technique to estimate the mean velocity of simulated ultrasound signals", Ultrasound Med. Biol., Vol. 19, pp. 727-740, 1993.
19. D.W. Rickey et al. "A velocity evaluation phantom for colour and pulsed Doppler instruments", Ultrasound Med. Biol., Vol. 18, pp. 479-494, 1992.
20. P.R. Hoskins, W.N. McDicken, "Colour ultrasound imaging of blood flow and tissue motion", British Journal of Radiology, Vol 70, pp. 878-890, 1997.
21. W.K. Kremkau, "Doppler ultrasound: principles and instruments", pp. 180-181, WB Saunders Company, 1995.
22. D.J. Phillips, F.M. Green Jr, Y. Langlois, G.O. Roederer, D.E. Strandness Jr., "Flow velocity patterns in the carotid bifurcation of young, presumed normal subjects", Ultrasound Med. Biol., Vol. 9, pp. 39-49, 1983.
23. J.M. Rubin, R.O. Bude, P.L. Carson, R.L. Bree, R.S. Adler, "Power Doppler US: a potentially useful alternative to mean frequency based color Doppler US", Radiology, Vol. 190, pp. 853-856, 1994.
24. W.N. Mc Dicken, G.R. Sutherland, C.M. Moran, L.N. Gordon, "Colour Doppler imaging of the myocardium", Ultrasound Med. Biol., Vol. 18, pp. 651-654, 1992.
25 A.D. Fleming, P. Palka, W.N. Mc Dicken, "Myocardial velocity gradients detected by Doppler imaging", British Journal of Radiology, Vol. 67, pp. 679-688, 1994.
26. D. Censor, V.L. Newhouse and T. Vontz "Theory of ultrasound Doppler-spectra velocimetry for arbitrary beam and flow configurations", IEEE Trans. on Biomedical Engineering, Vol. 35, pp. 740-751, 1988.
27. P. Tortoli, G. Guidi, F. Guidi, C. Atzeni: A review of experimental Transverse Doppler studies, IEEE Transactions on UFFC, Vol. 41, N. 1, pp. 84-89, 1994.
28. M. Postema, A. van Wamel, C. T. Lanc´ee, and N. De Jong, "Ultrasound-induced encapsulated microbubble phenomena," *Ultrasound Med. Biol.*, vol. 30, no. 6, pp. 827–840, 2004.

ULTRASOUND CONTRAST AGENTS

CHRISTIAN CACHARD* AND OLIVIER BASSET
CREATIS CNRS 5515, Inserm U630, Université Claude Bernard Lyon 1, 69621 Villeurbanne, France

Abstract. While the use of contrast agents in other imaging modalities (X ray, MRI, PET, ...) has been routinely accepted for many years, the development and commercialization of contrast agents designed specifically for ultrasound imaging has occurred only very recently. As in the other imaging modalities, the injection of contrast agents during an ultrasound examination is intended to facilitate the detection and diagnosis of specific pathologies. Contrast agents efficiency is based on the backscattering of ultrasound by microbubbles. These microparticules are intravenously injected in the blood flow. After an introduction and generalities on ultrasound contrast agents (UCA) the microbubble physics in an acoustic field will be developed. Second, physics characteristics of contrast agents will be compared (bubbles with or without shell, gas nature, size distribution). Influence of acoustic pressure on the behaviour of the microparticules (linear, non linear and destruction) will be discussed. Finally, a review of specific imaging adapted to contrast agent properties as harmonic imaging, pulse inversion imaging will be presented.

Keywords: ultrasound contrast agents; ultrasound imaging; backscattering; microbubbles; harmonic imaging; pulse inversion imaging.

1. Principle

From an acoustic point of view, blood and biological tissues are equivalent to liquids (ultrasound velocity, acoustic impedance, ...). With a simplistic assumption, ultrasound imaging depends on number of scatters. In blood,

* Christian Cachard, CREATIS CNRS 5515, Inserm U630, Université Claude Bernard Lyon 1, 69621 Villeurbanne, France ; e-mail : Christian.Cachard@creatis.insa-lyon.fr.

Y. Lemoigne et al. (eds.), Physics for Medical Imaging Applications, 137–150.

the acoustic wave is scattered by red blood cells. The contrast agent injected in the blood flow (bolus or perfusion) increases the number of scatterers, so imaging of blood vessels and perfused tissues is improved. The maximum efficiency is obtained with gaseous (resonant) particles. The reason of choosing such kind of particles is based on the principle of impedance mismatching between two mediums. This leads to a high reflexion coefficient. The additional diagnostic information obtained by using a contrast agent could potentially benefit a significant numbers of the ultrasound scans performed.

1.1. HISTORIC

In 1968, Gramiak and Shah[1] observed an enhancement of ultrasound echoes induced by injected small free air bubbles and published the first work about Ultrasound Contrast Agents (UCA). About thirty years later the first UCA are commercialized: Optison, Sonovue. During this research period, many contrast agents solutions were tested: hand shaken solutions, solids, liquids or gaseous particles. The superiority of gaseous particles was demonstrated. At the beginning, researches were turned on development of material particles to enhance images using grey-scale (broadband). In the 1990s, a major step was accomplished with the introduction of the harmonic imaging. Now, the development concerns specific UCA imaging and the echographic machines are adapted to contrast agents properties.

Simply defined, an ultrasound contrast agent intended to characterize blood flow through tissues, is a material that (1) has sound reflection properties that differs significantly from the surrounding media (tissue and blood), and (2) is composed of particles of a size that allows them to flow freely through, but remains within, the vessels of the microcirculatory system.

Today's ultrasonic contrast agents consist of encapsulated microbubbles on the order of 1 to 8 μm in diameter. They may be filled with air or with a lower solubility gas. Gasfilled bodies are highly echogenic due to the difference in the compressibility between the microbubbles and the surrounding medium (blood). The shell, designed to reduce gas diffusion into the blood, can be rather stiff (e.g. human serum albumin) or more flexible (e.g. phospholipids), and the shell thickness can vary from approximately 10 to 200 nm. A contrast agent is administered by intravenous injection, in bolus or in perfusion and, ideally, must cross the microcirculation of the pulmonary bed. Its physical properties should remain stable during the ultrasonic examination and provide a strong acoustic response. Unlike contrast agents used for Computed Tomography or Magnetic Resonance Imaging, ultrasound contrast agents do not diffuse into the extra cellular compartment. Thus, they are strictly blood pool agents. It should be non-toxic and harmlessly eliminated from the system after the ultrasonic examination. Preclinical and clinical studies have proven the efficacy and tolerance of the agent by patients[2-4].

1.2. THE LIFE TIME OF A BUBBLE

In this section, the bubble is studied without acoustic field. Persistence T is the life time of a bubble decreasing from its initial radius R_0 till its disappearance. Bubble persistence under saturated solutions can be calculated from the Epstein-Plesset formula, with neglecting surface tension coefficient[5]

$$T = \frac{\rho R_0^2}{2dC_S} \tag{1}$$

T persistence of the microbubble
ρ density of the gas
R_0 initial radius of the microbubble
d diffusivity of the gas
C_s saturation coefficient

TABLE 1. Persistence time of bubble as function of the gas and its radius.

Radius (μm)	Gas		
	Air	Sulfur hexafluoride	Dodecafluoropentane
1	0.2	0.1	3
3	1.7	1.2	33

With air the persistence time is too short to perform an examination, so gas with low diffusivity in blood as sulfur hexafluoride (Sonovue, Bracco) or Dodecafluoropentane (Echogen) are used and a shell has been added. On opposite, for some specific examination, recirculation of the contrast agents has to be avoided and longevity can be a drawback.

1.3. THE MECHANICAL INDEX

The Mechanical Index (*MI*) is proportional to the probability of seeing cavitation phenomenon to occur. It is used to quantify the emitted acoustic energy in a body rather than the time averaged intensity. This parameter is defined as the ratio between the maximum emitted impulse amplitude p in [MPa] and square root of the emitted pulse frequency f in [MHz].

$$MI = \frac{p}{\sqrt{f}} \tag{2}$$

TABLE 2. Corresponding pressure level with MI as function of the frequency.

Frequency (MHz)	1.6	1.6	1.6	1.6	3.2	3.2	3.2	3.2	5	5	5
Negative pick pressure (MPa)	0.05	0.2	0.4	1	0.05	0.2	0.4	1	0.05	0.2	0. 4
MI	0.04	0.16	0.32	0.79	0.03	0.11	0.22	0.56	0.02	0.09	0.18

1.4. THE MICROBUBBLE PHYSICS IN AN ACOUSTIC FIELD

In this section. a single bubble is considered and not the real contrast agent composed of a population of bubbles. In equations. the bubble is denoted by its radius R. In literature. the bubble is designed by its size. but the diameter (D = 2 R) is more relevant term.

1.4.1. *The scattering cross section*

The capacity of the particle to scattered an acoustic wave is described by the scattering cross section. The scattering cross section is defined as the scattered power divided by the incident intensity. For a spherical particle. the scattering cross section is:

$$\sigma_s = \frac{P_s}{I_i} = \left[\frac{4}{9}\pi R^2\left(\frac{2\pi f}{c}R\right)^4\right]\left[\left|\frac{\kappa_s - \kappa}{\kappa}\right|^2 + \frac{1}{3}\left|\frac{3(\rho_s - \rho)}{2\rho_s - \rho}\right|^2\right] \tag{3}$$

The subscript s is reported to the scatterer. Higher is the scattering cross section. better is the scatterer and so the contrast agent. The scattering cross section is proportional to the sixth power of the bubble radius (Eq. 3) so the larger the bubbles are. the more effective is the backscattered of the ultrasound wave. However. the limit is set by the necessity for transpulmonary passage; bubbles with a diameter higher than 7 μm (i.e larger than red blood size) are unable to pass through the pulmonary capillaries[6].

In Eq. (3) two parts may be distinguished. The first term is function of bubble radius R and frequency f of the incident pulse and is independent of materials particle and surrounding medium. The second term compares the compressibility and the density of the two mediums. If three same size spherical particles made of different materials in the same incident wave are compared (Table 3): (i) a gas particle (air or nitrogen bubble), (ii) a solid particle (iron bullet), (iii) a liquid particle (ether drop), the first term is constant and the scattering cross section is proportional to the change of the compressibility and the density of the two mediums. So in blood. materials having suitable sound reflection properties to be used as ultrasound contrast agents are gas-filled microbubbles as shown in table 3.

The size of the particles is much smaller than the wavelength of the acoustic field ($2\ R_0 << \lambda$). so the assumption of Rayleigh diffusion is valuable.

TABLE 3. Comparison of the scattering cross section of three spherical particles of 1.5 μm radius at a frequency of 3.75 MHz.

Materials	Compressibility (m^2N^{-1})	Density ($kg\ m^{-3}$)	Second term of Eq. (3)	σ_S (m^{-2})
Water	$4.5\ 10^{-10}$	998		
Microbubble (air)	$8.2\ 10^{-6}$	1.16	$3.3\ 10^{8}$	$3.4\ 10^{-10}$
Liquid solution (ether drop)	$1.8\ 10^{-11}$	792	10.1	$1.0\ 10^{-17}$
Solid particle (iron ball)	$3.7\ 10^{-2}$	7800	1.63	$1.7\ 10^{-18}$

1.4.2. *Bubble behaviour in the incident ultrasound field*

When the transmitted ultrasound wave hits a microbubble. a volume pulsation is initiated. The acoustic wave alternately compresses the microbubble on the positive pressure. and expands it on the negative pressure. As function of the pressure amplitude of the incident wave. the behaviours of agent contrast bubbles can be classified in three main categories:

- linear domain
- stationary non-linear domain
- transient non-linear domain

Each of these three domains corresponds to a range of acoustic pressure level which involves different responses from the contrast agent. For MI < 0.2. the response of micro-bubbles is linear. The response becomes stationary and non-linear in the MI range 0.2 and 0.4. One can observe that harmonic amplitudes become non-neglectable within this range of pressure level. Since the response is stationary, the phenomenon is reversible which means that the bubble integrity is not damaged. Beyond 0.4 bubbles collapse which gives a non-linear transient response[7] and bubbles are disrupted. In this case, theoretical modelisation of the contrast agent behaviour is rather complex and is dedicated to specific research.

1.4.2.a. Linear response

For small amplitudes of the ultrasound incident wave. the relative compression and expansion of the bubble is the same. Hence the bubble's walls vibration is said to be linear. To respect the linear assumption. the vibration amplitude must be neglectable compared to initial radius R_0. The backscattered wave is enhanced. resulting in an augmentation of the echo from blood compared to echoes from non perfused tissue. It was the behaviour originally envisaged by contrast agent manufacturers.

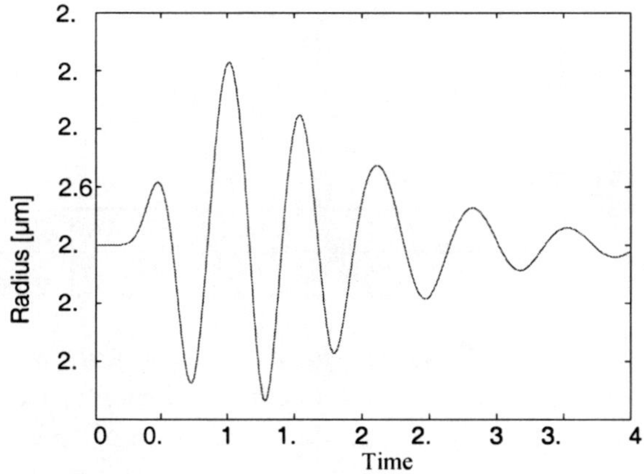

Figure (a)

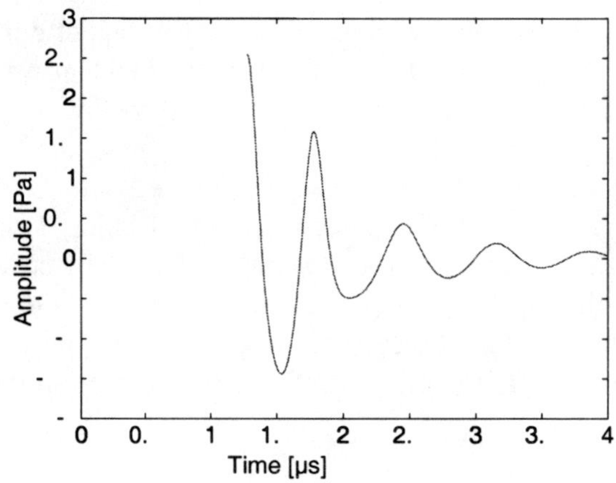

Figure (b)

Figure 1. Three sinus period 2 MHz driving pulse with 42 kPa inpinging a 2.5 µm free radius bubble in water: (a) radius oscillation and (b) radiated pressure.

It has been shown in Minnaert[8]. that a bubble behaves as mass-damper oscillator with its own resonance frequency. The expression of this resonance frequency for an adiabatic process is:

$$f_0 = \frac{1}{2\pi R_0}\sqrt{\frac{3\kappa P_0}{\rho_1}} \tag{4}$$

κ: polytropic constant
ρ_l: density of the surrounding medium (kg m^{-3})
R_0: stable radius of the bubble (m)
P_0: hydrostatic pressure (Pa)

By a fortunate (lucky) coincidence, the resonance frequency for bubbles of 1 to 6 μm diameter is within the range used for diagnostic ultrasound (Table 4). For an air bubble in water a simplified formula of resonance frequency is :

$$f_0 R_0 = f_0 D_0 /2 \approx 3.3 \text{ MHz μm} \tag{5}$$

TABLE 4. Resonance frequency (MHz) of air bubbles in water.

Diameter (μm)	1	2	4	6
Resonance frequency (MHz)	6.6	3.3	1.6	1.1

The overall effects of microbubble resonance are, therefore, an increase in the amplitude of the backscattered signal at its fundamental frequency (the frequency of the transmitted signal) and the generation of second harmonic (with frequency two times that of the fundamental) and higher multiples of that frequency.

In summary, the ultrasonic response of a contrast agent depends on the size of the bubble compared to the frequency of the incident wave. An ideal contrast agent would have dimensions at the resonant size (having a resonant frequency equal to that of the driving frequency) for optimal scattering of ultrasound.

The wavelength of the driving pressure signal being larger than the bubble's dimension one can assume that the bubble will scatter acoustic energy equally with respect to the direction. In that case the excitation is uniform for the entire bubble surface which implies a radial motion. Including the total damping coefficient, the efficient section of a radiating bubble[9] is:

$$\sigma_{SC}(f, R_0) = \frac{P_S}{I_i} = \frac{4\pi R_0^2}{\left[\frac{f_0^2}{f^2} - 1\right] + \delta_{tot}^2(f, R_0)} \tag{6}$$

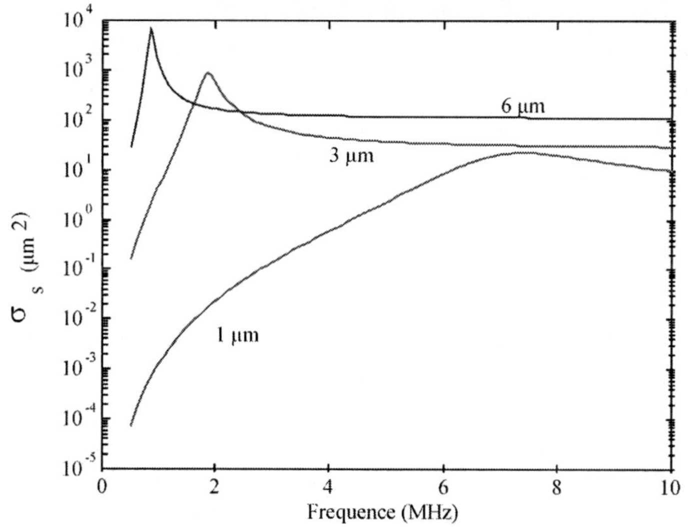

Figure 2. Scattering cross section versus frequency of three bubbles of diameter 1 μm, 3 μm and 6 μm.

1.4.2.b. Stationary non-linear harmonic response

Instead of producing a nice sinusoidal wave with a clean frequency spectrum. it produces an odd looking waveform with non symmetrical top and bottom. It is this asymmetry which produces harmonics. In Morgan[10], the authors present a modified Herring equation with shell terms that is solved for the time-dependent bubble radius and wall velocity, and these outputs are used to formulate the predicted echo from a single encapsulated bubble.

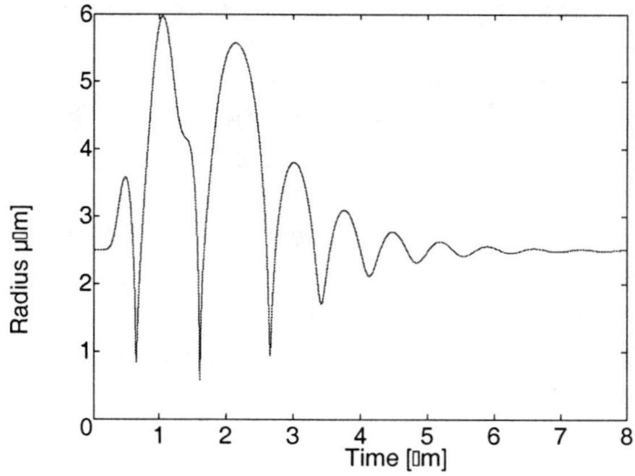

Figure 3. Three sinus period 2 MHz driving pulse with 420 kPa inpinging a 2.5 μm free radius bubble in water.

1.4.2.c. Transient non-linear response

As the amplitude of the acoustic wave is increased more, the scattering level of most of the contrast agents increases abruptly for a short time. This has been associated with bubble rupture and release of free gas bubble.

1.4.3. *The shell of the microbubble*

The shell has two major effects on the oscillation and scattering of sound from the microbubble. First, the shell makes the microbubble stiffer than a free gas bubble of equal size. This causes the resonant frequency of the shell-encapsulated microbubble to be at a higher incident ultrasonic frequency than for the free bubble and it also dampens the oscillation amplitude. Stiffer-shelled contrast agents have been found to rupture with less acoustic pressure than agents with more flexible shells[11]. The more elastic the shell, the greater its range of contraction and expansion prior to rupture. Flexible-shelled agents demonstrate stronger nonlinear response at a given ultrasound field acoustic pressure than their stiffer counterparts. The second important effect introduced by the shell is viscosity. The additional viscosity of the shell causes more of the energy in the incident ultrasonic pulse to be converted to heat instead of being reradiated. This reduces the scattering-to-attenuation ratio of the microbubble[12, 13].

The shell resistance to ultrasound exposure is proportional to its viscosity[14]. No threshold criterion described in the literature takes into account the shell's intrinsic properties in spite of the fact that they clearly play an important role in the acoustic pressure required for shell rupture[11]. Research on this relationship is underway that should help to link shell properties to microbubble vibration and rupture[15].

1.5. PARTICLE DISTRIBUTION IN AN ACOUSTIC FIELD

For an examination by contrast ultrasonography, a few millimetre of microbubble contrast agent is injected intravascular. It is known that the usual approach based on analyzing ultrasound effects of single scatterers in a liquid can be inadequate for appropriate modelling of a bubbles cloud. However, for very small and widely separated scatterers it is admissible to assume a linear behaviour. At high bubbles concentration, multiple scattering attenuation and shadowing cause a flattening of the curve and can even lead to a degradation of the scattering cross section of the cloud.

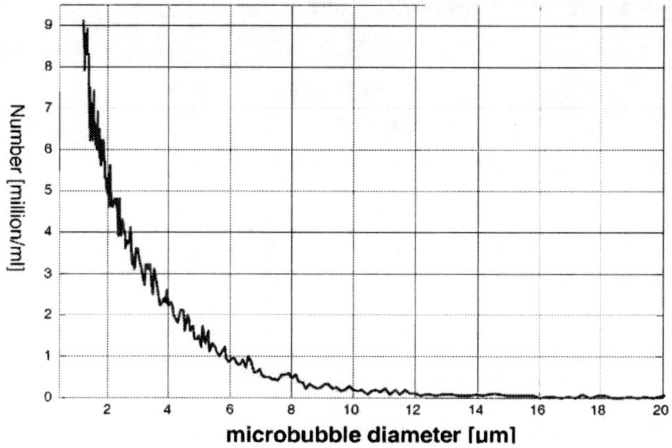

Figure (a)

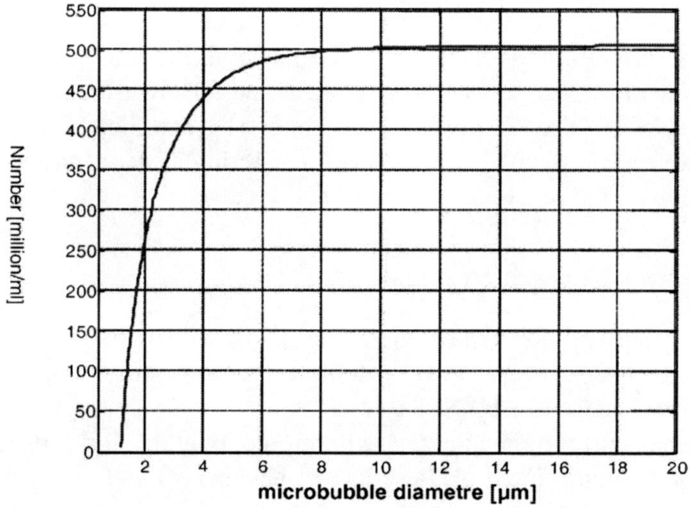

Figure (b)

Figure 4. Distribution of a sample of SonovueTM (a) with respect to bubble's diameter, (b) cumulated number with respect of the diameter.

In theory, on one hand backscattering of the contrast agent increase linearly while the attenuation due to the cloud increases exponentially with respect to the bubble concentration. Consequently reflected intensity must increases linearly until a given concentration. then decreases exponentially. The attenuation due to a cloud of bubbles cannot be neglected when bubble's concentration becomes high.

1.5.1. *Total backscattering coefficient*

The total backscattering coefficient can be defined as the sum of the individual bubble's responses belonging to a radius bubble distribution $n(R_0)$ for a unit of volume. This is valid under a certain bubble concentration. The backscattering coefficient $\eta(f)$ can be expressed in by:

$$\eta(f) = \frac{1}{4\pi}\sum_{i=1}^{N_b} n(R_0^i)\sigma_{cs}(f, R_0^i) \tag{7}$$

The backscattering coefficient can be seen as the equivalent of the frequency dependent transfer function of the contrast agent.

1.5.2. *Lungs filtering*

In a human body, the lungs have a direct influence on the bubble distribution function. Indeed they act like a filter by removing bubbles with large radius from the distribution function. This filtering function can be modelled[13] by removing 10% of the bubbles with radius between 3 and 4 μm, 50% with radius between 6 and 7 μm and 90% with radius larger than 10 μm. Notice that large bubbles have a great impact on the total backscattered energy. This imply that the filtering effect of the lungs in not neglectable on the resulting echographic image.

1.6. CONTRAST IMAGING

Contrast imagings are dedicated to the comportment of UCA. The main imaging techniques are now available in ultrasonic scanners. Their principles are presented by N. De Jong[16].

The ones currently available are harmonic imaging. and Pulse Inversion Harmonic Imaging. Each of these has its advantages and drawbacks and is useful in certain situations. Further investigations into the clinical utility of ultrasound contrast agents will help identify which imaging mode should be used in a given clinical situation. Also, as more is learned about the physics of microbubbles and their interaction with ultrasound, improvements will be made on these imaging methods. as well as new ones being developed.

1.6.1.a. Fundamental imaging

This technique corresponds to the classic mode B imaging which builds an image from the concatenation of a set of RF-signals that have been post processed.

1.6.1.b. Harmonic imaging

Harmonic imaging, or more correctly second harmonic imaging. is set up to transmit a signal at one frequency (the fundamental or first harmonic) and to receive echoes at twice that frequency (the second harmonic). Harmonic imaging is based on the higher non linearity of contrast agents compared to tissues.

This technique takes advantage of the non-linear properties of the bubbles that composes the contrast agent. Indeed when non-linearity occurs, the bubble vibrates at second and higher multiples of the transmitted frequency. The frequency content of the backscattered signal is then composed of this fundamental frequency and harmonic frequencies. This phenomenon can be neglected in tissue. This is why backscattered signal coming from the tissue can be discriminated from the backscattered signal coming from the contrast agent. This is achieved by filtering the fundamental frequency out by using a high-pass filter. The resulting signal is then processed for imaging. Unfortunately this technique is highly related to the frequency bandwidth of the transmitted signal. Indeed, for good results the bandwidth of the emitted signal must narrow in order to minimize the spectral overlap between the fundamental and the harmonics frequency content. This induces a loss of accuracy in axial resolution.

1.6.1.c. Pulse inversion imaging

In pulse inversion imaging, a sequence of two ultrasound waves is transmitted into tissue. The second wave is transmitted after a suitable delay and is an inverted replica of the first wave. For a linear medium the response of the second wave front is an inverted copy of the first wave front. The sum of the two responses is zero. For non-linear region the response of the second wave front is not an inverted copy. The sum is not zero and the resulting value is proportional to the degree of non-linearity.

1.6.1.d. Contrast Doppler Imaging

The techniques developed in imaging have been extended in Doppler mode and they are performing In the following we have tried to explain simply the properties of the contrast agent used in the imaging modalities.

Why Imaging is performing with contrast agent?

Strong (resonant) scatterers are increased in the perfused tissue

Why Harmonic Imaging is performing with contrast agent?

The non linearity is higher for contrast agent than the tissue one

Why Pulse Inversion Harmonic Imaging is performing with contrast agent?

Pulse inversion increase the signal to noise ratio

Why Pulse Inversion Harmonic Doppler Imaging is performing with contrast agent?

The Doppler technique images the decorrelation of ultrasound signals and is performing because:

- contrast agent is moving with the blood flow,
- contrast agent distribution is changing (time life of microbubbles),
- contrast agent spatial distribution is modified by the ultrasound field (radiation force).

1.7. BENEFITS AND DRAWBACKS

- Contrast agent enhance the echoes and so the grey level of cavity and perfused tissues
- Ultrasound contrast imaging is
  - contrast particle dependent
  - injection mode dependent (bolus or perfusion)
  - frequency dependent
  - imaging modality dependent
  - operator dependent
- The protocol have to be adapted to the organ.

References

1. Gramiak R. and Shah P.M. Echocardiography of the aortic root. investigated Radiology. 3: 356-366, 1968
2. Fritzsch T. Schartl M. and Siegert J. *Preclinical and clinical results with an ultrasonic contrast agent.* Invest Radiol. 23 Suppl 1: pp. S302-5, 1988
3. Geny B. et al. *Safety and efficacy of a new transpulmonary echo contrast agent in echocardiographic studies in patients.* J Am Coll Cardiol. 22(4): pp. 1193-8, 1993
4. Bilotta F. et al. *Pulmonary transit of echocontrast agents during mechanical ventilation: a clinical transthoracic echocardiographic study.* Echocardiography. 22(5): pp. 395-401, 2005
5. Quay S.C. Ulrasound contrast agent development: the role of gaz selection in microbubbles persistance. J. Ultras. Med. vol. 13. p. 59, 1994

6. Meltzer R.S. et al. *Transmission of ultrasonic contrast through the lungs.* Ultrasound Med. Biol. 7(4): pp. 377-84, 1981
7. De Jong N. and Frinking P. Characteristics of contrast agents an 2nd imaging. Proc. IEEE Ultrason. Symp. pp. 1449-1458, 1996
8. Minnaert M. On musical air bubbles and the sound of running water. Philosophical Magazine. 10. pp. 235-248, 1933
9. Medwin H. Counting Bubble Acoustically: a Review. Ultrasonics. pp. 7-13, 1977
10. Morgan K.E. Allen J.S. Dayton P.A. Chomas J.E. Klibanov A.L. and Ferrara K.W. *Experimental and theoretical evaluation of microbubble behaviour: effect of transmitted phase and bubble size.* IEEE Transactions on Ultrasonics. Ferroelectrics and Frequency Control (UFFC). 47(6):1494-1509, 2000
11. Stride E. and Saffari N. *On the destruction of microbubble ultrasound contrast agents.* Ultrasound Med Biol. 29(4): pp. 563-573, 2003
12. Bouakaz A. et al. *In vitro standardc acoustic parameters of ultrasound contrast agents: Definitions and calculations.* IEEE Ultrason Symp Proc. 96CH35993: pp. 1445-1448, 1996
13. Bouakaz A. De Jong N. and Cachard C. *Standard properties of ultrasound contrast agents.* Ultrasound Med Biol. 24(3): pp. 469-72, 1998
14. Hoff L. Sontum P.C. and Hovem J.M. *Oscillations of polymeric microbubbles: effect of the encapsulating shell.* J Acoust Soc Am. 107(4): pp. 2272-80, 2000
15. Sarkar K. et al. *Characterization of ultrasound contrast microbubbles using in vitro experiments and viscous and viscoelastic interface models for encapsulation.* J Acoust Soc Am. 118(1): pp. 539-50, 2005
16. De Jong N. Frinking P.J.A. Bouakaz A. and Ten Gate F.G. Detection procedures of ultrasound contrast agents. Ultrasonics. 38(1-8):87-92, 2000

SPECKLE FORMATION, ANALYSIS AND PROCESSING APPLIED TO ULTRASOUND TISSUE CHARACTERIZATION

JOHAN M. THIJSSEN*
833 Clinical Physics Laboratory, University Children's Centre, University Medical Centre Nijmegen, P.O. Box 9101, 6500 HB, Nijmegen, The Netherlands

Abstract. This paper describes the image formation in medical ultrasound for the case of scattering media. The texture statistics are dominated by speckle formation which results from random interference of backscattered echoes. The effects of spatial, fixed and adaptive, filtering, as well as, of grey scale encoding on the detection of lesions are analytically described and illustrated with representative images.

Keywords: ultrasound; imaging; speckle; texture; grey scale; filtering.

1. Introduction

Ultrasonic tissue characterization (UTC) is a well-established field of scientific research since the first publication on computer analysis of radiofrequency (rf)-signals by Mountford and Wells (1972). After another milestone, i.e. the first symposium at the National Bureau of Standards, Gaithersburg, MD, in 1975 (Linzer, 1976), followed by a similar European meeting in 1979 (Thijssen, 1980), an ever lasting stream of scientific papers has been devoted to this topic. The industrial interest has been relatively low, partly because of a few disappointments, partly because the technological demands of developed UTC methods were too high. Presently, modern digital scanners may be considered to be "ready" to accept these methods!

A definition of UTC could be: "The assessment by ultrasound of quantitative information about the characteristics of biological tissues, and

* Johan M. Thijssen, 833 Clinical Physics Laboratory, University Children's Centre, University Medical Centre Nijmegen, P.O. Box 9101, 6500 HB, Nijmegen, The Netherlands; e-mail: j.thijssen@cukz.umcn.nl.

Y. Lemoigne et al. (eds.), Physics for Medical Imaging Applications, 151–176.

pathological changes thereof''. This quantitative information is extracted from the echographic data at two different levels: first, the raw data, i.e. the rf-signals corresponding to single transmit–receive echo lines in an image after linear amplification; second, the two-dimensional (2D) echographic images, which are constructed by the scanner from the rf-signals. The scan lines generally have not only been demodulated but also, in addition, compressed and filtered.

A basic problem inherent in both ways of data analysis is the so-called beam diffraction: due to the limited transmit- and receive-apertures of the ultrasound transducer and additional transmit/receive focussing regimes the spatial and spectral beam characteristics are depth dependent. This means that even in an isotropic and homogeneous medium the backscattered signals are depth dependent. The position and width of the corresponding amplitude spectrum, but also the ''speckle'' pattern in images, changes from the transducer face to greater depth. Furthermore, amplitude decay due to attenuation in the examined tissues is changing the information contents and the amplitude of the signal with depth. The latter is compensated for by a time gain compensation (TGC) setting of the receiving amplifier. These problems can be accounted for by adequate calibration measurements when rf-signals are processed (e.g., Huisman and Thijssen, 1996). In case of image analysis, the derived UTC parameters have to be corrected or a posteriori calibrated (cf., Valckx et al., 1996).

Processing of rf-data is used for estimating acousto-spectrographic tissue characteristics, i.e. the frequency dependence of the attenuation and backscattering coefficients in the spectral domain. The attenuation is mainly due to absorption of the ultrasound. Absorption is caused by leakage of vibration energy into vibration and rotation modes of relaxation oscillation of biological (macro) molecules. In soft tissues this results in a linear with frequency increase of the absorption coefficient (e.g., Jongen et al., 1986). The backscattering of tissues is related to the presence of collagen structures, like microvasculature and elastic fibres. The backscattering characteristics are depending on the scatterer dimensions relative to the wavelength of the ultrasound (e.g., Nicholas, 1982).

Analysis of rf-data is also performed in ''elastography'' (cf., Ophir et al., 1991). Here the backscattered rf-signals are acquired prior to and after a slight compression of the tissues. The apparent time shift between equivalent windows of these rf-signals is estimated by a correlation technique and, by using the speed of sound of the tissue, the local displacement is obtained. Then, the relative compression, i.e. the tissue strain, is estimated vs. depth. This strain is related to the elasticity of the tissue (i.e. Young's modulus), hence the terms elastography, elasticity imaging, or sonoelasticity (Levinson, 1988). It may be obvious that elastography yields the estimate of a completely new tissue characteristic as compared to other methods.

Finally, it may be remarked that the rf-based techniques are used to estimate the discussed acoustic tissue parameters locally ("sliding window") and to produce so-called parametric images. This might be identified as the ultimate goal of innovation in medical ultrasound!

A second part of UTC is based on the analysis of the texture in 2D echographic images. It was shown by realistic simulations, that the statistical and speckle characteristics of echographic texture are dependent on the "number density" of the scattering sites within a medium (Oosterveld et al., 1985). In other words, the grey level statistics and the speckle size changed continuously when the number of scatterers, on average, within the resolution cell corresponding to the transmitted beam width and pulse length increased from 1 to approximately 10. In the latter case "fully developed" speckle occurs. This result shows that image analysis could in principle reveal the effective number density of tissues, as well as pathological changes of this parameter. The number density may be considered as a histological characteristic of tissues.

The literature is somewhat confused by the problem that the authors not always realize that the displayed echographic images are constructed after logarithmic compression. Some authors construct the images after rf-signal preprocessing (diffraction correction) and subsequent software demodulation, thus resulting in linear speckle production (Oosterveld et al., 1991, 1993; Romijn et al., 1991; Momenan et al., 1988; Schmitz and Ermert, 1999). Other authors just digitize the analog video format images and which means that the incorporated log-compression has completely changed the speckle statistics (Nicholas et al., 1986; Giessen et al., 1996).

Another method of texture analysis was devised to reveal eventual spatial regularity in the histology, e.g., the lobular structure of the liver parenchyma (Wagner et al., 1986). Based on second order statistical texture analysis some evidence for its usefulness could be shown. Further texture analysis methods are not making use of any 660 J.M. Thijssen / Pattern Recognition Letters 24 (2003) 659–675 a priori assumptions about tissue histology but merely use methods developed in other, generally technical, fields of image analysis (e.g., Oosterveld et al., 1991, 1993; Romijn et al., 1991; Giessen et al., 1996; Schmitz and Ermert, 1999; Valckx et al., 2000).

Speckle reduction filtering has been investigated recently by using adaptive filtering methods. The assumption behind it is that the detectability of local texture changes (so-called "lesions", or clinically "tumours") can be improved by diminishing the speckle noise (cf., Bamber and Daft, 1986; Loupas et al., 1989; Kotropoulos and Pitas, 1992; Verhoeven and Thijssen, 1993). The latter authors investigated the results of filtering by estimation of the "lesion detectability" and they could not show considerable improvements.

The final topic of the paper is concerned with grey scale manipulation, i.e. the effects of changing the gamma of the grey level encoding (Thijssen et al., 1988). This question was induced by literature on the effect of log-compression on the grey level statistics and by the wish to optimize the displayed information.

2. Physical backgrounds

2.1. ACOUSTIC TISSUE MODEL

A general model of biological tissues is shown in Fig. 1. The model, suitable for most parenchymal tissues (i.e. constituting organs like liver, spleen etc.), is assuming homogeneity and isotropy. Acoustically it is characterized by a fixed speed of sound (on the order of 1540 m/s in soft tissues) and by absorption and scattering. In case the whole "region-of-interest" (ROI) would not consist of the same tissue, the speed of sound might vary in depth. Although many attempts have been made to locally estimate the speed of sound (see for a survey Robinson et al., 1991), it has not been successful. This is partly due to the condition that at least two independent measurements (i.e. from different transducer positions) are necessary. Therefore, the UTC is sometimes applied to ROIs of limited size. Absorption is due to relaxation phenomena of translational and rotational vibration modes of biological macromolecules. Scattering is considered to obey the Born approximation (i.e. single scattering, undisturbed wave propagation) and is related to small inhomogeneities in acoustic impedance, which are randomly distributed in three-dimensional (3D) space. For instance, the microvasculature (arterioles), and the collagen matrix of parenchymal tissues are producing relatively strong scattering contributions as compared to individual cells (e.g., blood cells). The overall attenuation due to absorption and scattering is accessible for estimation. In terms of backscattered echo amplitude $e(t,z)$:

$$e(t,z) = e(t,0)e^{-\mu(f)2x} \tag{1}$$

where t is the time after ultrasound transmission, f is the frequency, z is the depth (i.e. $2z$ is the two-way travel path of scattered wave) and μ is the attenuation coefficient (nepers/cm).

In addition to the random (=diffuse) scattering, in some tissues small range order is present which is related to structural regularity on a millimetre scale, e.g., the triads of Kiernan constituting the portal system of the liver. So, in addition to diffuse scattering, a structural scattering component is present which is revealed in the texture of 2D echographic images (Jacobs and Thijssen, 1991).

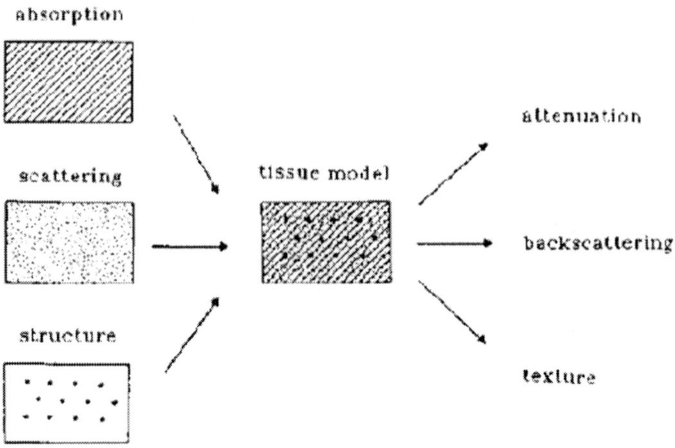

Figure 1. Block diagram of "acoustic tissue model" and kind of information that can be derived with ultrasonic tissue characterization.

2.2. PREPROCESSING

The characteristics of the ultrasound beam are depth dependent, both in continuous wave and in pulsed mode. The limited dimensions of the transducer cause this. The resulting intensity distribution is called the "beam diffraction pattern", or "directivity function", which applies both to the transmission and to the reception mode of the transducer employment (reciprocity principle). Generally, the directivity effects and the additional depth effects of transmit- and receive-focusing are summarized in the same term: beam diffraction. Furthermore, it was shown experimentally (Thijssen et al., 1981; Cloostermans and Thijssen, 1983) that the received echographic spectra, calculated from the rf-data, are highly depth dependent and, therefore, influence the estimates of the frequency dependence of the attenuation- and backscattering- coefficients (see also Fink and Cardoso, 1984; Robinson et al., 1984; Verhoef et al., 1985). Finally, it was shown by experiments and simulations (Oosterveld et al., 1985) that also the texture features (i.e. speckle characteristics) estimated from echographic images display a significant depth dependence (see also Section 4).

The tissue attenuation yields both an amplitude decay of the echographic signals, as well as an effective downshift of the spectral information. The latter effect is not shown in echographic images, due to the amplitude demodulation. It is, however, used in acousto-spectrographic approaches to estimate the frequency dependence of the attenuation coefficient (Kuc et al., 1976; Cloostermans et al., 1986; Dines and Kak, 1970; Fink et al., 1983).

It can be concluded that prior to UTC, the rf-signals have to be corrected, in the frequency domain, for the influence of beam diffraction (cf., Thijssen et al., 1981; Cloostermans and Thijssen, 1983; Fink and Cardoso, 1984; Verhoef et al., 1985; Romijn et al., 1989, 1991; Oosterveld et al., 1991; Huisman and Thijssen, 1996) and, in addition, for the employed depth dependent amplification TGC.

Eq. (1) is obviously too simple to describe the real world conditions in echography. After transforming Eq. (1) to the frequency domain and insertion of transducer performance, the backscattered spectrum $E(f,z)$ becomes:

$$E(f,z) = P^2(f)\, D^2(f,z) T^2(f,z)\, S(f), \tag{2}$$

where $P(f)$ is the electro-acoustic transfer function of transducer, $D(f;z)$ is the beam diffraction transfer function (spectrogram), $T(f;z)$ is the tissue transfer function (attenuation), and $S(f)$ is the backscatter transfer function.

In Eq. (2) it is assumed that the TGC amplification has been corrected for. The diffraction correction consists of the estimation of PðfÞ from a reflection in focus in a medium with known acoustic characteristics and of $D^2(f;z)$from the average backscatter vs. depth of a medium with known backscatter characteristics (e.g., Romijn et al., 1989; Huisman and Thijssen, 1996). The next step is to analyse the corrected spectra vs. depth:

$$E_c(f,z) = T^2(f,z)\, S(f), \tag{3}$$

By taking the logarithmic spectra (in dB) and differentiating with respect to depth z:

$$\frac{\partial}{\partial z}\{\log_{10}[Ec(f,z)]\} = 2\,\frac{\partial}{\partial z}\{\log_{10}[T(f,z)]\} \tag{4}$$

In this way, the tissue transfer function (i.e. attenuation coefficient $\mu(f)$) is obtained, which is then used to compensate the spectrogram for the effects of attenuation and resulting is:

$$E_c^*(f,z) = S(f), \tag{3}$$

So, the homogeneous backscattering vs. frequency is obtained over the whole depth range that is being analysed.

Finally, the attenuation-compensated data can be software demodulated and a corrected echographic image can be constructed ready for texture analysis (cf., Romijn et al., 1991; Oosterveld et al., 1991).

When rf-signals are not available, a pragmatic approach for preprocessing the video images prior to analysis is the use of a reference measurement on a tissue-mimicking phantom with known acoustic characteristics. This method allows for correction of the diffraction and attenuation effects in the axial direction, i.e. along the scan lines, and for the equipment settings of overall gain and TGC (cf., Valckx et al., 1996, 2000). In the past, many authors neglected position dependencies of the areas in the image selected for analyse, or they restricted the analysis to a single ROI in the focal transmit zone of the transducer.

3. Texture

3.1. SPECKLE FORMATION

The scattering due to small (subwavelength) inhomogeneities yields an image of "soft" tissues. Since, the echoes from nearby scatterers arrive at the transducer face with small time differences interference occurs (cf., Fig. 2). The rf A-lines constituting a 2D image are displaying random amplitude vs. time, as well as small phase jumps. The interference may be described as a random walk in 2D space. If the number of scatterers within the resolution cell of the transducer is high (>10) then it follows from the central limit Fig. 2. Backscattered echoes from within a ''resolution cell'' are arriving with small time differences at transducer face. Due to linear summation an interference pattern results, corresponding to one rf scan line. theorem, that the analytical signal has a 2D Gaussian probability distribution function (pdf). When considering amplitude demodulation as estimation of the envelope of the analytical rf-signals, it can be shown that the magnitude (envelope) has a Rayleigh pdf and the phase has a uniform pdf (cf., Papoulis, 1965).

At lower scatterer number densities, grey levels are distributed according to the so-called K-distribution (Jakeman and Tough, 1987), which degrades to the Rayleigh pdf at higher densities.

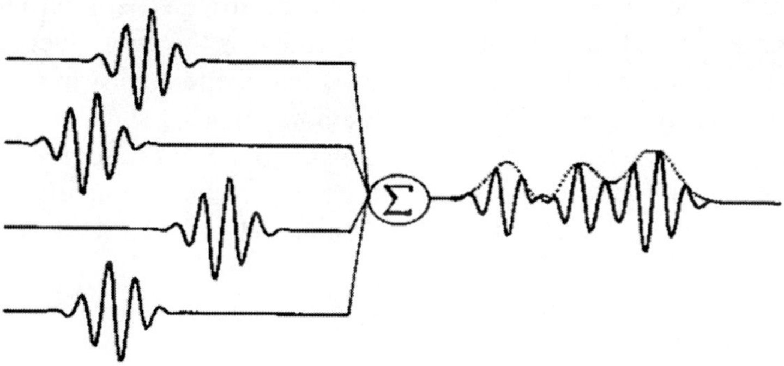

Figure 2. Backscattered echoes from within a "resolution cell" are arriving with small time differences at transducer face. Due to linear summation an interference pattern results, corresponding to one rf scan line.

In 2D imaging, the sound beam is moved, which means a gradual decorrelation of the axial speckle pattern. This gradual change results in a pattern of ‘‘banana’’ shaped speckles (cf., Fig. 3). Each speckle has some similarity with the 2D image of a single point reflector (i.e. the point-spreadfunction (PSF)). It should be realised, however, that the lateral speckle size is highly dependent on the distance to the transducer and not a direct image of any structure!

Figure 3. Top row: echographic "speckle" images from homogeneously scattering medium. Depth increases from zero (transducer) to 16 cm, from left to right. Bottom row: PSF obtained by scanning a point reflector at depths from 1 to 15 cm.

3.2. TEXTURE STATISTICS

Texture can be defined as the arrangement of grey levels in an image. The most simple quantification of this arrangement is the grey level histogram, which is a first order statistic. This means that the texture is measured pixel wise, without considering the relations among pixels.

For large scatterer number density the histogram of echographic images, when the amplitude of the echo strength is processed linearly, obeys the Rayleigh pdf (e.g., Burckhardt, 1978; Wagner et al., 1983; Oosterveld et al., 1985):

$$p(\hat{e}) = \frac{\hat{e}}{\sigma^2} e^{-\frac{1}{2}\frac{\hat{e}^2}{\sigma^2}}, \tag{6}$$

where $\hat{e}$ is the demodulated signal ("envelope"), and σ^2 is the signal power.

It can be shown (cf., Papoulis, 1965) that mean, μ_e, and variance, σ_e^2, are:

$$\mu_e = \sqrt{\frac{\pi\sigma^2}{2}}, \tag{7}$$

$$\sigma_e^2 = \sigma^2 (2 - \frac{\pi}{2}) \tag{8}$$

Hence, the so-called "signal-to-noise" ratio (SNR) becomes:

$$SNR = \frac{\mu}{\sigma} = \sqrt{\frac{\pi}{4-\pi}}, \tag{9}$$

The general expression (irrespective of scatterer density) for the pdf of the signal intensity I is given by (Jakeman, 1984):

$$p(I) = \frac{1}{2}\int_0^\infty u du J_0 (u\sqrt{I} \langle J_0^N (u\beta)\rangle_{N,\beta}, \tag{9}$$

where J_0 is the Bessel function of zero order, first kind, β is the scattering amplitude of individual scatterers, and N is the number of scatterers within resolution cell.

This equation can not be analytically solved, but expressions were obtained for the moments of the pdf. For large N (i.e. $N > 10$) this pdf turns into the exponential pdf. It was shown (Jakeman, 1984) that the "SNR" equals:

$$SNR_I = \frac{\mu_I}{\sigma_I} = \left(1 + \frac{\langle \beta^4 \rangle}{N \langle \beta^2 \rangle^2}\right)^{-1/2} \tag{10}$$

Taking the scatterer number density n and the volume corresponding to the transmitted ultrasound pulse (in focus) equal to V, then $N = nV$. The number density can then be estimated from Eq. (11) if β is constant.. In soft biological tissues $\beta \approx 3$ (cf., Sleefe and Lele, 1988).

More recently, it was shown that the pdf of the envelope amplitude for low (effective) scatterer density is given by the K-distribution (Jakeman and Tough, 1987):

$$p(\hat{e}) = \frac{2b}{\Gamma(\alpha)} \left(\frac{b\hat{e}}{2}\right)^{\alpha} K_{\alpha-1}(b\hat{e}), \tag{11}$$

where $b = 2\sqrt{\alpha / E(\hat{e}^2)}$, $\Gamma()$ is the gamma function, $\alpha = (\mu+1)N$ is the effective number of scatterers within resolution cell, μ is the constant, depending on scatterer and transducer characteristics, and K_x is the modified Bessel function of second kind, and order x.

The moments of the K-distribution have been analytically derived and may be used to calculate the SNR of the envelope (Dutt and Greenleaf, 1995):

$$SNR_{\hat{e}} = \frac{\sqrt{\pi}\Gamma(\alpha + \frac{1}{2})}{\sqrt{4\Gamma(\alpha+1) - \pi \cdot \Gamma^2(\alpha + \frac{1}{2})}}, \tag{12}$$

It can be shown that in the limit case, where α goes to infinity (in practice $\alpha \geq 10$) the K-distribution approaches the Rayleigh pdf (''fully developed speckle'').

It should be realised, that in general terminology of echographic scanners a 'linear' processing means that a linear look-up table is used to encode the image grey levels. The systems use nevertheless a logarithmic compression amplifier prior to digitisation, which yields an overall histogram that is considerably different from the Rayleigh pdf (Kaplan and Ma, 1994; Dutt and Greenleaf, 1996). This pdf of the logarithmic echo signals yields a fixed standard deviation (i.e. additive, rather than multiplicative, speckle noise).

Recently, it has been argued that when a structural component is present in the location of backscattering sites in tissue (i.e. a more or less regular pattern, like in the distribution of liver triads), the Rayleigh pdf should be replaced by the Rician pdf (e.g., Goodman, 1975; Insana et al., 1986; Thijssen, 1992).

A method has been developed to unravel the contributions of the diffuse and structural scattering components (Wagner et al., 1986; Insana et al., 1986). The characteristic distance of the structural scatterers is revealed in the power spectrum by spectral peaks superimposed on the spectrum corresponding to the diffuse scattering (cf., Fig. 4).

Taking the characteristic structural distance Δz_s, i.e. a time difference

$$2\Delta\tau = \frac{2\Delta z_s}{c}, \tag{13}$$

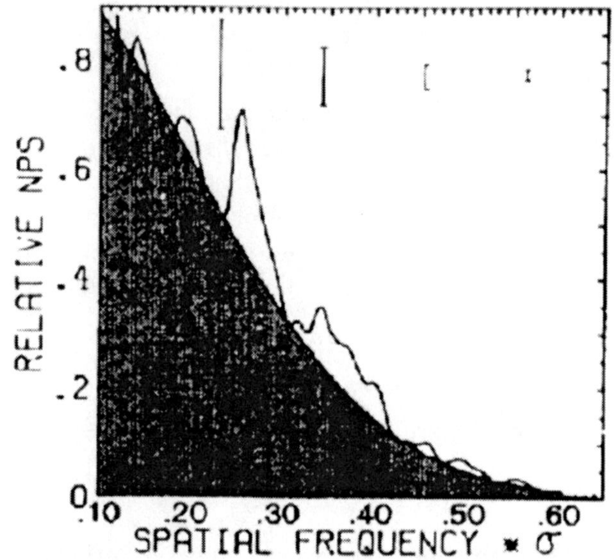

Figure 4. Example of power spectrum obtained from echographic image of liver tissue. Shaded area corresponds to power spectrum of transmitted ultrasound pulse and is due to ''diffuse scattering''. Superimposed spectral peaks are related to structural scattering component (triads of Kiernan) (Insana et al., 1986).

which equals the two-way travel time in pulse-echo mode, then it can be shown, that *F(f)*, the backscattered power spectrum (in the temporal frequency domain, one-dimensional (1D) case), is given by (Mesdag, 1985; Jacobs and Thijssen, 1991):

$$|F(f)| = \left|\frac{\sin(N2\pi f\tau}{\sin(2\pi f\tau}\right| |S(f)|, \tag{14}$$

where N is the number of scatterers within duration of transmitted ultrasound pulse, τ, and S is the power spectrum, corresponding to the envelope of diffusely backscattered echo signal.

The spectral peaks can be analysed by estimating the first peak at $f > 0$, or more sophisticated methods like for instance cepstral analysis (Varghese and Donehue, 1995) could be applied. Till now the analysis, initiated by Fellingham and Sommer (1984), has not been highly successful in clinical applications. The simulation study by Jacobs and Thijssen (1991) indicated that if the structural distance variations throughout the tissue volume are larger that 10% of the mean, the distance estimate becomes inaccurate.

4. Texture analysis

4.1. GREY LEVEL STATISTICS

The first order statistics are quantified by calculating the moments and derived parameters. A useful derived histogram parameter is the SNR (Eq. (9)). This parameter was analysed as an UTC parameter (Oosterveld et al., 1985; Dutt and Greenleaf, 1995) and depends on the density of scatterers (Eqs. (11) and (13)). The SNR was also used for parametric imaging (Verhoeven et al., 1991).

The second order statistics are used to describe and quantify the relations between neighbouring pixels. At first, the autocovariance function (ACVF) was introduced (Wagner et al., 1983; Oosterveld et al., 1985) (cf., Fig. 5) which can be considered as a measure of the average size of the speckles in echographic images:

$$\mathrm{ACVF}(x,z) = \iint_{ROI} d\xi d\zeta \{\hat{e}(\xi + x, \zeta + z) - \langle \hat{e} \rangle\}\{\hat{e}(\xi,\zeta) - \langle \hat{e} \rangle\} \qquad (15)$$

where ROI is the "region-of-interest" area, $\langle \hat{e} \rangle$ is the mean of signal envelope in ROI, x, z, and ξ, ζ is the lateral, axial coordinates, respectively.

The average size of the speckles in the focal zone of the transducer (i.e. plane wave conditions) can be quantified by the full-width-at-half-maximum (FWHM) of the ACVF (Wagner et al., 1983; Oosterveld et al., 1985):

$$FWHM_x = 0.80 \frac{\lambda f}{D'}, \qquad (16)$$

$$FWHM_z = \frac{0.61}{\Delta f} \qquad (17)$$

where $\lambda = c/f_c$ is the wavelength of transmitted ultrasound at centre frequency, F is the focal distance, D' is the effective diameter of transducer *(=D/1.08),* and Δf is the half width (_6 dB width) of spectrum corresponding to transmitted pulse.

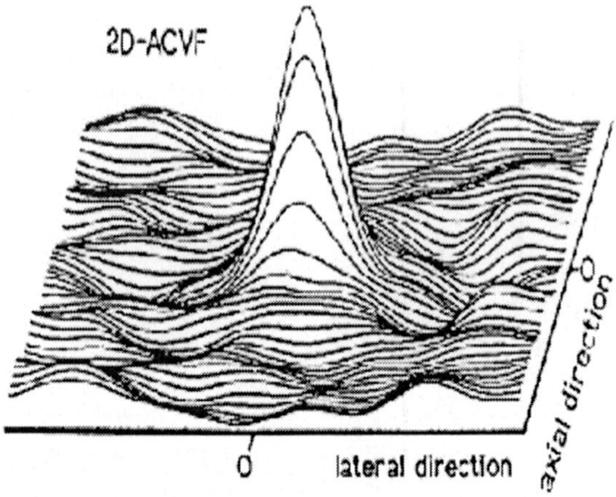

Figure 5. Two-dimensional graph of the autocorrelation function of an echographic image of a homogeneous, isotropic scattering.

Eqs. (17) and (18) apply in case of fully developed speckle (i.e. high scatterer number density) in the focal zone. It may be remarked, that the single scatterer case, corresponding to the PSF of the used transducer can be written in similar formulas, and yield 10–20% larger values (Thijssen, 1992).

The texture analysis based on a tissue model containing both diffuse and structural scattering (cf., Fig. 1) was undertaken by Wagner et al. (1986) and Insana et al. (1986). These authors, among other parameters, estimated the "Rician" variance ($\sigma_r^2(I)$) of the backscattered intensity from the Fourier transform of the ACVF. This is equivalent to the area under the curve fitted through the minima of the estimated power spectrum. This curve is modelled (or measured) as the power spectrum derived form a plane reflector in focus. Subtraction of the Rician power from the total area under the spectrum yields the variance of the structurally backscattered power, $\sigma_S^2(I)$.

These variances can be used as tissue parameters, or to calculate the backscattered diffuse and structural intensities. The latter procedure (Wagner et al., 1986) contains a square root operation, which yielded algebraic instabilities in a study of diffuse liver diseases (Oosterveld et al., 1991).

The speckle size is not only dependent on the transducer characteristics, but varies with the scatterer number density as well (Oosterveld et al., 1985). A serious problem with the ACVF as a texture parameter is that in particular the lateral size is strongly dependent on the depth and also on the attenuation within the tissue (Oosterveld et al., 1985; Huisman and Thijssen, 1998b). These dependencies have often been overlooked, or neglected, in literature. The only way out seems to be to analyse the second order statistics in the axial direction exclusively. The problem remains that even after the diffraction and attenuation corrections (see Section 2.2) and software demodulation, the lateral speckle size is still strongly depth dependent (cf., Fig. 3).

Further methods of second order statistics analysis have been taken from the general image processing literature. An often used statistic is the co-occurrence matrix (Haralick et al., 1973). This is actually a second order (i.e. two-dimensional)- histogram approximating the probability of occurrence of all combinations of two grey levels for a set of two pixels at a certain spatial distance (or "pixel displacement"). Many parameters can be derived from this matrix, some of which are often used in UTC: contrast, correlation, entropy and angular second moment (cf., Thijssen et al., 1993; Valckx and Thijssen, 1997).

Several other second order statistics applied to UTC can be found in literature of the last 10 years. It is questionable whether these contribute in essentially different ways as compared to the ACVF and the co-occurrence matrix. In any case, at the scale of the size of the speckles (i.e. small displacements), all methods will be equivalent. At a larger scale (supra-speckle scale) macroscopic changes in tissue structure should be present which can produce a kind of modulation of (local-mean)- grey levels, e.g., as is found in a fibrotic liver. In that case, the displacement used in the estimation of the second order statistic should be much larger than most authors have considered so far. An alternative to these methods has recently been presented by Huisman and Thijssen (1998a) in the way of an adaptive feature extraction algorithm based on higher order neural networks (HONN). This approach does not make assumptions about the scale and it can be based on first and/or second order statistics without further restrictions. Their results, as worked out for the first order case, are rather promising.

4.2. CLASSIFICATION

This is the final goal of UTC: labelling an analysed window, or a ROI, as normal or, alternatively, as diseased. The number of estimated parameters used for the classification can be relatively high, but the number of independently proven cases of a disease is however very often relatively low. This means, that learning the ''classifier'', i.e. the algorithm to decide, is based on a limited set of clinical cases. The classifier can be e.g., a discriminate function (Bayesian classifier; Oosterveld et al., 1993; Wagner et al., 1995), a neural network (Klein Gebbink et al., 1993; Huisman and Thijssen, 1998a; Schmitz and Ermert, 1999), or a decision tree (Giessen et al., 1996). In any case, it has been shown by Foley (1972) that the number of parameters that can be taken into account in a linear classification algorithm should not exceed the smallest number of cases, in any of the classes, divided by three (or more, in case of non-linear classifiers). This rule is often not considered in literature!

5. Speckle reduction

5.1. INTRODUCTION

Since speckle is a 2D noise in echographic images from scattering media, it has been attempted for a long time already to reduce this image noise. One approach is to scan the same tissue region several times, either from different directions, or from the same direction while using different transducer performance characteristics. Both methods yield a set of images from the same region and speckle reduction is obtained by averaging the images (''compounding''). Scanning a region from different directions is feasible if the beam produced by a linear array transducer is angled and shifted along the array simultaneously. A modern version of this old technique (Trahey et al., 1986b) is now called ''Sono CT'', i.e. ultrasound computed tomography, marketed by Philips-ATL. The improvement of SNR is inversely proportional to the square root of the number of independent views. This number will be much less than the number of used views if an array transducer is considered.

Frequency compounding breaks up the overall bandwidth of the received echo signals into many sub bands by filtering. The filter outputs are then processed (AM-demodulation, compression) and finally accumulated into a single image (cf., Magnin et al., 1982; Trahey et al., 1986a; Gehlbach and Sommer, 1987). Although, it has been claimed that enormous improvement of speckle SNR could be achieved (Gehlbach and Sommer, 1987), it appeared from consideration of the lesion detectability (cf., Section 5.2 below) that this improvement is counter balanced by an equivalent loss in spatial resolution. The latter effect follows from Eq. (18); since the Df is decreased by for instance a factor of four, the axial speckle size increases by the same factor. The axial resolution therefore decreases also by this factor (cf., Trahey et al., 1986a). This loss together with the partial correlation of the frequency compounded images may even lead to a decreased lesion detectability!

5.2. LESION DETECTABILITY

The effects of speckle reduction filtering can only be revealed if an adequate measure of performance (improvement) is available. The conditions are: a circular lesion, of diameter D, with a reflectivity contrast relative to the surrounding scattering medium. Defining the background grey level amplitude $\hat{e}_1$ and of the lesion $\hat{e}_2$, then a "lesion SNR" is defined as:

$$SNR_L = \frac{\langle g(\hat{e}_2)\rangle_D - \langle g(\hat{e}_1)\rangle_D}{(\sqrt{\sigma_2^2 + \sigma_1^2})_D}, \tag{18}$$

where $g(\ )$ is the monotonous function of echo amplitude $\hat{e}$, $\langle \hat{e} \rangle$ is the ensemble average of mean grey level of a lesion, and σ^2 is the variance of grey level averaged over lesion area.

This SNR_L exactly specifies the receiver operating characteristic (ROC)-measure d'_a (Swets and Pickett, 1982), when the noise is Gaussian. In terms of Eq. (19) this means that the number of speckles covered by the lesion area should be in the order of 10 or more, to reach the Gaussian limit of the sum of independent variables (central limit theorem).

Rather than producing a large number of test images to estimate the SNR_L from Eq. (19), it might be attractive to assess it from the grey level "pixel statistics", i.e. the histogram parameters. It can be shown (Smith et al., 1983; Thijssen et al., 1988; Verhoeven and Thijssen, 1993):

$$SNR_L = C_L SNR_P \sqrt{M}, \tag{19}$$

where $C_L = (m_2 - m_1) / ((1/2)(m_2 + m_1))$ *is* the lesion contrast, (m_1, m_2) is the mean grey level in lesion and surroundings, respectively,

$$SNR_P = \frac{\frac{1}{2}(m_2 + m_1)_P}{(\sqrt{\sigma_2^2 + \sigma_1^2})_P} = \text{pixel SNR},$$

where σ_1^2, σ_2^2 is the pixel variance of lesion and surroundings respectively,

$$M = \frac{A_L}{A_C} = \frac{\pi D^2 / 4}{\delta_x \delta_z}$$

is the number of speckles in the lesion area, A_L and A_C are lesion area and speckle correlation size (two-dimensional), and δ_x, δ_z are speckle correlation size in axial and depth direction, respectively.

It is clear from the above equation for M that this parameter defines the "information grains" of the speckle image. Also, any effect due to filtering which changes the effective size of the lesion area, or of the speckles will change the SNR_L. Verhoeven and Thijssen (1993) used the 2D ACVF to estimate the speckle size and the lesion size before and after filtering.

5.3. FIXED FILTERS

The filtering is performed by applying some function (operator) $g(\,)$ to the pixels within a moving window and replacing the central pixel value by the result of the operator. Defining the window $N = (n_x\,,\, n_y)$, with $n_x\,,\, n_y$ odd, then the filter F is applied as:

$$E(\xi,\eta) = F(g(\hat{e})).{}_N\,, \tag{20}$$

where ξ,η are the coordinates of the central pixel of the window. The mean filter then follows as:

$$E(\xi,\eta) = \frac{1}{n_x n_y} \sum_{(n,m)\in N} .g(\hat{e}(n,m))\,, \tag{21}$$

The median filter is given by

$$E(\xi,\eta) = \text{med}\, g(\hat{e}(n,m))_N\,. \tag{22}$$

Another non-linear filter that was investigated is called the L_2-mean filter

$$E(\xi,\eta) = \text{SQRT}\left[\frac{1}{n_x n_y} \sum_{(n,m)\in N} .(\hat{e}(n,m))^2\right] \tag{23}$$

It has been shown by Kotropoulos and Pitas (1992) that the L_2-mean filter is equivalent to the maximum likelihood estimation in case of (multiplicative) Rayleigh pdf as in echographic speckle.

5.4. ADAPTIVE FILTERS

In adaptive filtering, the action of a fixed filter is adapted to some characteristics defined within the filtered window. So:

$$F'(\xi,n) = \sum_{i=1}^{L} K_i(n,m)F_i(n,m)\,, \tag{24}$$

where F' is the adaptive filter, L is the total number of applied filters i, and K_i is the result of local pixel analysis, with $\sum_{i=1}^{L} K_i(n,m) = 1$.

An adaptive filter applied to echographic speckle images is the edge preserving smoothing (unsharp masking) filter, e.g., the adaptive mean filter (cf., Bamber and Daft, 1986; Verhoeven and Thijssen, 1993):

$$F'(\xi,n) = K(n,m)g(\hat{e}(n,m)) + (1 - K(n,m)) \times \frac{1}{n_x n_y} \sum_{(n,m)\in N} \cdot g(\hat{e}(n,m)), \quad (25)$$

Where

$$K(n,m) = \gamma - \frac{\langle g(\hat{e}(n,m))\rangle_N}{\mathrm{var}[g(\hat{e}(n,m))]_N} \frac{\mathrm{var}[g(\hat{e}(n,m))]_R}{\langle g(\hat{e}(n,m))\rangle_R}$$

$$-\gamma - \frac{SN^2R_N}{SN^2R_R}$$

with γ is the scaling factor (≈ 1) to keep $K(x,y) \in (0,1)$, SN^2R_N is the pixel signal to-(noise)^2 ratio of some window *N*, and SN^2R_R is the same for image obtained from reference medium.

From the definition of *K(n,m)* it follows that in windows containing contours, i.e. SNR_N low, $K \approx 1$, hence the original pixel value will be preserved.

5.5. RESULTS OF FILTERING

The aforementioned fixed and adaptive filters were applied to a set of images containing lesions of exactly known size and contrast, which were produced by a 3D simulation program (Oosterveld et al., 1985). Fully developed speckle was obtained and the lesion (diameter: 1 cm) was centred around the focus of the employed transducer (f_c = 3 MHz, Δf = 1.55 MHz, focus at 8 cm). Lesion contrast with respect to background area was +3 dB (or +5 dB).

An example of the effects of window shape, square and rectangular (aspect ratio 1:2), on the filtering by a fixed mean filter is shown in Fig. 6. Also the size ($n_x \times n_y$) of the window was investigated by Verhoeven and Thijssen (1993). It appeared that the optimal window size is dependent on the lesion size.

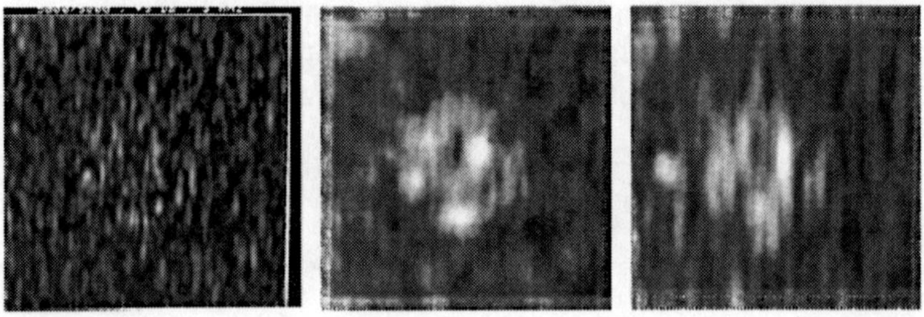

Figure 6. Left: original, simulated, image containing a central lesion (+5 dB contrast). Centre: fixed mean filter applied, $(n_x, n_y) = (15, 15)$ pixels. Right: same filter, $(n_x, n_y) = (10, 25)$ pixels (Verhoeven and Thijssen, 1993).

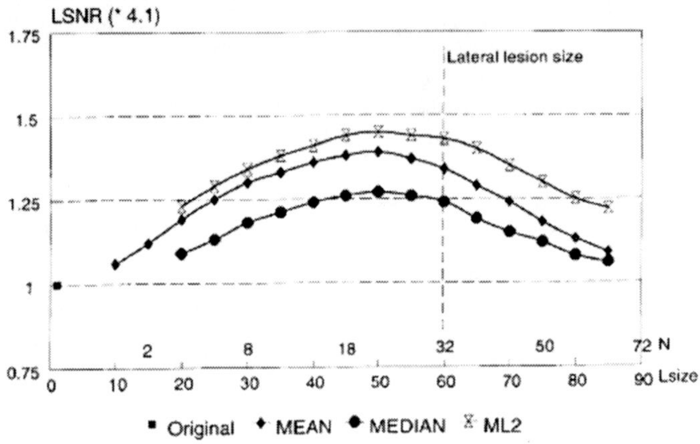

Figure 7. Fixed mean-, median- and L_2-mean filtering. SNR_L vs. size of filter window. L_{size} is the window area in $(pixels)^2$, N is the area of window in (speckles) (Verhoeven and Thijssen, 1993).

Furthermore, at optimal window size the lesion detectability (SNRL) was increased by approximately 45% (i.e. $\approx \sqrt{2!}$) by the fixed L_2-mean filter. The median and mean fixed filters yielded 25% and 35% improvement, respectively (Fig. 7). Logarithmic grey scale compression yielded somewhat less improvements (35%, 20%, and 25% respectively). The adaptive filter results were disappointing in a sense, because no significant difference with the results from fixed filters was found. However, the relatively low contrast (+3 dB, +5 dB) lesions obviously did not produce clear contours and edge preserving action was not involved.

The edge preserving effect of the adaptive mean filter is, however, evident in Fig. 8. For this reason, the adaptive version of the L_2-mean filter might be considered optimal for echographic images (after linear amplification). The question remains whether non-linear (e.g., logarithmic) grey scale en coding yields different results. This is discussed in Section 6.

5.6. OTHER SPECKLE REDUCTION FILTERING APPROACHES

In recent literature several wavelet transform based strategies have been proposed (Zong et al., 1998; Hao et al., 1999; Rakotomamonjy et al., 2000). These methods are using some form of thresholding the wavelet coefficients or of filtering coefficients. A quantitative assessment of performance of these methods is still lacking.

6. Grey scale encoding (''gamma'')

6.1. LOGARITHMIC (COMPRESSION) ENCODING

The log-compression is incorporated in the preprocessing of medical echographic equipment in order to adjust the large echo dynamic range (60–100 dB) to the 8 bits of the digitization in the scan converter. This makes it interesting to investigate the effects of this compression on the grey level statistics and the resulting lesion detection (Thijssen et al., 1988).

For fully developed speckles, i.e. the Rayleigh pdf applies, it can be shown that the Fisher–Tippett pdf results (Thijssen et al., 1988; Kaplan and Ma, 1994; Dutt and Greenleaf, 1996):

$$f(\log\langle\hat{e}+1\rangle = \frac{\log\langle\hat{e}+1\rangle}{\sigma^2} e^{-2\log\langle\hat{e}+1\rangle} \exp\left\{\frac{-e^{-2\log\langle\hat{e}+1\rangle}}{2\sigma^2}\right\} \tag{26}$$

The pixel SNR could be shown to become:

$$SNR_P \approx \frac{\ln(\sigma^2\pi/2)-(4/\pi-1)}{2\sqrt{4/\pi-1}} \tag{27}$$

It should be mentioned that after log-compression σ_p becomes a constant ($\approx\sqrt{4/\pi-1}$), hence, additive noise is present. The lesion SNR for log compressed data could not be derived completely, because an analytic expression for the ACVF of the speckle could not be derived. So, simulations were performed and the ACVF was calculated. The other components (SNR_p, CL) were both calculated and estimated (Thijssen et al., 1988). The results are shown in Fig. 9, for the pure log-compression, as well as for the case of combined log-compression and power law $(\ln(\hat{e}+1))^n$.

Although the pixel SNR_p is increased due to log-compression, the SNR_L appears to be lower than for the linear case.

6.2. POWER LAW ENCODING

The power law is being used in echographic equipment, although often after log-compression (cf., Section 5.2). From theoretical point of view, it can be shown, that the square of the echo amplitude is optimal for lesion detectability (Smith et al., 1983). The first order statistics can be derived for fully developed speckle (Thijssen et al., 1988):

$$f(\hat{e}^n) = \frac{(\hat{e})^{2-n}}{n\sigma^2} \exp\left\{-\frac{\hat{e}^2}{2\sigma^2}\right\}, \tag{28}$$

where n is power.

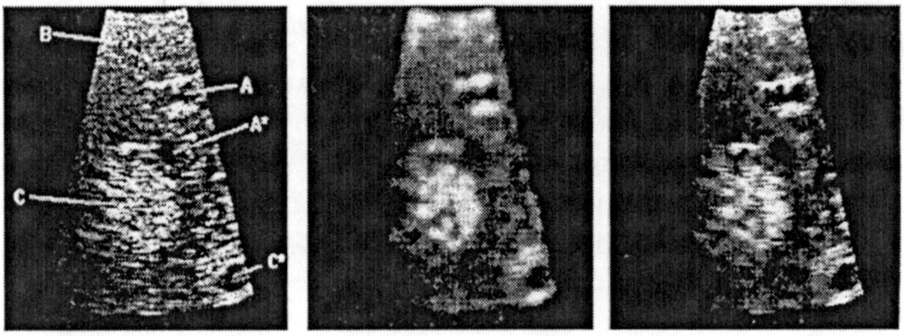

Figure 8. Left: image of liver; A is the artery, A0 is the vein, B is the inhomogeneous region, C, C' is tumour. Centre: fixed mean filter. Right: adaptive mean filter. (Verhoeven and Thijssen, 1993).

It can also be shown, that the pixel SNR becomes:

$$SNR_P = \frac{\Gamma(n/2+1)}{\left\{\Gamma(n+1) - \Gamma^2(n/2+1)\right\}^{1/2}}, \tag{29}$$

where $\Gamma(\ )$ is the gamma function.

The autocorrelation function of the speckle can be derived as well (cf., Middleton, 1960):

$$ACF = E\langle(\hat{e}(x))^n, (\hat{e}(x+\Delta x))^n\rangle$$

$$= (2\sigma_{\hat{e}}^2)^n \Gamma^2(\frac{n}{2}+1) 2F_1(-\frac{n}{2}, -\frac{n}{2}, -1; |k_0|^2), \tag{30}$$

where $2F_1$ is the hyper geometric function.

For this case the effective number of speckles within the lesion area M (cf., Eq. (20)) could therefore be calculated and the lesion SNR_L be obtained as a function of power n. As for the log compression with power law, the pure power law yields a strong decrease of the SNR_p vs. n, whereas, the SNR_L increases from n = 1/8 through n = 2, yielding a maximum detectability at $n = 2$, and a decrease at higher power values (Fig. 9).

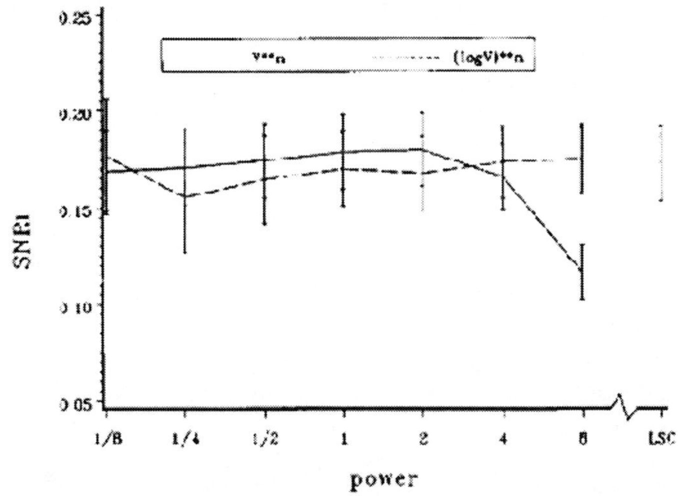

Figure 9. Lesion signal-to-noise ratio SNR_L vs. magnitude of power law, n = 1/8 till n = 8. Drawn line connects mean values (±sd) for pure power law, dashed line for power of log law (Thijssen et al., 1988).

This result is interesting from the point of view of optimal design of TV monitors for echographic equipment since, the PAL standard ‘‘gamma’’ is 2.8 and the NTSC standard gamma is 2.2!

7. Conclusions

The tendency of Industry to incorporate high performance and very powerful computing hardware into the echographic equipment might lead very soon to feasibility studies in which UTC is not only on-line, but even in real time available to the diagnostician in a form of “computer aided diagnosis”. The author believes these will be the conditions that may induce clinical interest and thereby will bring about more and more clinical evidence and, in the end, broad acceptance of various techniques available for improved echographic diagnosis.

One of the major challenges still is to find the optimal set of UTC parameters to be used in “multiparametric imaging” (e.g., Schmitz and Ermert, 1999; Valckx et al., 1996, 2000). Indeed this set has to be established by analyzing a high number of clinical cases and by comparison with some kind of gold standard.

The effects brought about by speckle reduction and grey scale encoding are not very impressive, but consistent improvements could be shown. The

L_2-mean filter yielding a square root of 2 (45%) improvement of the lesion detectability. This result is consistent with the improvement found for power law encoding of the grey scale. The logarithmic grey scale compression incorporated in most medical ultrasound equipment appeared to be less favourable than the power law encoding. However, the combined log-power encoding, with $n = 2$, was found to be acceptable.

Further progress is to be expected from new techniques which are being introduced such as: elastography, contrast echography and second harmonic imaging. The introduction of these techniques has proceeded to a more or less clinical phase of explorations. Applications are proceeding towards the fields of tissue differentiation and tumour diagnosis. The elastography by assessing the hardness of tissues, the contrast echography and second harmonic imaging by improved contrast of (neo)vascularization of tissues and by dynamic flow and perfusion studies of tissues.

The future of ultrasound is still bright, but, as before, we can state that the future has begun!

ACKNOWLEDGEMENTS

The author wants to thank The European Scientific Institute (Archamps, France), for preparing the word version of the manuscript.

References

Bamber, J.C., Daft, C., 1986. Adaptive filtering for reduction of speckle in ultrasonic pulse-echo systems. Ultrasonics 24, 41– 44.

Burckhardt, C.B., 1978. Speckle in ultrasound B-mode scans. IEEE Trans. Son. Ultrason. 25, 1–6.

Cloostermans, M.J.T.M., Mol, H., Verhoef, J.A., Thijssen, J.M., 1986. In vitro estimation of acoustic parameters of the liver and correlations with histology. Ultrasound Med. Biol. 12, 39–51.

Cloostermans, M.J.T.M., Thijssen, J.M., 1983. A beam corrected estimation of the frequency dependent attenuation of biological tissues from backscattered ultrasound. Ultrason. Imag. 5, 136–147.

Dines, K.A., Kak, A.C., 1970. Ultrasonic attenuation tomography of soft tissues. Ultrason. Imag. 1, 16–33.

Dutt, V., Greenleaf, J.F., 1995. Speckle analysis using signal to noise ratios based on fractional order moments. Ultrason. Imag. 17, 251–268.

Dutt, V., Greenleaf, J.F., 1996. Statistics of log-compressed echo envelope. J. Acoust. Soc. Am. 99, 3817–3825.

Fellingham, L.L., Sommer, F.G., 1984. Ultrasonic characterization of tissue structure in the in vivo human liver and spleen. IEEE Trans. Son. Ultrason. 31, 418–428.

Fink, M.A., Cardoso, J.F., 1984. Diffraction effects in pulseecho measurements. IEEE Trans. Son. Ultrason. 31, 313– 329.

Fink, M., Hottier, F., Cardoso, J.F., 1983. Ultrasonic signal processing for in vitro attenuation measurement: Short time Fourier analysis. Ultrason. Imag. 5, 117–135.

Foley, D., 1972. Considerations of sample and feature size. IEEE Trans. Inform. Theory 18, 618–626.

Gehlbach, S.M., Sommer, F.G., 1987. Speckle reduction processing. In: Proc. SPIE Int. Symp. Pattern Recogn. Acoust. Imag., Vol. 768.

Giessen, R.J.B., Huynen, A.L., Aarnink, R.G., de la Rosette, J.J.M.C.H., Debruyne, F.M.J., Wijkstra, H., 1996. Construction and application of hierarchical decision tree for classification of ultrasonic prostate images. Med. Biol. Eng. Comp. 15, 105–109.

Goodman, J.W., 1975. Statistical properties of laser speckle. In: Dainty, J.C. (Ed.), Laser Speckle and Related Phenomena. Springer Verlag, Berlin, pp. 9–75.

Hao, X., Gao, S., Gao, X., 1999. A novel multiscale nonlinear thresholding method for ultrasonic speckle suppression. IEEE Trans. Med. Imag. 18, 787–794.

Haralick, R.M., Shanmugam, K., Dinastein, I., 1973. Textural features for image classification. IEEE Trans. Syst. Man Cyber. 3, 614–624.

Huisman, H.J., Thijssen, J.M., 1996. Precision and accuracy of acoustospectrographic parameters. Ultrasound Med. Biol. 22, 855–871.

Huisman, H.J., Thijssen, J.M., 1998a. Adaptive feature extraction with application to ultrasonic image analysis. Ultrason. Imag. 20, 132–148.

Huisman, H.J., Thijssen, J.M., 1998b. An in vivo model of liver parenchyma. IEEE Trans. Ultrason. Ferroel. Fre. Contr. 45, 739–750.

Insana, M.F., Wagner, R.F., Garra, B.S., Brown, D.G., Shawker, T.H., 1986. Analysis of ultrasound image texture via generalized Rician statistics. Opt. Eng. 25, 743–748.

Jacobs, E.M.G.P., Thijssen, J.M., 1991. A simulation study of echographic imaging of structurally scattering media. Ultrason. Imag. 13, 316–333.

Jakeman, E., 1984. Speckle statistics with a small number of scatterers. Opt. Eng. 23, 453–461.

Jakeman, E., Tough, R.J.A., 1987. Generalized k-distribution: A statistical model for weak scattering. J. Opt. Soc. Am. 4, 1764–1772.

Jongen, H.A.H., Thijssen, J.M., van der Aarssen, M., Verhoef, W.A., 1986. A general model for the absorption of ultrasound by biological tissues. J. Acoust. Soc. Am. 79, 535–540.

Kaplan, D., Ma, Q., 1994. On the statistical characterization of log-compressed Rayleigh signals. In: IEEE Ultrasonics Symposium.

Klein Gebbink, M.S., Verhoeven, J.T.M., Thijssen, J.M., Schouten, T.E., 1993. Application of neural networks for the classification of diffuse liver disease by quantitative echography. Ultrason. Imag. 15, 205–217.

Kotropoulos, C., Pitas, I., 1992. Optimum nonlinear signal detection and estimation in the presence of ultrasonic speckle. Ultrason. Imag. 14, 249–275.

Kuc, R., Schwartz, M., von Micsky, L., 1976. Parametric estimation of the acoustic attenuation coefficient slope for soft tissue. In: IEEE Ultrason. Symp. Proc., no. 76 CH11200-SSU.

Levinson, S.F., 1988. The ultrasonic investigation of dynamic muscle elasticity in vivo. J. Ultrasound Med. 7, 127 (abstract).

Linzer, M., 1976. Ultrasonic tissue characterization I. NBS Special Publication, 453, Gaithersburg.

Loupas, A., McDicken, W.N., Allan, P.L., 1989. An adaptive weighted median filter for speckle suppression in medical ultrasound images. IEEE Trans. Circ. Syst. 36, 129– 135.

Magnin, P.A., Ramm, O.T., Thurstone, F.L., 1982. Frequency compounding for speckle contrast reduction in phased array images. Ultrason. Imag. 4, 267–281.

Mesdag, P.R., 1985. Estimation of medium parameters by acoustic echo measurements. Ph.D. Thesis, Technological University Delft, The Netherlands.

Middleton, D., 1960. Introduction to Statistical Communication Theory. McGraw-Hill, New York.

Momenan, R., Insana, M.F., Wagner, R.F., Garra, B.S., Brown, D.G., 1988. Application of cluster analysis and unsupervised learning to multivariate tissue characterization. J. Clin. Eng. 13, 455–461.

Mountford, R.A., Wells, P.N.T., 1972. Ultrasonic liver scanning: The A-scan in the normal and cirrhosis. Phys. Med. Biol. 17, 261–269.

Nicholas, D., 1982. Evaluation of backscattering coefficients for excised tissues: Results, interpretation and associated measurements. Ultrasound Med. Biol. 8, 17–28.

Nicholas, D., Nassiri, D.K., Garbutt, P., Hill, C.R., 1986. Tissue characterization from ultrasound B-scan data. Ultrasound Med. Biol. 12, 135–143.

Oosterveld, B.J., Thijssen, J.M., Hartman, P.C., Romijn, R.L., Rosenbusch, G.J.E., 1991. Ultrasound attenuation in B-mode texture analysis of diffuse liver disease: Methods and preliminary results. Phys. Med. Biol. 36, 1031– 1064.

Oosterveld, B.J., Thijssen, J.M., Hartman, P.C., Rosenbusch, G.J.E., 1993. Detection of diffuse liver disease by quantitative echography: Dependence on a priori choice of parameters. Ultrasound Med. Biol. 19, 21–25.

Oosterveld, B.J., Thijssen, J.M., Verhoef, W.A., 1985. Texture in B-mode echograms: 3-D simulations and experiments of the effects of diffraction and scatterer density. Ultrason. Imag. 7, 142–160.

Ophir, J., Cespedes, I., Ponnekanti, H., Yazdi, Y., Li, X., 1991. Elastography: A method for imaging the elasticity in biological tissues. Ultrason. Imag. 13, 111–134.

Papoulis, A., 1965. Probability, Random Variables and Stochastic Processes. McGraw Hill, New York.

Rakotomamonjy, A., Deforge, P., March_e, P., 2000. Waveletbased speckle noise reduction in ultrasound B-scan images. Ultrason. Imag. 22, 73–94.

Robinson, D.E., Ophir, J., Wilson, L.S., Chen, C.F., 1991. Pulse-echo ultra-sound speed measurement: Progress and prospects. Ultrasound Med. Biol. 17, 633–646.

Robinson, D.E., Wilson, L.S., Bianchi, T., 1984. Beam (diffraction) pattern correction for ultrasonic attenuation measurement. Ultrason. Imag. 6, 293–303.

Romijn, R.L., Thijssen, J.M., Delft, J.L., Wolff-Rouendaal, D., Best, J., Oosterhuis, J.A., 1989. In vivo ultrasound backscattering estimation for tumor diagnosis: An animal study. Ultrasound Med. Biol. 15, 471–479.

Romijn, R.L., Thijssen, J.M., Oosterveld, B.J., Verbeek, A.M., 1991. Ultrasonic differentiation of intracular melanoma: Parameters and estimation methods. Ultrason. Imag. 13, 27–55.

Schmitz, G., Ermert, H., 1999. Tissue characterization of the prostate using radiofrequency ultrasonic signals. IEEE Trans. Ultrason. Ferroel. Fre. Contr. 46, 126–138.

Sleefe, G.E., Lele, P.P., 1988. Tissue characterization based on scatterer number density estimation. IEEE Trans. Ultrason. Ferroel. Fre. Contr. 35, 749–757.

Smith, S.W., Wagner, R.F., Sandrik, J.M., Lopez, H., 1983. Low contrast detectability and contrast/detail analysis in medical ultrasound. IEEE Trans. Son. Ultrason. 30, 164–173.

Swets, J.A., Pickett, R.M., 1982. Evaluation of Diagnostic Systems: Methods from Signal Detection Theory. Academic Press, New York.

Thijssen, J.M., 1980. Ultrasonic Tissue Characterization: Clinical Achievements and Technological Potentials. Stafleu, Alphen aan den Rijn.

Thijssen, J.M., 1992. Echographic image processing. In: Hawkins, P.W. (Ed.), Advances in Electronics and Electron Physics, Vol. 84. Academic Press, Boston, pp. 317–349.

Thijssen, J.M., Cloostermans, M.J.T.M., Verhoef, W.A., 1981. Measurement of ultrasonic attenuation in tissues from scattering reflections: In vitro assessment of applicability. In:

Thijssen, J.M., Verbeek, A.M. (Eds.), Ultrasonography in Ophthalmology 8. Dr.W. Junk Publishers, The Hague, pp. 431–439.

Thijssen, J.M., Oosterveld, B.J., Hartman, P.C., Rosenbusch, G.J.E., 1993. Correlations between acoustic and texture parameters from rf- and B-mode liver echograms. Ultrasound Med. Biol. 19, 13–23.

Thijssen, J.M., Oosterveld, B.J., Wagner, R.F., 1988. Grey level transforms and lesion detectability in echographic images. Utrason. Imag. 10, 171–195.

Trahey, G.E., Allison, J.W., Smith, S.W., Ramm, O.T., 1986a. A quantitative approach to speckle reduction via frequency compounding. Ultrason. Imag. 8, 151–164.

Trahey, G.E., Smith, S.W., Ramm, O.T., 1986b. Speckle pattern correlation with lateral translation. IEEE Trans. Ultrason. Ferroel. Fre. Control 33, 257–264.

Valckx, F.M.J., Thijssen, J.M., 1997. Characterization of echographic image texture by co-occurrence matrix parameters. Ultrasound Med. Biol. 23, 559–571.

Valckx, F.M.J., Geemen, A.J., Thijssen, J.M., 1996. Online parametric ultrasound imaging. Ultrason. Imag. 18, 51.

Valckx, F.M.J., Thijssen, J.M., Geemen, A.J., Rotteveel, J.J., 2000. Calibrated parametric ultrasound imaging. Ultrason. Imag. 22, 57–72.

Varghese, T., Donehue, K.D., 1995. Estimating mean scatterer spacing with the frequency-smoothed spectral autocorrelation method. IEEE Trans. Ultrason. Ferroel. Fre. Contr. 42, 451–463.

Verhoeven, J.T.M., Thijssen, J.M., 1993. Improvement of lesion detectability by speckle reduction filtering: A quantitative study. Ultrason. Imag. 15, 181–204.

Verhoeven, J.T.M., Thijssen, J.M., Teeuwes, A.G.M., 1991. Improvement of lesion detection by echographic image processing: Signal-to-noise ratio imaging. Ultrason. Imag. 13, 238–251.

Verhoef, W.A., Cloostermans, M.J.T.M., Thijssen, J.M., 1985. Diffraction and dispersion effects on the estimation of ultrasound attenuation and velocity in biological tissues. IEEE Trans. Biomed. Eng. 32, 521–529.

Wagner, R.F., Insana, M.F., Brown, D.G., 1986. Unified approach to the detection and classification of speckle texture in diagnostic ultrasound. Opt. Eng. 25, 738–742.

Wagner, R.F., Smith, S.W., Sandrik, J.M., Lopez, H., 1983. Statistics of speckle in ultrasound B-scans. IEEE Trans. Son. Ultrason. 30, 156–163.

Wagner, R.F., Wear, K.A., Perez, J.E., McGill, J.B., Schechtman, K.B., Miller, J.G., 1995. Quantitative assessment of myocardial ultrasound tissue characterization through receiver operating characteristic analysis of Bayesian classifiers. J. Am. Coll. Cardiol. 25, 1706–1711.

Zong, X., Laine, A.F., Geisser, F.A., 1998. Speckle reduction and contrast enhancement of echocardiograms via multiscale non-linear processing. IEEE Trans. Med. Imag. 17, 532–540.

PERFORMANCE TESTING OF MEDICAL ECHO/DOPPLER EQUIPMENT

J.M. THIJSSEN[*], M.C. VAN WIJK, AND M.H.M. CUYPERS
435 Clinical Physics Laboratory, University Children's Hospital and Children's Heart Centre, P.O. Box 9101, 6500 HB Nijmegen, The Netherlands

Abstract. This paper describes the methods applied in a software package developed by the authors for use in a performance testing protocol for medical ultrasound equipment. The history of performance testing of medical ultrasound equipment is briefly reviewed. This paper is confined to the testing of performance of usage aspects, i.e. imaging performance and Doppler velocity estimation performance. Simple test objects are used which have a long life expectancy. The tests performed both in fundamental and in (tissue) harmonic modes when applicable are spatial resolution, contrast sensitivity, and clutter. The concept of a computational observer is used to define the lesion signal to-noise ratio and the tissue-to-clutter ratio. Further imaging performance features are penetration depth, slice thickness and geometric conformity of display. Pulsed Doppler velocity measurement features tested are: sensitivity, depth and 3D size of the sample volume, velocity measurement, channel separation. The whole performance measurement protocol as well as the quantitative measurements in the digitized images are implemented in software, together with the graphs and data obtained from the measurements.

Keywords: performance testing; echography; Doppler velocity measurement; harmonic imaging; ultrasound.

---

[*] J. M. Thijssen, 435 Clinical Physics Laboratory, University Children's Hospital and Children's Heart Centre, P.O. Box 9101, 6500 HB Nijmegen, The Netherlands; e-mail: j.thijssen@cukz.umcn.nl.

Y. Lemoigne et al. (eds.), Physics for Medical Imaging Applications, 177–195.

1. Introduction

Many national and international organizations have been involved for some time already in the definition and methods for the assessment of the quality of medical ultrasound equipment. Examples are the American Institute of Ultrasound in Medicine (AIUM, 1974), the American Association of Physicists in Medicine (AAPM) (Carson and Zagzebski, 1977) and the International Electrotechnical Committee (Hill, 1977). These efforts have not led to internationally accepted quality standards yet. One of the reasons is the ongoing rapid evolution of equipment features and performance. Another reason might be the ambiguity of these committees in defining a clear end point: should it be a technical standard, or a quality assessment from the user's point of view. Another limitation of the methods involved in quality standards so far is the involvement of subjective assessments in case of image quality. This limitation also holds for a recent AAPM report (Goodsitt et al., 1998), although it has evidently been based on the end point of quality of use, i.e. imaging performance and for the paper by Dudley et al. (2001).

In this paper, the approach is followed that rather than "subjective" assessment, the performance measurements are carried out on echographic images which are digitally acquired in a computer. This allows for adequate calibrations and for more "objective" assessments of performance. A similar approach is followed in this paper for the assessment of Doppler flow velocity (cf. Hoskins et al., 1994).

A recent technical development of medical ultrasound equipment is "second harmonic imaging". This feature is used to image the information related to the double of the transmitted frequency which is caused by finite amplitude effects ("tissue harmonic imaging"), or to nonlinear resonance of small gas bubbles in contrast media ("contrast imaging"). Harmonic imaging became feasible with the improved transducer technology (e.g. composite materials) which combines the production of short pulses (100% relative bandwidth) with a high efficiency. The equipment enables the full bandwidth to be split into three sub-bands, which can be either used in both transmission and reception (conventional imaging), or the lowest one in transmission and the highest one in reception (harmonic imaging). It is not yet clear how the general performance of the equipment is changed in harmonic mode and this is one of the topics of this paper.

The advantage of Tissue Harmonic imaging is said to be clutter reduction. In other words, the disparity between the transducer's directivity function in transmission (e.g. at 2 MHz) and in reception (e.g. at 4 MHz) yields a suppression of a certain class of image artefacts, i.e. those related to multiple reflections, beam width, side lobes and eventually to grating lobes.

2. Test objects and equipment

The imaging performance testing was performed by using a tissue mimicking ‘‘phantom’’ (test object) consisting of urethane rubber base material (ATS Inc., Bridgeport, CT 06608, USA). This phantom (Fig. 1) contains several thin wires arranged in special patterns, cylindrical objects of known scattering contrast as compared to the surroundings and cylinders void of scatterers (‘‘cysts’’) which are imaged as voids in a scattering environment. The backscattering and the attenuation of the base material are similar to those of parenchymal soft tissues. The nominal contrast values of the ‘‘solid lesions’’ with respect to the background grey level are specified by the manufacturer for the frequency range 2.25"/7.5 MHz The speed of sound is however 7% lower than in the human body. The manufacturer has adapted the geometry in the axial direction as to compensate for image distortion. Nevertheless, the focussing of array transducers will be somewhat disturbed, leading to an underestimation of the resolution measurements (i.e. an overestimation of the size of the point-spread-function). The authors have the opinion that the long preservation time of the phantom far outweighs the minor effect on resolution.

The Doppler flow velocity measurements were carried out with a ‘‘string’’ test object (JJ&A Instr., Duvall, WA 98019, USA), as shown in Fig. 2. A thin rubber string (diameter: 1 mm) is rotated over small wheels by a stepping motor. The linear speed of the string can be electronically varied over a range from 10 to 200 cm/s.

The measurements were based on images from the medical equipment taken from the analogue video output. The images were digitally grabbed in RGB format (3#/8 bits) by a ‘‘frame’’ grabber (MRT Videoport, PCMCIA) and transferred to the hard disk of a personal computer (PIII, 550 MHz). Image resolution was (790#/510) pixels, grey level range (0"/255). The analysis software was written with MATLAB (release 12). Recently C-code was generated from the MATLAB scripts and programmed.

3. Imaging quality assurance

3.1. STANDARD EQUIPMENT SETTINGS

Although not part of the image analysis, the TV monitor of the equipment should be adjusted until the grey wedge usually displayed in 16, or 32 levels, is visually set optimally by using the ‘‘intensity’’ and ‘‘contrast’’ controls. In this setting (to be marked), a test signal (image) generator, generally used in TV repair shops, is used to measure the screen luminance at full screen black and, maximum, white levels (in: [cd/m2]). These numbers can be used to quantify the display "dynamic range".

In the chosen echographic examination mode and transducer (frequency and mode), the transmitted output power should be fixed by using either the displayed "thermal index" (TI), or "mechanical index" (MI) value (aium/nema, 1992).

The preprocessing control functions when available should be set as used in clinical practice. These settings must be reproducible, i.e. the readout numbers should be noted. If no read-out is available, it is recommended to switch the "rejection" control to the lowest level (maximum displayed echoes). The time-gain-compensation (TGC) in most equipment is anticipating tissue attenuation and it will yield a reasonably homogeneous grey level over depth when a tissue mimicking phantom is used. If not, the user has to adjust the TGC slide potentiometers until the image from the zone in the phantom without objects is seen as a homogeneous average grey level over a maximum attainable depth. The settings of the TGC should be marked or noted. The overall gain is displayed on the screen of most medical equipment, either as a number, or in decibel (dB). If gain is not displayed, the control knob should be marked at the selected position.

The post processing comprises the "look-up table" (LUT) used to select the coding of stored echo levels (i.e. after log compression and digitization) for display at the screen. For testing, a linear LUT is to be preferred because then a pure logarithmic relation exists between echo level and display voltage. This relation is needed for some of the performance measurements. The LUT is estimated from the grey level wedge by inserting a rectangular measurement box in the redisplayed image, while using the developed software. The grey levels versus position are then fitted with a linear regression line. Any deviation from a linear relation can visually be noted and the best fitting LUT can be found. Both the displayed dynamic range and the code of the LUT should be noted for future reference.

3.2. IMAGING QUALITY MEASUREMENTS

3.2.1. *Geometric conformity*

The axial and lateral length measurements (or more general: calipers) can be calibrated by using the horizontal row and vertical column of wires in one image (Fig. 1).

So, when properly calibrated the equipment should indicate the correct distances. After "freezing" the image, the user is asked to position calipers in the centre of the wires imaged on top and bottom, and furthest left and right, successively. The distances indicated by the equipment are noted and compared to the nominal values of the phantom (Fig. 2). Finally, the computer program estimates from these data the software ruler (mm/pixel) to be used in subsequent measurements (resolution, penetration depth).

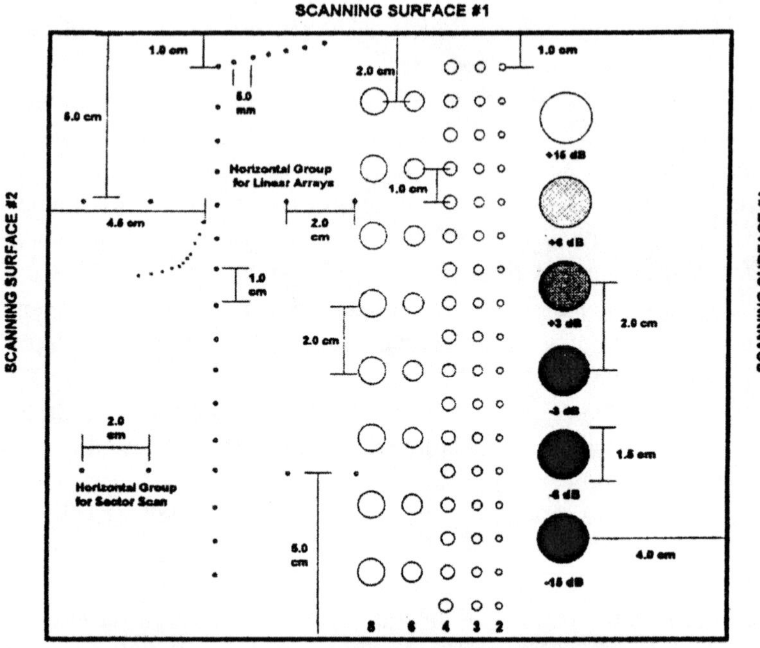

(a)

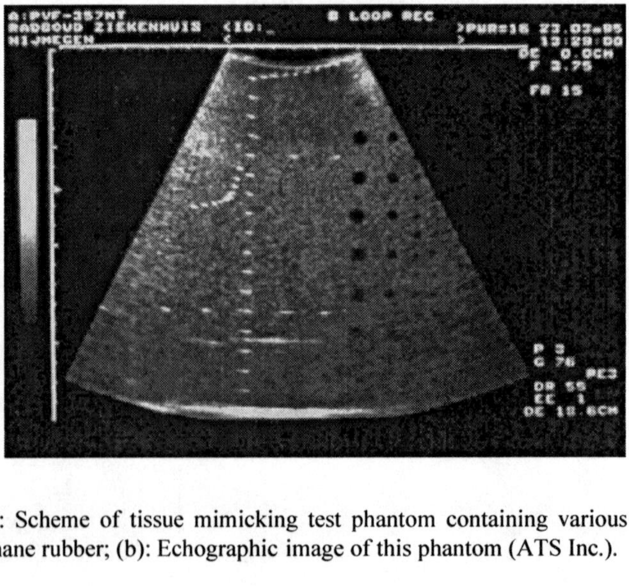

(b)

Figure 1. (a): Scheme of tissue mimicking test phantom containing various objects, base material urethane rubber; (b): Echographic image of this phantom (ATS Inc.).

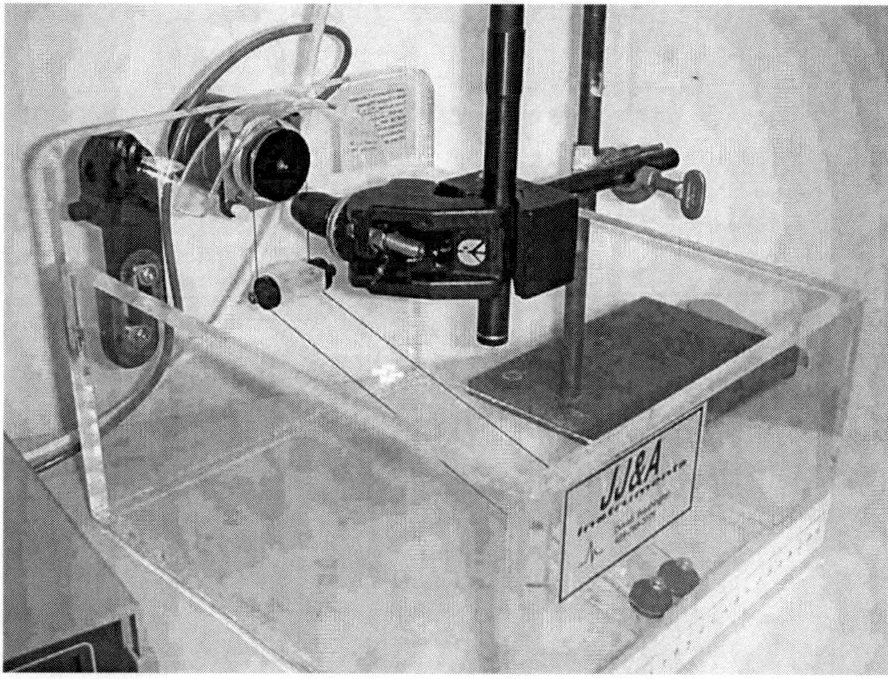

Figure 2. String phantom for assessment of Doppler velocity measurement (JJ&A Instr.).

3.2.2. *Slice thickness/ elevation focus*

In general, array transducers have a fixed focus cylindrical lens at their surface which yields a narrowing of the transmitted sound field. This focussing action is in the direction perpendicular to the scanning plane, hence the term elevation focus, or slice thickness reduction. A special phantom is used to measure the depth where the elevation focus is located (ATS, model 538N). The scheme of the phantom is shown in Fig. 3. It contains a plane thin scattering sheet at a 758 angle with the horizontal plane. A scan along the plane yields a horizontal bar image, whose thickness is used to calculate the slice thickness at the fixed elevation focus.

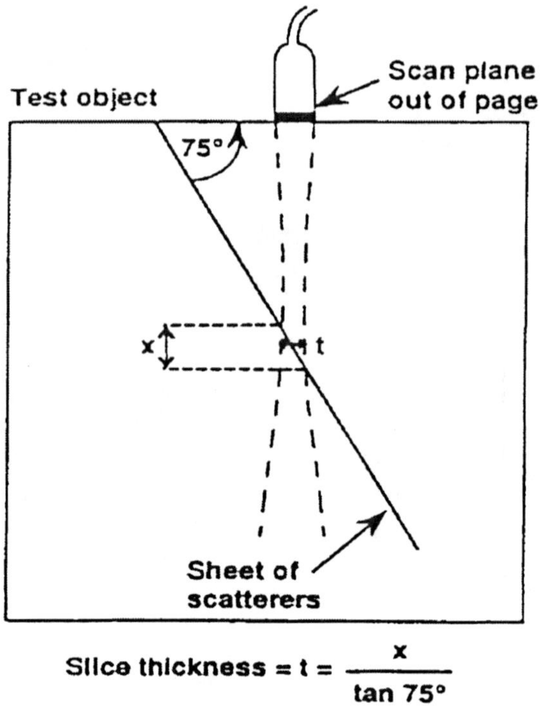

Figure 3. Scheme of slice thickness phantom, beam width (-6 dB) in elevation direction is obtained by using known angle (75°) of scattering layer.

The examiner moves the transducer over the top of the phantom to estimate the position where the bar width is minimum. The resulting image is stored then for further analysis. The depth at which the bar is located, i.e. the elevation focal depth, is used to install the (software controlled) in-plane focus (or azimuth focus) for the following measurements.

The software algorithm permits location of arectangular box over the bar (Fig. 4). The averaged full-width-at-half-maximum (FWHM) thickness (x) is then estimated by using the echo level-to grey level ratio (dB per level, i.e. the ‘‘gamma’’, assessed in the section ‘‘contrast resolution’’ measurement) to convert the grey levels into dB. The slice thickness t (at $/6 dB) is then calculated from: t%/x/(tan 75).

3.2.3. *Image contrast resolution/ dynamic range*

The contrast resolution is defined as: the smallest contrast of a lesion of a certain size with respect to the surrounding tissue that can be detected with a

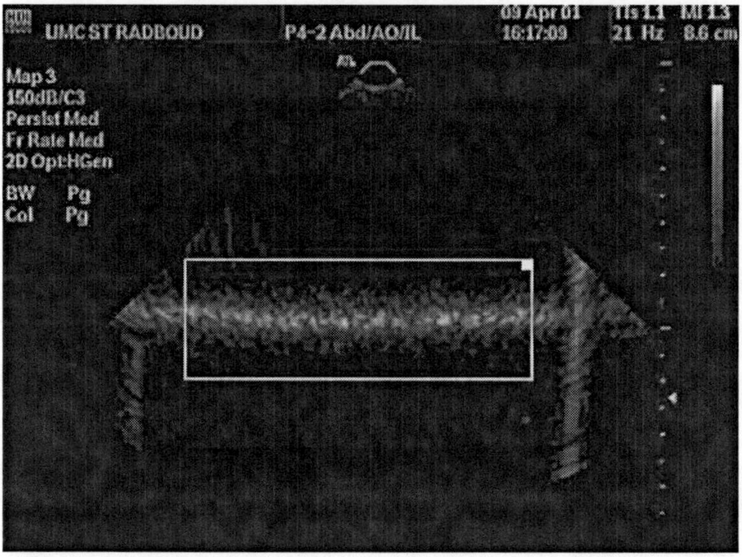

Figure 4. Image obtained from slice thickness phantom, showing the measurement box. Width of the bar is defined as -6dB width averaged.

certain probability (e.g. 75%). In practice it is useful to employ the ''lesion signal-to noise ratio'', SNRL (Smith et al., 1983; Wagner et al., 1986; Thijssen et al., 1988), sometimes called ''Mahalanobis distance'' (cf. McLachlan, 1992):

$$SNR_L = \frac{|<\mu_2> - <\mu_1>|}{\sqrt{\sigma^2_{\mu_2} + \sigma^2_{\mu_1}}} \qquad (1)$$

where $<\mu_2>$ and $<\mu_1>$ are mean grey level within a circular area of surrounding (background) tissue (2), and of lesion (1), averaged over the ensemble of images from independent scans. $\sigma^2_{\mu_2}$ and $\sigma^2_{\mu_1}$ are variance of mean grey level of area of background (1) and of lesion (2). This SNRL can be considered as the performance measure of a ''computational'' observer performing a detection experiment under ''SKE'' conditions (i.e. ''signal known-exactly'', cf. Swets and Pickett, 1982).

In addition, the displayed dynamic range and the gamma can be estimated. For this purpose, the six contrast cylinders are scanned five times each at different positions, and also a homogeneous region at the same depth is scanned five times. The mean and standard deviation of the grey levels within a fixed circular area fitting into the displayed discs is estimated for each object. This circle is also used for selecting the background discs. These values are plotted against the nominal contrast values in dB, as shown in Figs. 5 and 6. The estimated contrast resolution (i.e. gamma), the dynamic range and SNRL values for fundamental and harmonic imaging modes are then estimated.

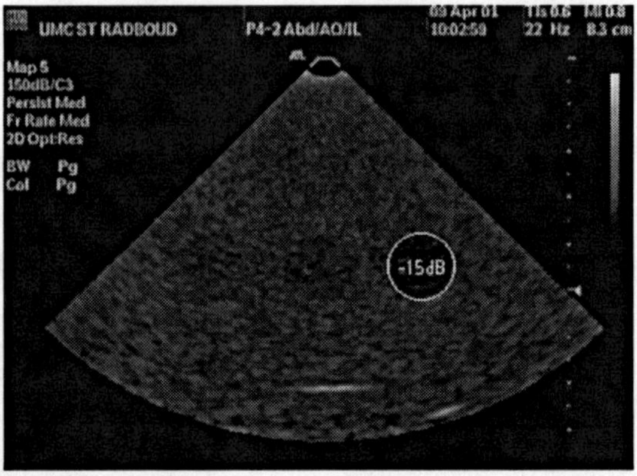

Figure 5. Echographic image obtained from scattering ''lesions'' and measurement disc inserted in one of them.

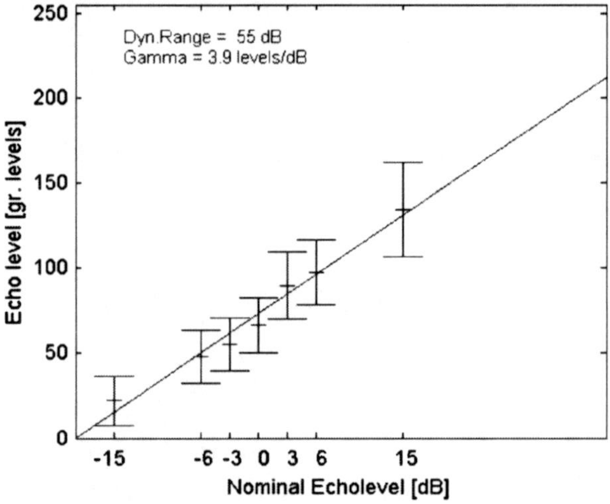

Figure 6. Plot of nominal contrast values of scattering lesions vs. mean pixel of lesions (average of five measurements each and standard deviation). Linear regression yields "Gamma" (pixel level/dB) and dynamic range (number of dB for full range (0.255)).

3.2.4. *Spatial resolution (2D point-spread-function, PSF)*

The geometrical resolution is defined as the axial, or lateral, FWHM, i.e. $6 dB width, of the image of a small object at the focus (i.e. both elevation and azimuth foci). For this purpose, the vertical column of wires is scanned and the image of the wire nearest to the focus is analyzed (Fig. 7, left). The gain has to be reduced generally because the reflectivity of the wires is much higher than of the scattering background medium. This reduction ensures that saturation of the PSF and the possible influence of ''ringing'' of the wire echoes, are avoided. The software estimates the position of the maximum grey level occurring in the measurement box and by using the gamma (see Section 3.2.3) the grey level corresponding to $6 dB is calculated. The dB"grey level profiles in axial and lateral direction through the maximum are then used to obtain the FWHM (Fig. 7, right). Again these measurements were carried out in fundamental and in harmonic mode.

3.2.5. *Penetration depth/sensitivity*

The overall sensitivity of the transducer and equipment for imaging can be assessed by estimating the maximum depth at which echoes from a scattering medium can be imaged. It should be noted that this measurement is depending on the actual setting of the TGC. The output level and the overall gain do not influence the method if saturation of the grey values is avoided. Given a reproducible TGC setting, i.e. yielding the maximum depth range of identical echo level in the phantom image (i.e. in a region without any objects), the software estimates the mean echo level over a user defined depth range. For this purpose a rectangular box is inserted in the central position of the image, spanning the depth range in which ‘‘speckle’’ resulting from the scattering medium can be seen. The grey level profile versus depth is then calculated and averaged in lateral direction. The mean echo level in the range where these echoes are still approximately constant (i.e. from a few centimetres below the transducer to a few centimetres beyond the focus) is estimated. The depth at which this mean level has dropped by 6 dB is then estimated (Fig. 8). A simple sensitivity test for daily use might be the documentation (by photograph) of the number of ‘‘ring-down’’ stripes of a transducer held in air, while using the default settings of the equipment (Dudley et al., 2001). When this number has decreased over time, it might be assumed that the performance, i.e. overall sensitivity of the transducer has deteriorated.

(a)
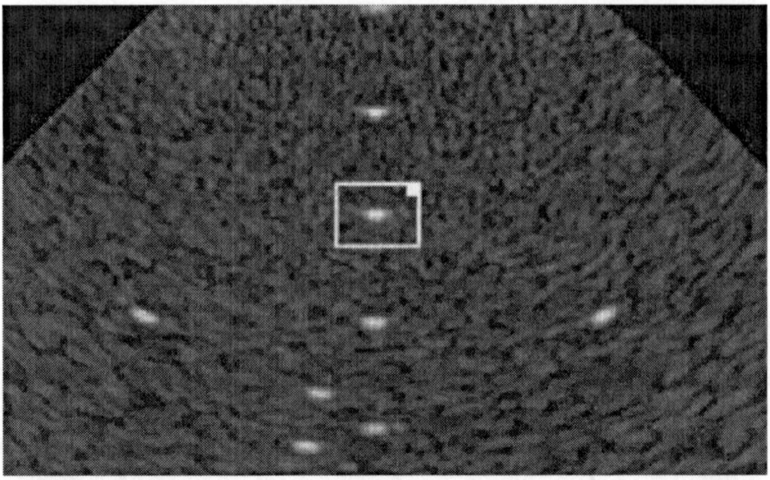

(b)
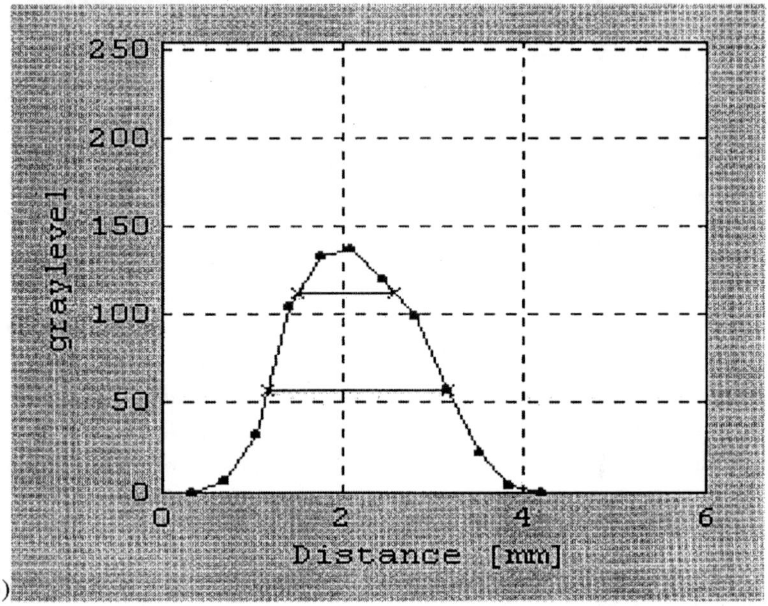

Figure 7. (a): Echographic image zoomed in on the wires in the focal zone. Box indicates area selected for the measurement of the geometrical resolution, i.e. size of point spread function. (b): Example of grey level profile through centre of wire image. FWHM, or -6dB width is estimated by using "Gamma".

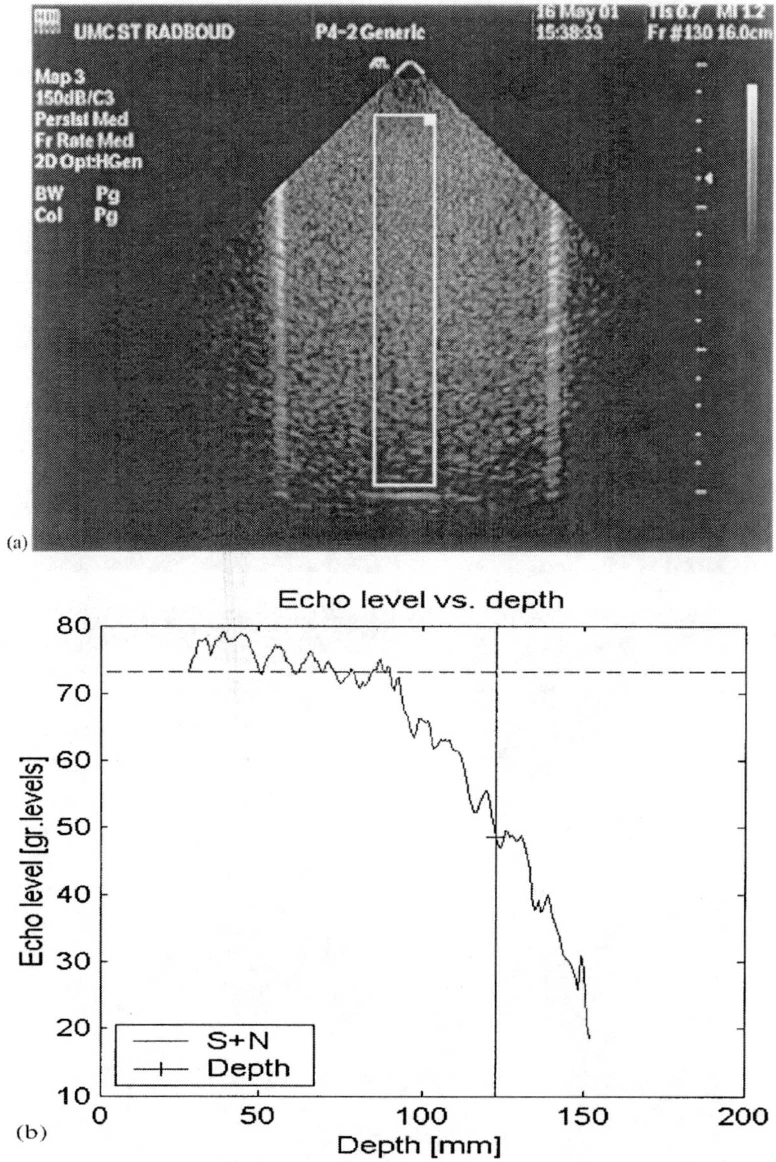

Figure 8. (a): Image for estimation of penetration depth (overall sensitivity). TGC is eventually adjusted to obtain homogeneous grey level with depth, measurement box inserted by software. (b): Mean grey level vs. depth. Estimate of depth at which grey level has dropped to -6 dB of mean in near zone.

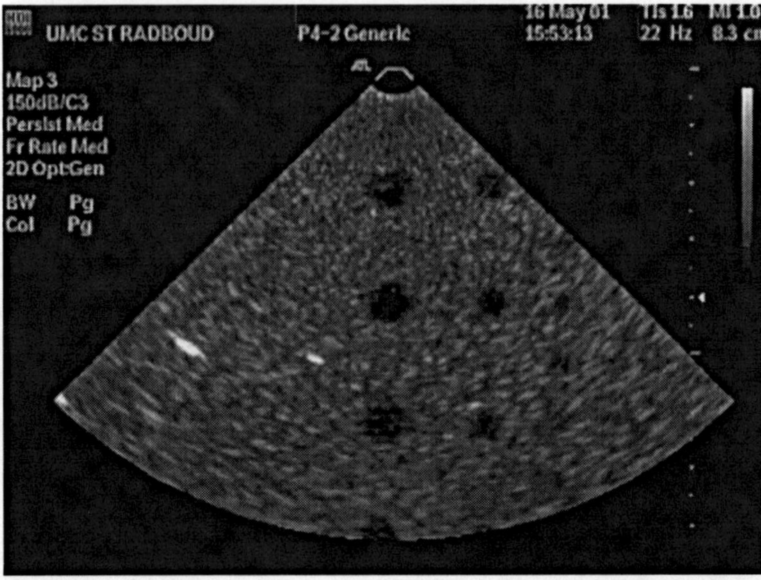

(a)

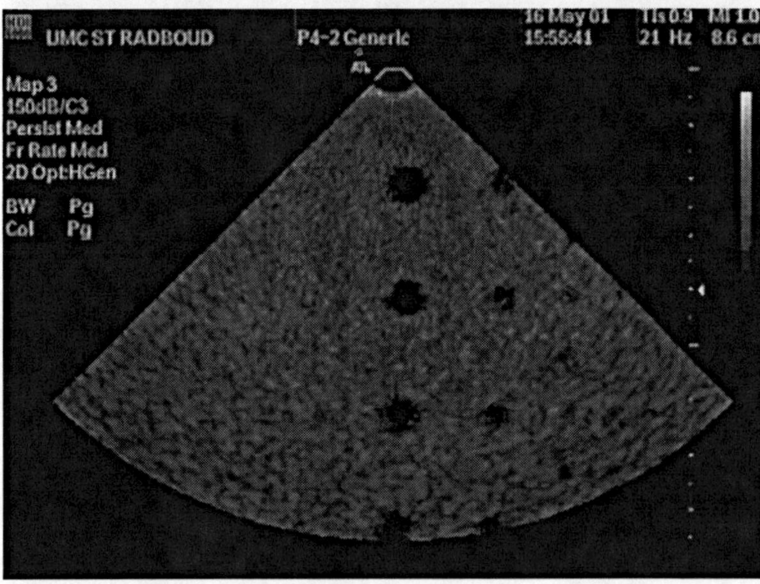

(b)

Figure 9. (a): image of "voids" (cysts) at depths 2, 4, and 6 cm in fundamental imaging made (f=2-4 MHz). (b): same image in harmonic modes (f=4 MHz).

3.2.6. *Tissue-to-clutter ratio (TCR)*

Ideally this aspect of echographic imaging should be tested by imaging liquid filled spherical objects (''cysts'') in the neighbourhood of strongly reflecting structures. The authors used the liquid filled cylinders in the ATS phantom for this purpose (Fig. 1). After storing a set of images in the fundamental mode, the equipment was switched to harmonic mode and the output power as indicated by the MI, cf. AIUM/NEMA (1992), was adjusted to the one used in fundamental mode. Five images were stored in both modes (Fig. 9) and the mean echo level was estimated over an enclosed disc area within the cystic object and the same disc placed in the surrounding medium at equal depth. The "tissue-to-clutter" ratio (TCR) can be used to quantify the clutter reduction:

$$TCR = 20\log_{10}\frac{\mu_T}{\mu_C} \tag{2}$$

where: μ_C is the mean grey level in "cyst" (cylindrical area filled with fluid, diameter: 7.5 mm), and μ_T is the mean grey level of surrounding medium, estimated at the same depth as cyst and averaged over the same disc area.

The mean TCR was estimated for objects at 2, 4 and 6 cm depth (Fig. 10) It might be noted that at a depth of 20 mm the TCR in the harmonic mode is improved as compared to the one of the fundamental mode.

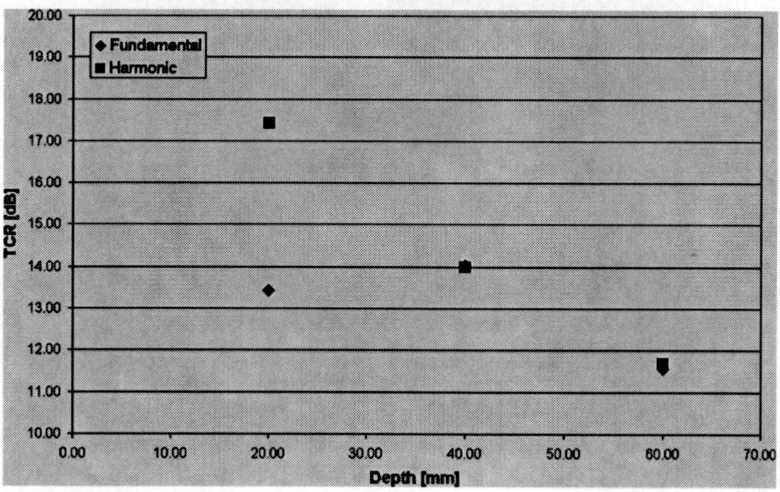

Figure 10. TCR for fundamentals (♦) and harmonic (■) from data in Figure 9 (cf. van Wijk and Thijssen, 2002).

3.3. DOPPLER BLOOD FLOW VELOCITY

3.3.1. *Introduction*

Performance testing of Doppler systems has many aspects which are directly, or indirectly of importance to the interpretation of the obtained data (Hoeks et al., 1984; Hoskins et al., 1994; Evans and McDicken, 2000). The characteristic to be tested without doubt is the flow velocity. In particular, in pulsed Doppler systems it is of importance to assess the conformity of the sample volume size and location, and the angle between scan line and core line of the blood vessel. Further characteristics to be tested are the overall sensitivity, and channel separation (bi-directional systems). As the 2-D colour Doppler imaging is generally interpreted in a qualitative sense, performance testing of these modes might be omitted.

Although several advantages can be listed for Doppler flow test objects (Hoskins et al., 1994; Teirlinck et al., 1998), for instance, containing tubing with circulating blood mimicking fluid, the authors consider the disadvantages crucial for the acceptance in hospital routine testing. The string test object used in this study (Fig. 2) was slightly modified: the silk string (with knot, disturbing the measurements) was replaced by a 1 mm diameter rubber string which could be seamlessly glued to make a ring. The reflectivity level of this material is low enough to enable a simple sensitivity measurement, as well as accurate velocity calibration measurements (Fig. 11).

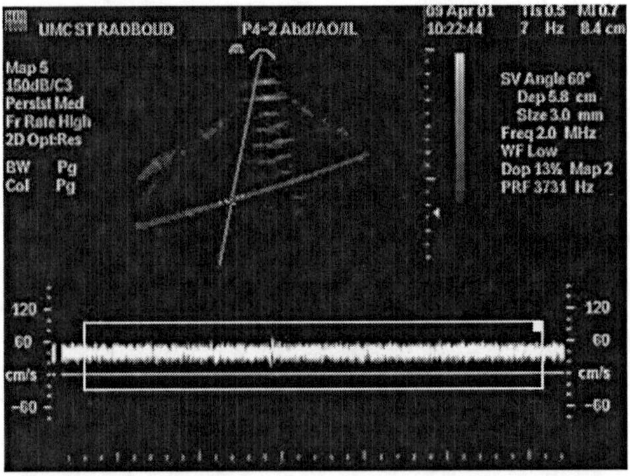

Figure 11. Example of pulsed Doppler spectrogram obtained from string phantom (Figure 2). Measurement box for software estimation of mean velocity value.

Finally, due to its small effective reflection cross section diameter (0.3 mm), it is well suited for the assessment of the dimensions of the pulsed Doppler sample volume.

3.3.2. *Doppler sensitivity*

The (sector, linear) array transducer is positioned to obtain a longitudinal image of the string. A Doppler gate is then installed at the depth of the elevation focus (see Section 3.2.2) and is just enclosing the string. A convenient angle is selected by choosing the Doppler measurement line yielding an angle of 308. The transducer is further slightly angled, or shifted horizontally, to obtain the maximum Doppler signal. The Doppler gain should be gradually decreased to improve the sensitivity of this positioning action. The transducer is fixed at the maximum sensitivity (signal) position and the gain is further lowered until the Doppler signal just disappears. The gain is slightly increased then till the signal becomes just visible again. The used gain at this point is defined as the "Doppler" sensitivity and is noted for further reference. The gain is then increased by 20 dB (when the gain is not indicated on the screen, the whole procedure could be performed by using the MI, while changing the output power rather than the amplifier gain). The obtained setting of the +20 dB gain is then used throughout the Dopplertesting procedure.

3.3.3. *Sample volume depth*

When the image quality has been tested (cf. Section 3.2), the accuracy of the calipers has already been assessed. Therefore, the indicated depth of the sample volume can be simply compared with a caliper measurement made along the scan line along which the Doppler velocity measurement is being made. Velocity measurements should preferably be made at a depth near to the depth of the elevation and in plane foci. As can be seen in Fig. 12, this position yields an unbiased velocity estimate.

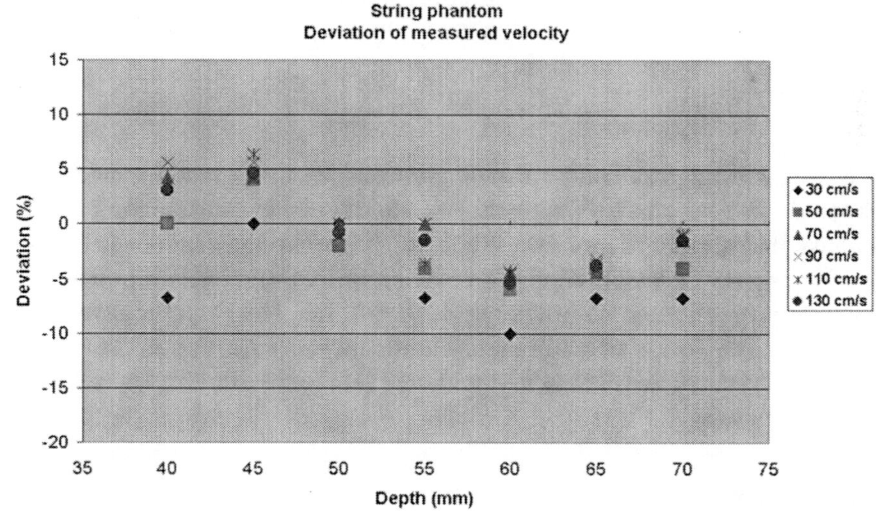

Figure 12. Deviation of measured velocity from string velocity vs. depth of sample volume. Optimal accuracy in focus (50 mm).

A possible explanation might be found in the paper by Barannik (2001), who by theoretical analysis of flow lines come to the conclusion that "the modal Doppler frequency changes with flow line displacement in the illuminating field except if the sample volume centre and the beam focus coincide".

3.3.4. *Sample volume dimensions*

Although the effective size in three dimensions is dependent not only on the settings of the equipment, but also on the reflectivity of the blood cells, a useful indication can be obtained with the string phantom while using the !20 dB setting of the Doppler gain. It requires an XYZ-positioning system, with adequate read-out. The simplest procedure would be to move the transducer along any of the coordinates and note the positions where the Doppler signal just disappears. The authors do not consider this assessment to be part of a routine quality assurance procedure in the hospital.

3.3.5. *Velocity measurement*

The transducer is positioned such that the string is again in the centre of the sample volume at a depth corresponding to the depth of the elevation and azimuth foci. The Doppler spectrogram is registered (Fig. 11), acquired by the frame grabber and transferred to the PC. A rectangular box is interactively positioned around the displayed spectral band and the software then estimates the position in vertical direction of the averaged maximum grey level. The vertical scale calibration is taken from the Doppler display. The thus estimated string velocity is compared to the string velocity installed with the read-out of the test object.

3.3.6. *Channel separation*

Strictly speaking, this feature should be tested by using either an electronic testing device, or a test object enabling simultaneous measurement of flow velocity in opposite directions. When using the string phantom (Fig. 2), a simple test can be made after the velocity calibration, i.e. while using the Doppler gain setting estimated as in Section 3.3.2. The Doppler gain is now increased until just a velocity signal is displayed (Fig. 11) at the velocity direction opposite to the used direction. This gain position is noted for future reference.

4. Conclusions

The performance testing as described in this paper can be carried out with a minimum number of test objects. Moreover, these test objects have a long

life expectancy (15 years) and are simple in their use. The authors for this reason expect wide acceptance of these methods in the larger hospitals. The software developed by the authors is well documented, has an extensive Handbook and is easy to use. The original Matlab Source code is still available, but the stand alone version (compiled C^{++}) is more suitable for distribution and does not need the relatively expensive Matlab license. The software is available on request and it can be tested at the Website of the authors
(http://www.umcn.nl/scientist/Departments/Paediatrics/index.html).

The authors anticipate that the standardization activities of national and international organizations will continue. However, as long as no distinction is made between technical (performance) testing, e.g. of bandwidth, pulse shape, beam profile, noise figure, etc., and performance of use (image quality, Doppler velocity) it may take much time to produce useful standards. In the meantime, new equipment developments are still becoming available faster than standards. This paper will hopefully contribute to a wider introduction of performance testing in hospitals, which at least in the Netherlands is expected to become part of the institutional certification of quality assurance.

References

American Institute of Ultrasound in Medicine. Standard for real-time display of thermal and mechanical acoustic output indices on diagnostic ultrasound equipment. Rockville/ Washington: AIUM/NEMA; 1992.

American Institute of Ultrasound in Medicine. The AIUM 100 mm test object and recommended procedures for its use. Rockville, MD: American Institute of Ultrasound in Medicine; 1974.

Barannik EA. Pulsed Dopple flow-line spectrum for focussed transducers with apodized apertures. Ultrasonics 2001;39:311-/7.

Carson DL, and Zagzebski JA. Pulse echo ultrasound imaging systems: performance tests and criteria. New York: American Institute of Physics, 1977.

Dudley NJ, Griffith K, Houldsworth G, Holloway M, and Dunn MA. A review of two alternative ultrasound quality assurance programs. Eur J Ultrasound 2001;12:233-/45.

Evans DH, and McDicken WN. Doppler Ultrasound, 2nd ed.. Chichester: Wiley, 2000.

Goodsitt MM, Carson PL, Witt S, Hykes DL, and Kofler JM. Real time B-mode ultrasound quality control test procedures. Med Phys 1998;25:1385-/406.

Hill CR. Methods of measuring the performance of ultrasonic pulse-echo diagnostic equipment (Discussion document: IEC Working Group). Ultrasound Med Biol 1977;2:343-/s50.

Hoeks APG, Ruissen CJ, Hick P, and Reneman RS. Methods to evaluate the sample volume of pulsed Doppler systems. Ultrasound Med Biol 1984;10:427-/35.

Hoskins PR, Sherriff SB, and Evans JA. Testing of ultrasound equipment. Institute of Physical Sciences in Medicine, Report no. 71; 1994.

McLachlan GJ. Discriminant analysis and statistical pattern recognition. New York: Wiley, 1992.

Smith SW, Wagner RF, Sandrik JM, and Lopez H. How contrast detectability and contrast/detail analysis in medical ultrasound. IEEE Trans Sonics Ultrasonics 1983;SU-30:164"/73.

Swets JA, Pickett RM. Evaluation of diagnostic systems. New York: Academic Press, 1982.

Teirlinck CJ, Bezemer RA, Kollman C, Lubbers J, Hoskins PR, et al. Development of an example test object and comparison of test objects in various laboratories. Ultrasonics 1998;36:653-/60.

Thijssen JM, Oosterveld BJ, and Wagner RF. Gray level transforms and lesion detectability in echographic images. Ultrasonic Imag 1988;10:171-/95.

Wagner RF, Insana MF, and Brown DG. Unified approach to the detection and classification of speckle texture in diagnostic ultrasound. Opt Eng 1986;25:738-/42.

Van Wijk MC, and Thijssen JM. Performance testing of medical ultrasound equipment: fundamental vs. harmonic mode. Ultrasonics 2002;40:585-91.

ULTRASONIC ELASTOGRAPHY

RÉMI SOUCHON*
INSERM UMR 556,151 rue Albert Thomas, F-69429 Lyon CEDEX 04 Lyon, France

Abstract. Elastography is a new ultrasound-based imaging technique that provides images (called elastograms) of internal strain in soft tissues under a static compression. The strain is related to the stiffness of the tissues, which is in turn related to the pathological state of tissues. For example, it has been known for long that breast and prostate cancer are stiffer than normal tissues, and palpation is a standard medical practice. Elastography is able to provide new quantitative information about the stiffness of the tissues that would not be obtainable with existing imaging modalities. Elastograms are created by digitizing ultrasonic signals before and after applying a small static compression (usually about 1%) to the tissues being scanned. For small compressions, the displacement and the compression of the tissues induce a displacement and a temporal compression of the ultrasonic signal. These displacements are tracked by locating the position of the maximum of the cross-correlation function between the pre- and post-compression signals. Strain is obtained by taking the gradient of the displacements, and a strain image is formed. Applications include breast, prostate, kidney and intravascular elastography, as well as thermal therapy follow-up. Although elastography is still only used in research today, it is expected that this imaging modality may become available for standard practice within a few years.

Keywords: elastograms; strain; cancer; static compression.

1. Elasticity imaging

Over the past 20 years there have been numerous investigations conducted to characterize the mechanical properties of biological tissue systems, which have often been idealized as homogeneous, isotropic and linear

* Rémi Souchon, INSERM UMR, 151 rue Albert Thomas, F-69424 Lyon CEDEX, Lyon, France; email: souchon@lyon.inserm.fr.

Y. Lemoigne et al. (eds.), Physics for Medical Imaging Applications, 197–209.

elastic materials. These mechanical attributes may include the shear or elastic modulus (Young's modulus), the Poisson's ratio, or any of the longitudinal or shear strains that occur in tissues as a response to an applied load. In general, lesions may or may not possess echogenic properties that would make them ultrasonically detectable. Since the echogenicity and the stiffness of tissue are generally uncorrelated, it is expected that imaging tissue stiffness – or strain – will provide new information related to pathological tissue structure (Ophir et al. 1999). Krouskop et al. (1998) showed that a stiffness contrast exists between malignant and normal tissues in the breast and in the prostate (Table 1).

TABLE 1. Stiffness of malignant and normal prostate tissues, measured during cyclic loading at 1 Hz (Krouskop et al. 1998).

Tissue type	Elastic modulus at 2% precompression (kPa)	Elastic modulus at 4% precompression (kPa)
Normal (anterior)	62 ± 17	63 ± 18
Normal (posterior)	69 ± 17	70 ± 14
Benign prostate hyperplasia (BPH)	36 ± 9	36 ± 11
Cancer	100 ± 20	221 ± 32

A review of selected elasticity imaging techniques was recently published (Greenleaf et al. 2003). Basically, tissue elasticity imaging methods based on ultrasonics fall into three main groups:

- Methods where a quasi-static compression is applied to the tissue and the resulting components of displacement or of the strain tensor are estimated. These methods include elastography, based on a global compression of the medium (Ophir et al. 1991, O'Donnell et al. 1994, Ophir et al. 1999), and acoustic radiation force imaging (ARFI) that uses localized compression (Nightingale et al. 2001).
- Methods based on a monochromatic low-frequency vibration such as sonoelasticity (Krouskop et al. 1987, Lerner et al. 1990, Yamakoshi et al. 1990) which uses Doppler signals to estimate tissue displacement, and vibro-acoustography (Fatemi et al. 1998) which uses ultrasound-stimulated acoustic emission. Stationary waves inherent to monochromatic excitation have to be avoided using short bursts, or have to be accounted for when estimating the elastic properties.
- Transient elastography (Sandrin et al. 1999) relies on the observation of the propagation of a transient (pulsed) shear wave to determine the visco-elastic properties of the tissues. This method offers the advantage of producing a spatially and temporally localized excitation, independent of boundary conditions, that propagates through the

medium so that a whole volume can be scanned rapidly. The solution of the inverse problem (identification of the mechanical parameters) is greatly simplified in this case. The implementation of the technique requires the use of an ultrafast ultrasound scanner to track the propagation of the shear wave. The ultrafast scanner is based on an unfocused plane wave illumination of the medium instead of adjacent focused beams, resulting in a reduction in resolution.

As an alternative to ultrasound, magnetic resonance elastography (MRE) was shown to be able to measure tissue displacements in 3D with accuracy on the order of 200 nm (Muthupillai et al. 1995). MRE can measure all three components of the velocity of a low-frequency shear waves, and hence provide local estimates (and images) of the shear modulus.

2. Principles of ultrasonic elastography

Elastography is an ultrasonic imaging modality, capable of displaying the internal strains induced in a soft material undergoing an axial stress (Ophir et al. 1991, Ophir et al. 1999). Its principle, methods, and characteristics are explained in this chapter.

Ultrasonic elastography relies on radio-frequency (RF) ultrasonic signals acquired using standard ultrasound scanners. When a stress is applied to the tissues, these are being displaced and compressed – or elongated, depending on whether the stress is compressive or tensile. The local displacement of each particle along the direction of propagation of the ultrasonic beam induces a shift in the time domain of the corresponding echo in the backscattered signal, due to changes in the time of flight. Assuming a constant speed of sound, the induced time shift τ is directly proportional to the local displacement d. This process is described in Figure 1:

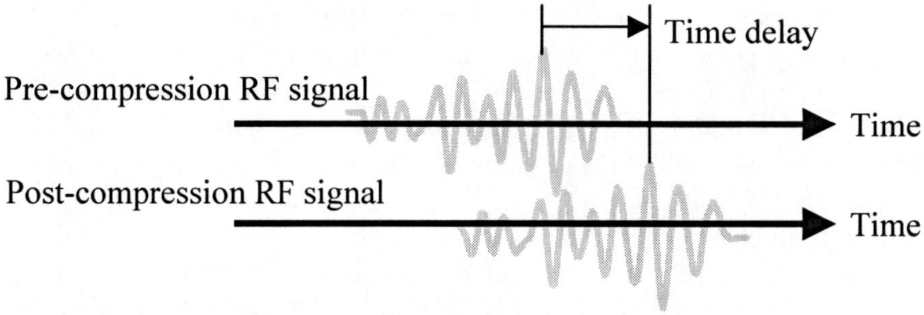

Figure 1. Pre- and post-compression RF signals. The displacement of the tissues induced a time shift.

Figure 2. Compression in the time domain of the RF signal.

The compression (or elongation) also induces a compression (or respectively stretching) in the time domain of the ultrasonic signal (Céspedes et al. 1993), similar to a frequency modulation, as shown below:

The process of elastographic acquisition is as follows:

1. Acquire pre-compression RF data
2. Apply a small compression
3. Acquire post-compression RF data
4. Estimate strain from the acquired signals.

The following paragraph shows how the axial strain can be estimated from the pre- and post-compression RF signals.

3. Strain imaging techniques

3.1. GRADIENT OF THE TIME DELAYS

The pre- and post-compression RF lines are segmented into short windows, and the time delay τ between corresponding windows is estimated. If the displacement of the particles inside the window is d, and assuming a constant speed of sound c, the time shift τ is given by the variation in the round-trip time of flight of the ultrasonic signal:

$$d = c\tau / 2 \tag{1}$$

The time delay is usually obtained from the position of the maximum of the cross-correlation function (which is a quantitative measurement of the similarity between two delayed signals) between the corresponding pre- and the post-compression windows. The maximum value of the normalized cross-correlation function is an indicator of the quality of the time delay estimate. In order to obtain sub-sample precision, a parabolic or cosine interpolation is performed around the maximum of the function (Céspedes et al. 1995). A cyclic bias may appear, depending on the chosen interpolation method, and may even induce known artifacts (Céspedes et al. 1995, Alam et al. 2000). However the error would be much larger if no interpolation were used.

Other time-efficient estimators can be used to obtain the time delays, using the position of the minimum of the sum of absolute differences between the corresponding segments (Zhu et al. 2002), or the root of the phase of the cross-correlation of the analytic signals (Pesavento et al. 1999). The process is repeated for each window, and a profile of the axial displacement along the RF signal is obtained. Axial strain ε is given by the gradient of the axial displacements in the axial direction z (where $z = ct / 2$):

$$\varepsilon = \frac{\partial d}{\partial z} = \frac{\partial \tau}{\partial t} \tag{2}$$

Practically, the gradient is obtained using two consecutive windows separated by a window shift ΔT. The axial strain measured between the two windows is given by the variation of the time delays divided by the window shift (figure 3):

$$\varepsilon = \frac{\tau_2 - \tau_1}{\Delta T} \tag{3}$$

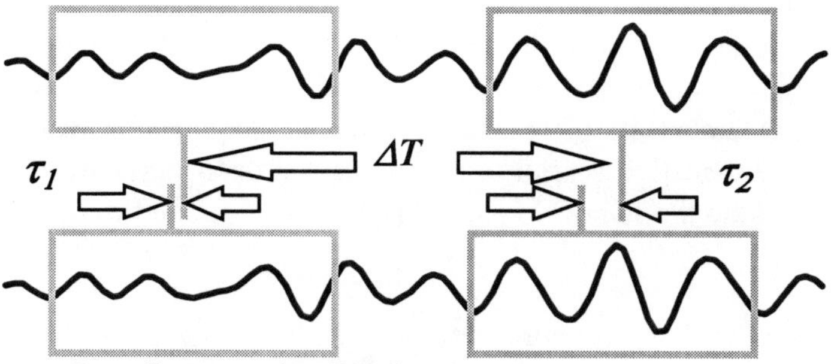

Figure 3. The strain is estimated using the time delays in two consecutive windows (courtesy of Pr. Jonathan Ophir, available online at www.elastography.com).

The process is repeated along multiple RF lines in order to form an image of the strain (elastogram). The major disadvantage of this method is the noise induced by the gradient operation. Various methods have been proposed to decrease the noise using filtering, either explicitly (O'Donnell et al. 1994) or implicitly by means of a least-squares estimator (Kallel and Ophir 1997a) or of a staggered strain estimator (Srinivasan et al. 2002).

As a result of the ultrasonic signal undergoing a compression (or elongation) in the time domain, the peak cross-correlation decreases with increasing strain, resulting in an increased noise in time delay estimation (Céspedes et al. 1999). Temporal stretching of the post-compression RF signal prior to time delay estimation improves the accuracy of the time delay measurements (Céspedes et al. 1993, Varghese et al. 1996).

3.2. STRETCHING FACTOR

Using an adequate temporal stretching of the post-compression signal improves the cross-correlation between the pre- and post-compression segments. Instead of looking for time delays, the stretching factor itself can be used as a direct measurement of strain. The pre- and post-compression signal can be modeled as (Alam et al. 1998):

$$r_1(t) = s(t) * p(t) + n_1(t) \tag{4}$$

$$r_2(t) = s(t/\alpha - t_0) * p(t) + n_2(t) \tag{5}$$

where $s(t)$ is the one-dimensional scattering distribution of the elastic target, $p(t)$ is the impulse response of the ultrasonic system, $n_1(t)$ and $n_2(t)$ are uncorrelated renditions of random noise, α is a temporal stretching factor close to unity because the applied strain is generally small, t_0 is a time delay, and * denotes convolution.

An iterative approach is used to find the stretching factor α that maximizes the normalized cross-correlation function between the two signals (Alam et al. 1998, Brusseau et al. 2001). This can be implement using an exhaustive search, a binary search or a hierarchical search. If the strain ε inside the window is small (i.e. $\varepsilon \ll 1$), it can be approximated by:

$$\varepsilon \approx 1 - \alpha \tag{6}$$

Because adaptive stretching does not contain any inter-window operation (gradient), it does not suffer from this type of degradation and it results in lower noise in strain estimates than gradient-based methods. A limitation of this method however is its computational complexity.

3.2.1. *Methods in the frequency domain*

The compression of the RF signal in the time domain is analogous to a frequency modulation and corresponds to a shift of the spectrum in the frequency domain. Spectral estimators have been proposed for elastography. The strain (or the stretching factor) can be estimated from an estimation of the frequency shift, using either the centroid (Konofagou et al. 1999) or spectral cross-correlation (Varghese et al. 2000). These estimators were shown to be less precise than correlation-based estimator for low strains, but more robust to decorrelation. As such, they represent an alternative for strain estimation in noisy environment (for example for low sonographic signal-to-noise ratio, or for signal decorrelation induced by excessive strain).

4. Image quality

Performances of elastographic imaging systems can be quantified using signal to noise ratio, contrast to noise ratio, sensitivity, dynamic range, and resolution. These characteristics are described in the following paragraphs.

4.1. STANDARD DEVIATION OF TIME DELAY ESTIMATES

The fundamental limitation in elastography is the smallest achievable standard deviation in time delay estimates. A theoretical expression of such limit is given by the modified Ziv-Zakai lower bound (ZZLB), whose expression was derived by Weinstein and Weiss (1984). The smallest variance is achieved for high sonographic signal to noise ratio (SNRs). In this case, the lower bound of the standard deviation of the time delay estimate is equal to the Cramér-Rao lower bound (σ^2_{CRLB}) and was given by Walker and Trahey (1995):

$$\sigma^2_{CRLB} \cong \frac{3}{2\pi^2 T(B^3 + 12Bf_o^2)}\left[\frac{1}{\rho^2}\left(1+\frac{1}{SNRs^2}\right)^2 - 1\right] \tag{7}$$

T is the window length, B the absolute bandwidth (fractional bandwidth $b = B/f_0$), f_0 is the central frequency, ρ is the normalized cross-correlation coefficient between the two signals, and the signal to noise ratio SNRs is the ratio of root mean square (rms) amplitudes.

The Cramér-Rao lower bound may be achieved only if the BTSNRc product is greater than a threshold η given by:

$$\eta = \frac{6}{\pi^2}\left(\frac{f_0}{B}\right)^2\left[\varphi - 1\left(\frac{B^2}{24 f_0^2}\right)\right]^2 \tag{8}$$

where $\phi(y) = \frac{1}{\sqrt{2\pi}}\int_y^{+\infty} e^{-\mu^2/2}d\mu$, and SNRc is a composite SNR that includes the effect of signal decorrelation (Weinstein and Weiss 1984):

$$SNRc = \frac{SNRs.SNR\rho}{1 + SNRs + SNR\rho} \tag{9}$$

In (9) the correlation coefficient ρ is converted into a $SNR\rho$ measure (Céspedes 1993):

$$SNR\rho = \frac{\rho}{1-\rho} \tag{10}$$

Below the threshold (i.e. for BTSNRc<η), phase ambiguities limit the precision of time delay estimates. Time delay estimates can still be obtained from the envelope of the signals, but with a lower precision defined by the Barankin bound (Weinstein and Weiss 1984).

There exist two types of errors: jitter and false peaks. False peaks occur when a secondary correlation peak exceeds the true peak, they are relatively large in magnitude (typically the error is a multiple of the wavelength, i.e. $\sim k\lambda$) and from our experience are usually related to poor correlation between signals. They appear as discontinuities and can be removed through non-linear processing, for example using a median filter on the estimated time delay. Jitter errors occur in the CRLB region when signal decorrelation, noise, and finite window length cause a slight displacement of the true correlation peak. They are small, but they cannot be removed and they pose a fundamental limit to the performance of a time delay estimator.

4.2. SIGNAL TO NOISE RATIO (SNRE)

The elastographic signal to noise ratio (SNRe) is a quantitative indication of the noise level in the strain images. It is defined in an area *where strain is supposed to be uniform* as the mean to standard deviation ratio (Céspedes et al. 1993):

$$SNRe = \frac{\mu_\varepsilon}{\sigma_\varepsilon} \tag{11}$$

4.3. THE STRAIN FILTER

Meunier and Bertrand (1995) derived the cross-correlation coefficient ρ as a function of the axial and lateral deformation:

$$\rho \approx \frac{2\sqrt{\alpha\beta}}{\sqrt{(\alpha^2+1)(\beta^2+1)}} \exp - \frac{f_0^{\ 2}}{2\sigma_f^2} \frac{(\alpha-1)^2}{\alpha^2+1} \tag{12}$$

where $\alpha=1-\varepsilon$, $\alpha\beta=1$ with the incompressibility condition, f_0 is the central frequency, and σ_f is the standard deviation of the Gaussian envelope of the signal spectrum.

This result was used to derive the strain filter, i.e. the upper bound of the SNRe as a function of strain (Varghese et al. 1997). The strain filter for a single compression experiment has a typical "band-pass" shape with maximum SNRe around 1% strain (figure 4). For low strains, the difference in time delay between consecutive windows is small and hidden by the noise in time delay estimates, resulting in high correlation but noisy strain estimates due to jitter in time delay estimates (Walker et al. 1995). For high strains, there exist an abrupt strain threshold above which the shape of the post-compression signal changes and decorrelation noise occurs, resulting in false peaks and the elastogram exhibits a typical "salt and pepper" decorrelation noise. Using a multi-compression sequence, the strain filter can be extended toward high strains and has a "high-pass" characteristic (Varghese et al. 1996).

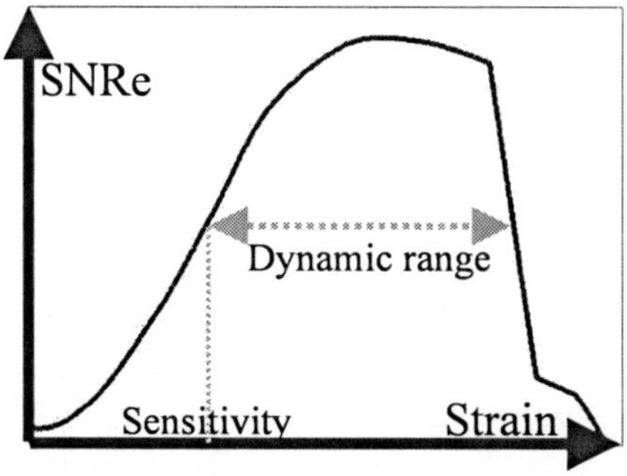

Figure 4. Typical shape of a strain filter.

The SNRe in the CRLB region is given by:

$$SNRe = \pi\varepsilon T\sqrt{\Delta T}\sqrt{\frac{(B^3 + 12Bf_o^2)}{3\left[\frac{1}{\rho^2}\left(1 + \frac{1}{SNRs^2}\right)^2 - 1\right]}} \tag{13}$$

Two important characteristics are obtained from the strain filter: the **sensitivity** ε_{min} is the smallest measurable strain at a given SNRe level (usually at half the maximum SNRe), and the **dynamic range** *DR* is the range of strains that can be reliably estimated using the elastogram. The dynamic range is defined using the minimum and maximum measurable strains at a given *SNRe* level as (Varghese et al. 1997):

$$DR = 20\log\left(\frac{\varepsilon_{\max}}{\varepsilon_{\min}}\right) \tag{14}$$

4.4. CONTRAST TO NOISE RATIO (CNRE)

The contrast-to-noise ratio is a quantitative measurement of the detectability of a target of a given size inside a homogeneous background and is defined as (Bilgen et al. 1999):

$$CNRe = 2\frac{(\varepsilon_t - \varepsilon_b)^2}{\sigma_t^2 + \sigma_b^2} \tag{15}$$

where ε_t, ε_b are the average strain in the target and in the background, and σ_t^2, σ_b^2 are the variances of the strain estimates in the target and in the background.

4.5. AXIAL AND LATERAL RESOLUTION

Elastographic axial resolution R_a was shown to be proportional to the ultrasonic wavelength λ, and inversely proportional to the bandwidth of the US system used to acquire the data (Righetti et al. 2002). Axial resolution is approximately equal to $k.(Z+\Delta Z)$ where $Z=cT/2$ is the spatial window length, $\Delta Z=c\Delta T/2$ is the spatial window shift, k is a constant that depends on how resolution is defined (typically $0.5 \le k \le 1$) the criteria that is used to define the resolution. However the best achievable resolution is limited by the spatial duration of the ultrasonic pulse, which is given by the wavelength λ:

$$R_a \approx \max(\, k.(Z+\Delta Z),\, k_a.\lambda\,) \quad \text{with } 0.5 \le k \le 1 \text{ and } k_a \approx 1.5 \tag{16}$$

Lateral resolution R_l is proportional to the beam width d as long as the pitch between consecutive RF lines is smaller than the beam width, with a proportionality factor k_l that depends on the criteria used to define resolution, but is on the order of 0.7 (Righetti et al. 2003):

$$R_l = k_l.d \tag{17}$$

4.6. INFLUENCE OF 3D MOTION

Mechanics is essentially a 3D problem, and in practice displacement are not purely axial, but also have components in the lateral and elevational directions, referring respectively to in-plane and out-of-plane motion in the image. Scatterers move out of the ultrasonic beam, other scatterers move in, resulting in a change in the shape of the RF signal. The correlation between pre- and post-compression signal decreases because we no longer image the same tissues (Kallel et al. 1997b). As a consequence elastograms become noisy, and eventually decorrelation noise dominates, resulting in a majority of false peaks seen as "salt and pepper" noise in the elastograms. Because the beam width of the US system used to acquire the data is usually small, elastography is sensitive to small lateral and/or elevational motion. Hence high acquisition frame rates are needed to reduce undesirable inter-frame motion in clinical applications – especially when scanning with hand-held probes.

2D or 3D compounding methods have been proposed to minimize decorrelation induced by such motion, or even to use it to measure lateral displacements, lateral strains, and Poisson's ratio (Insana et al. 1997, Chaturdevi et al. 1998, Konofagou et al. 1998, Bai et al. 2002). These methods do not restrict the search area to corresponding segments of the same RF line when looking for time delay estimates, but also search adjacent ultrasonic signals. Lateral search can be performed using a standard ultrasound scanner, but 3D scanning is required to track out-of-plane motion. These methods were shown to improve the axial elastogram in the presence of lateral or elevational motion. However, because of the geometry of the ultrasonic beam, the precision of the lateral or elevational components of the displacement is poor, and lateral strain elastograms are noisier than axial strain elastograms.

References

Alam SK, Ophir J, Konofagou E. An adaptive strain estimator for elastography. IEEE Trans Ultrason Ferroelectr Freq Control, 1998, Vol. 45, no. 2, pp. 461–472.

Alam SK, Ophir J. The effect of nonlinear signal transformations on bias errors in elastography. IEEE Trans Ultrason Ferroelectr Freq Control, 2000, vol. 47, no. 1, pp. 297–303.

Bilgen M. Target detectability in acoustic elastography. IEEE Trans Ultrason Ferroelectr Freq Control, 1999, vol. 46, no. 5, pp. 1128–1133.

Brusseau E, Fromageau J, Finet G, Delachartre P, Vray D. Axial strain imaging of intravascular data : Results on polyvinyl alcohol cryogel phantoms and carotid artery . Ultrasound Med Biol, 2001, vol. 27, no. 12, pp. 1631–1642.

Cespedes I, Ophir J. Reduction of image noise in elastography. Ultrasonic Imaging, 1993, vol. 15, pp. 89–102.

Cespedes I, Huang Y, Ophir J, Spratt S. Methods for estimation of subsample time delays of digitized echo signals. Ultrasonic Imaging, 1995, vol. 17, pp. 142–171.

Cespedes I, Ophir J, Alam SK. The combined effect of signal decorrelation and random noise on the variance of time delay estimation. IEEE Trans Ultrason Ferroelectr Freq Control, 1997, vol. 44, no. 1, pp. 220–225.

Fatemi M, Greenleaf JF. Ultrasound-stimulated vibro-acoustic spectrography. Science, 1998, vol. 280, pp. 82–85.

Greenleaf JF, Fatemi M, Insana M. Selected methods for imaging elastic properties of biological tissues. Annu Rev Biomed Eng, 2003, vol. 5, pp. 57–78.

Kallel F, Ophir J. A least-squares strain estimator for elastography. Ultrasonic Imaging, 1997a, vol. 19, pp. 195–208.

Kallel F, Varghese T, Ophir J, Bilgen M. The nonstationary strain filter in elastography, Part II Lateral and elevational decorrelation. Ultrasound Med Biol, 1997b, vol. 23, no. 9, pp. 1357–1370.

Konofagou E, Varghese T, Ophir J, Alam SK. Power spectral strain estimators in elastography. Ultrasound Med Biol, 1999, vol. 25, no. 7, pp. 1115–1129.

Krouskop TA, Dougherty DR, Vinson FS. A pulsed Doppler ultrasonic system for making noninvasive measurements of the mechanical properties of soft tissues. Journal of Rehabilitation Research, 1987, vol. 24, no. 2, pp. 1–8.

Krouskop TA, Wheeler TM, Kallel F, Garra BS, Hall T. Elastic moduli of breast and prostate tissues under compression. Ultrason Imaging, 1998, vol. 20, pp. 260–274.

Lerner RM, Huang SR, ParkeR KJ. Sonoelasticity images derived from ultrasound signals in mechanically vibrated tissues. Ultrasound Med Biol, 1990, vol. 16, no. 3, pp. 231–239.

Meunier J, Bertrand M. Ultrasonic texture motion analysis: theory and simulation. IEEE Trans Medical Imaging, 1995, vol. 14, no. 2, pp. 293-300.

Nightingale KR, Palmeri ML, Nightingale RW, Trahey GE. On the feasibility of remote palpation using acoustic radiation force. J. Acoust. Soc. Am., 2001, vol. 110, no. 1, pp. 625-634.

O'Donnell M, Skovoroda AR, Shapo BM, Emelianov SY. Internal displacement and strain imaging using ultrasonic speckle tracking. IEEE Trans Ultrason Ferroelectr Freq Control, 1994, vol. 41, no. 3, pp. 314–325.

Ophir J, Cespedes I, Ponnekanti H, Yazdi Y, LI X. Elastography: a quantitative method for imaging the elasticity of biological tissues. Ultrason Imaging, 1991, vol. 13, pp. 111–134.

Ophir J, Alam SK, Garra B, Kallel F, Konofagou E, Krouskop T, Varghese T. Elastography: ultrasonic estimation and imaging of the elastic properties of tissues. Proc Inst Mech Eng [H], 1999, vol. 213, pp. 203–233.

Pesavento A, Perrey C, Krueger M, Ermert H. A time-efficient and accurate strain estimation concept for ultrasonic elastography using iterative phase zero estimation. IEEE Trans Ultrason Ferroelectr Freq Control, 1999, vol. 46, no. 5, pp. 1057–1067.

Righetti R, Ophir J, Ktonas P. Axial resolution in elastography. Ultrasound Med Biol, 2002, vol. 28, no. 1, pp. 101–113.

Righetti R, Srinivasan S, Ophir J. Lateral resolution in elastography. Ultrasound Med Biol, 2003, vol. 29, no. 5, pp. 695–704.

Sandrin L, Catheline S, Tanter M, Hennequin X, Fink M. Time resolved pulsed elastography with ultrafast ultrasonic imaging. Ultrason Imaging, 1999, vol. 21, no. 4, pp. 259–272.

Srinivasan S, Ophir J, Alam SK. Elastographic imaging using staggered strain estimates. Ultrasonic Imaging, 2002, vol. 24, pp. 229–245.

Varghese T, Ophir J, Cespedes I. Noise reduction in elastograms using temporal stretching with multicompression averaging. Ultrasound Med Biol, 1996, vol. 22, no. 8, pp. 1043–1052.

Varghese T, Ophir J. A theoretical framework for performance characterization of elastography: The strain filter. IEEE Trans Ultrason Ferroelectr Freq Control, 1997, vol. 44, no. 1, pp. 164–172.

Varghese T, Konofagou EE, Ophir J, Alam SK, Bilgen M. Direct strain estimation in elastography using spectral cross-correlation. Ultrasound Med Biol, 2000, vol. 26, no. 9, pp. 1525–1537.

Walker WF, Trahey GR. A fundamental limit on delay estimation using partially correlated speckle signals. IEEE Trans Ultrason Ferroelectr Freq Control, 1995, vol. 42, no. 2, pp. 301–308.

Weinstein E, Weiss AJ. Fundamental limitations in passive time delay estimation - Part II: Wide-band systems. IEEE Trans Acoustics Speech Signal Processing, 1984, vol. 32, no. 5, pp. 1064–1078.

YamakoshI Y, Sato J, Sato T. Ultrasonic imaging of internal vibration of soft tissue under forced vibration. IEEE Trans Ultrason Ferroelectr Freq Control, 1990, vol. 37, no. 2, pp. 45–53.

Zhu Y, Hall TJ. A modified block matching method for real-time freehand strain imaging. Ultrasonic Imaging, 2002, vol. 24, no. 3, pp. 161–176.

ULTRASOUND AND THERAPY

CYRIL LAFON*
INSERM U281, 151 Cours Albert Thomas, F-69424 Lyon CEDEX 4, France

Abstract. This paper begins with an overview and a description of the interactions between ultrasound and biological tissues encountered during treatment protocols. In a second part of this seminar, two clinical applications of therapeutic ultrasound will be described in details: -Kidney stone destruction by ultrasound (lithotripsy) and High Intensity Focused Ultrasound for treating prostate cancer (HIFU).

Keywords: therapeutic ultrasound; shock waves; lithotripsy; high intensity focused ultrasound.

1. Introduction

Since the Langevin sonar trials during the First World War and the publication of evidence relating to ultrasound's potential for tissue destruction, ultrasound therapy had been somewhat forgotten; for a long time, the principal clinical use of ultrasound was purely diagnostic. However, with the development of new transducer, guidance and monitoring technologies, the therapeutic applications of ultrasound have seen renewed interest over the last two decades. Two applications in particular are routinely used in clinics today. The first, called extracorporeal lithotripsy, uses the extracorporeal application of shock waves to break kidney stones. The second aims to destroy tumors using the heat generated by high-intensity focused ultrasound (HIFU). These two applications enable us to examine the different effects that can be induced in tissue using ultrasound. These biological effects can be divided into two broad categories: thermal effects and non-thermal effects (Figure 1). Cavitation induces mechanical constraints in the tissue, and can also induce chemical effects. It should also be mentioned that, most of the time, biological effects are the result of a combination of different effects.

* C. Lafon, INSERM U281, 151 Cours Albert Thomas, F-69424 Lyon CEDEX 4, France; e-mail: lafon@lyon.inserm.fr

Y. Lemoigne et al. (eds.), Physics for Medical Imaging Applications, 211–226.

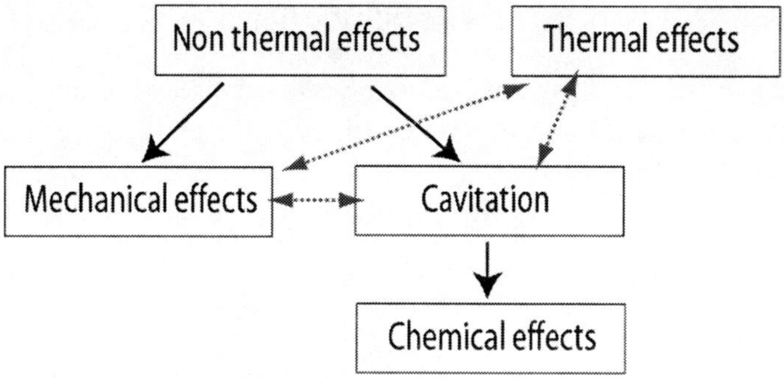

Figure 1. The different biological effects of ultrasound generally occur simultaneously during therapeutic applications.

2. Shock waves and extracorporeal lithotripsy

2.1. BACKGROUND

The disease pathophysiology of renal calculi was the subject of intensive research for many years. During this period, knowledge of the disease's etiology has improved considerably. However, no suitable solution was found for this disease, known as kidney stones. The revolution concerning the treatment of kidney stones is due principally to the emergence of treatment by shock waves, known as ESWL (extracorporeal shock wave lithotripsy). Since the first successes obtained by Chaussy in 1980 (Chaussy 1980), ESWL has become the preferred treatment for urinary calculi. 85% of urinary calculi are treated by ESWL (Sauerbruch 1991). Currently, over 800 extracorporeal lithotripsy centers exist worldwide, and several million patients have been able to benefit successfully from this technique. The treatment requires the direction of very high-amplitude, focused shock waves, generated outside the patient's body, towards the stone to be destroyed, the aim being to make it explode. Generally, this technique requires no anesthetic or, at the most, a simple analgesic. Many clinical details regarding this technique can be found in general publications on the subject (Chaussy 1989, Griffiths 1990, Eisenberger 1991).

Treatment by extracorporeal lithotripsy requires:

- a shock-wave generator;
- a system for localizing the stone to be destroyed;
- alignment of the stone with the focal point of the shock-wave generator.

2.2. DIFFERENT SOURCES OF SHOCK WAVES

A shock wave (Figure 2) is defined as a violent increase in pressure, over a very short time period (of the order of 100 ns) and a return to normal according to an exponential decay. The total duration of the impulse can be measured in microseconds; the pressure generated is around 100 MPa (1 MPa = 10 bar).

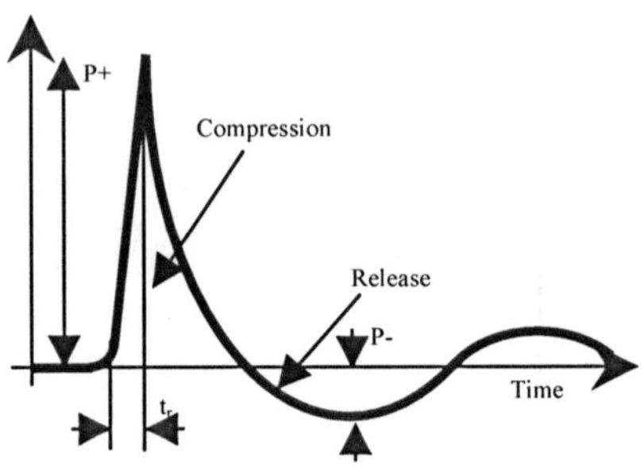

Figure 2. The shock wave is characterized by a compression phase with a brief rise time (in the order of 10 ns) followed by an expansion phase that can last several microseconds.

2.2.1. *Electrohydraulic and electroconductive generators*

This type of generator uses a point source. A violent electrical discharge is produced between two electrodes placed under water (Figure 3), which generates a bubble of steam. The high dilatation speed of this bubble creates a spherical shock wave. As the inter-electrode space is situated at the first focal point of a semi-ellipsoid reflector, the shock wave generated is partially reflected and focused at the reflector's second focal point. The electrical energy is supplied by a capacitor whose capacitance is generally close to 0.1 μF and which is regularly recharged by a high-voltage power supply of between 12 and 20 kV. Electrical switch is achieved using a gas-filled spark gap.

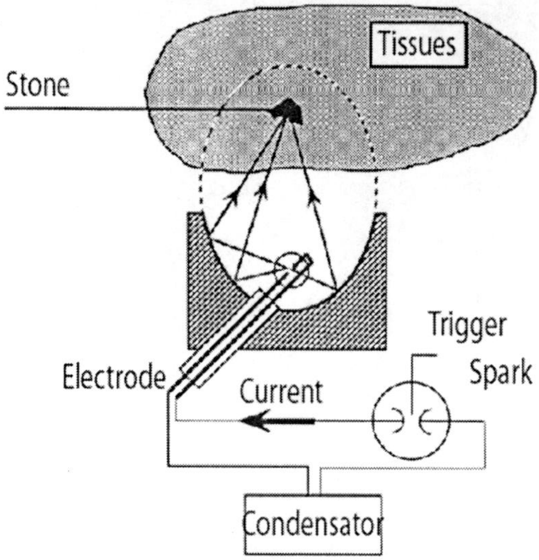

Figure 3. Principle of the electrohydraulic generator. The electrical discharge between the two electrodes gives rise a shock wave which, after reflection off the semi-ellipsoid wall, reaches a second focal point, where the stone to be destroyed is located.

The shock wave is transmitted between the generator and the patient via a coupling medium composed of degassed water (this can also be used with other types of generator) in order to avoid the formation of bubbles that would absorb and diffuse a large part of the shock wave. The pressure obtained at the focal point is of the order of several tens of MPa.

In degassed water, plasma formation is a non-reproducible phenomenon that can cause very large variations in the pressure impulse. There is a random lag time between the closure of the changeover switch and the appearance of the electric arc. Furthermore, the passage of current between the two electrodes does not take place over the whole surface, but between two randomly located points. A substantial improvement in these generators was proposed by Broyer (1996): this involved replacing the electric arc – a very unstable phenomenon – by an electrical discharge through a liquid conductor (Figure 4). This generator, also called an electroconductive generator, allows for a pressure-wave reproducibility rate of around 1%. Furthermore, as the discharge takes place along an infinite number of current lines spread over the electrodes' whole section, the lifespan of the electrodes is greatly increased. Finally, the constant conductivity in the electrolyte has made it possible to use a transmission line and move the shock-wave generator from the electrical discharge circuit.

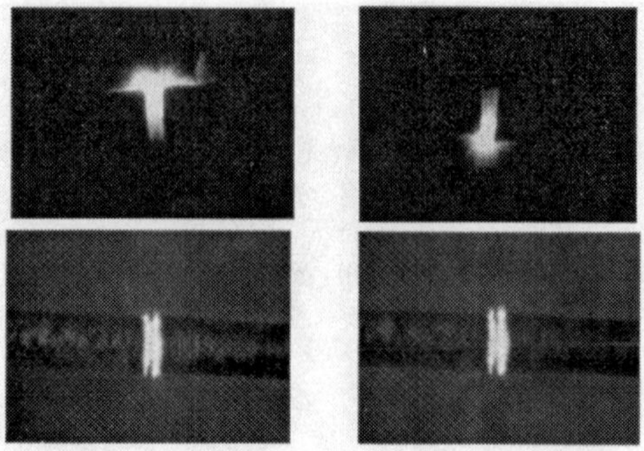

Figure 4. The place where the arc forms varies from one shock to another in the case of the electrohydraulic generator (a), whereas it is fixed in the case of the electroconductive generator (b).

2.2.2. *Electromagnetic generators*

This type of generator uses an extended source consisting of a metallic membrane. The principle is essentially that of a loudspeaker (Figure 5). A charged capacitance of between 16 and 22 kV is violently discharged through an electromagnetic coil in under 5 μs. The current within the coil induces an intense magnetic field that repels the metallic membrane situated opposite. The movement of the membrane creates a pressure wave which, spreading out inside a cylinder, becomes a shock wave. The wave is then focused using an acoustic lens. A variation also exists, where the membrane is cylindrical. The pressure wave is generated in a radial fashion and is focused by reflection off the surface of a semi-paraboloid.

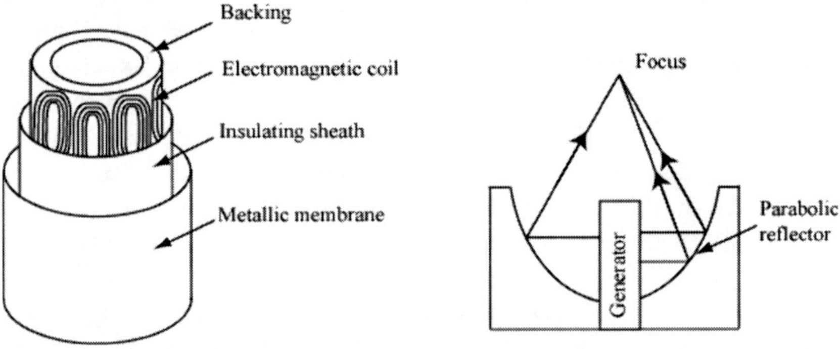

Figure 5. The principle of the electromagnetic generator is similar to that of the loudspeaker. The flat or cylindrical membrane is attracted very quickly, creating an ultrasound wave. This wave is then focused using a lens or parabolic reflector.

2.2.3. *Piezoelectric generators*

Piezoelectric sources consist of a two-dimensional matrix, composed of many piezoelectric transducer elements. These transducers are laid out on the internal face of a spherical shell (Figure 6). When subjected to an electric discharge, they expand violently and then return to their initial state after a certain number of oscillations. The sound wave generated is naturally focused at the center of the sphere. Because of the low pressure generated at the surface of the spherical shell, there needs to be a large transmitting surface, making for very cumbersome generators. To date, the efficiency of these generators has always been less than that of their rivals and, for some stones, two treatment sessions are necessary.

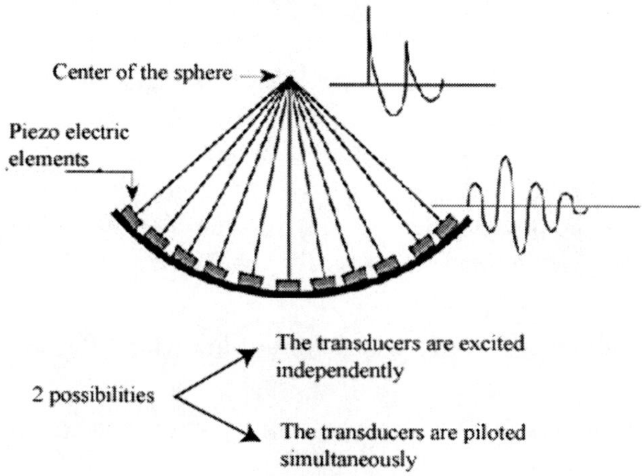

Figure 6. The piezoelectric generators are composed of a two-dimensional matrix of transducer elements. The wave focuses naturally at the focal point of the shell.

2.2.4. *Generator/patient coupling*

The optimum coupling is that which allows as little energy loss as possible between the generator and the stones that are to be destroyed. Of course, any interface between two media of different types will attenuate the ultrasound waves and must therefore be avoided. Two solutions have been proposed:

1. Immersion of the patient's trunk. This solution was used for the world's first machine, developed by the Dornier Company. This solution was complicated, since the patient had to be able to be moved in all three directions in order to bring the stone in line with the focal point of the shock waves.
2. Coupling via a membrane.

Although the first solution allowed for minimal shock-wave energy loss, it proved not only uncomfortable for the patient, but also cumbersome and inflexible with regard to the treatment of stones using a different approach.

The use of a membrane has proved far more flexible and ergonomic. The production of membranes whose acoustic impedance is close to that of water has enabled losses to be kept low. Original solutions have had to be found for changing electrodes easily (in the case of electrohydraulic and electroconductive generators) and for the avoidance of microbubble deposits on the membrane. Currently, energy losses are less than 5% and all manufacturers have opted for this technique.

2.2.5. *Characterization of shock waves produced by the different generators*

The sound waves produced by the different generators can be compared using the time curve of the shock wave generated at the focal point, as well as spatial distributions at the focal point and the skin (Coleman 1989, Buizza 1995). After passing through the coupling medium and being focused, the shape of the shock wave still resembles that shown in Figure 1. The generally recognized criteria for characterizing this shock wave include:

a. the maximum positive pressure value, known as the P+ compression wave,

b. the maximum negative pressure value, known as the P– expansion wave,

c. the width at mid-height t_w,

d. rise time t_r,

e. impulse duration t_t.

The exact value of all these parameters is quite difficult to determine and significant variations have been reported by different authors, linked to the fact that the hydrophones used are limited in terms of bandwidth and, above all, the fact that their active surface – which is too large (in the order of mm^2) – causes considerable distortion of the measured wave. The maximum pressure value is around 100 MPa. The negative pressure wavelength value (–10 MPa) is roughly the same, regardless of the type of machine. This is due to the fact that beyond a certain pressure, the negative pressure waves (expansion waves) are no longer transmitted; the water undergoes decohesion. The rise times are of the order of tens of nanoseconds for the electrohydraulic and piezoelectric machines, and hundreds of nanoseconds for the electromagnetic machines. This higher rise time is due to the fact that, since the frequency of the wave generated is lower, the wave does not have time to fully impact on its trajectory. The width at mid-height, around 350 ns, is the same for all generators. The pressure generated by a lithotripter can vary from one impact to another.

Reproducibility, on the other hand, is good for the piezoelectric, electromagnetic and electroconductive generators. For the electrohydraulic machines, the standard deviations of the pressure measurements are approximately 25 to 30%, depending on the machine and electrode wear. The energy is not distributed in one point at the geometric focus: it spreads out in the shape of an ellipsoid whose long dimension is determined by the direction of propagation of ultrasound. The volume depends to a large extent on the quality of focusing and on shock-wave intensity. Poor focusing and very high energy will increase the focal volume. The focal spot is greatest in the case of the electrohydraulic generators (long axis: approx. 6 to 8 cm; small axis: approx. 2 to 3 cm), and smallest in the case of the piezoelectric generators (long axis: approx. 3 to 4 cm; small axis: approx. 0.5 to 1 cm). From knowledge of the spatial distribution of the pressure, it is possible to calculate the energy density (energy per unit area) and/or the energy passing through the focal zone. Energy and energy density are around four times less in the case of piezoelectric generators. This point is extremely important, as the destruction of stones is essentially linked to the energy of the wave as soon as the energy threshold for destruction is reached. The aperture is characterized by the angle at the apex of the wave, moving towards the focal point. This parameter depends on the geometric characteristics of the generator. The greater this angle, the more the energy is diluted before focalization, and therefore the weaker the pressure will be when it reaches the patient, and the less pain the patient will feel. It is around 2 to 3 MPa for piezoelectric generators and two to three times stronger for the other types.

2.3. STONE IMAGING

The stone to be destroyed is currently localized using two different and complementary methods, namely X-rays and ultrasound. In the first case, the imager takes two images of the area concerned from two different angles, so as to determine the exact position of the stone (bearing in mind that X-ray imagery shows only the shadow cast by the object being examined). This remains a costly method causing exposure to radiation. It is particularly effective for kidney stones (except those that are radiolucent). Ultrasound localization involves performing an ultrasound scan of the area concerned using an ultrasound probe that is either fixed to the generator itself or situated at the end of an articulated arm, each joint of which is equipped with a position encoder. This technique is totally non-invasive, but requires more experience. The stone to be destroyed and the focal point of the shock waves are aligned by moving the patient or the generator in relation to each other.

2.4. FRAGMENTATION MECHANISMS

The fragmentation study was largely carried out using artificial stones. We can distinguish between direct effects, which result from the interaction of the shock wave with the stone, and indirect effects, which are due to the medium immediately surrounding the stone (Coleman 1993).

2.4.1. *Direct effects*

Owing to the small differences in acoustic impedance between tissue and water, the shock wave propagates inside the human body with only minimal refraction and reflection. However, the acoustic impedance of a kidney stone is 5 to 10 times greater than that of the tissue. When the shock wave hits the surface of the stone, part of the energy is reflected, creating a compression force at the surface, on the front of the stone. A compression wave then propagates through the stone, creating stresses all along its trajectory. When it reaches the rear of the stone, the compression wave is reflected, creating an expansion wave, which propagates in the opposite direction inside the stone (Hopkinson effect). Because of the often heterogeneous nature of the stone, these multiple stresses lead to fragmentation from the outside towards the center; it is this that starts the disintegration process (Figure 7). These direct effects are generally called "spalling" (Eisenberger 1991). A second effect, called “squeezing”, has also been reported. This effect is due to the fact that the velocity of the wave inside the stone is greater than its velocity in the surrounding medium. This causes almost static circumferential pressures which are responsible for fractures along meridians or parallels.

2.4.2. *Indirect effects*

The causes of stone fragmentation – other than the direct effects – also include secondary phenomena linked to the unstable cavitation that is created around the stone. Cavitation occurs because of the negative wave that immediately follows the shock wave (Figure 1). These expansion waves are also produced by the reflection of shock waves at interfaces, where the compression waves turn into expansion waves. If the amplitude of the expansion wave is sufficient, as is the case with lithotripsy, this causes an unstable cavitation in the liquid surrounding the stone (water, blood, urine, bile), as well as the collapse of microbubbles. Such collapses generate secondary shock waves and microjets which lead to a progressive erosion of the stone’s surface. A clear demonstration of the importance of this indirect effect was provided by Sass (1991), who photographed, at high speed, the progressive destruction of gallstones.

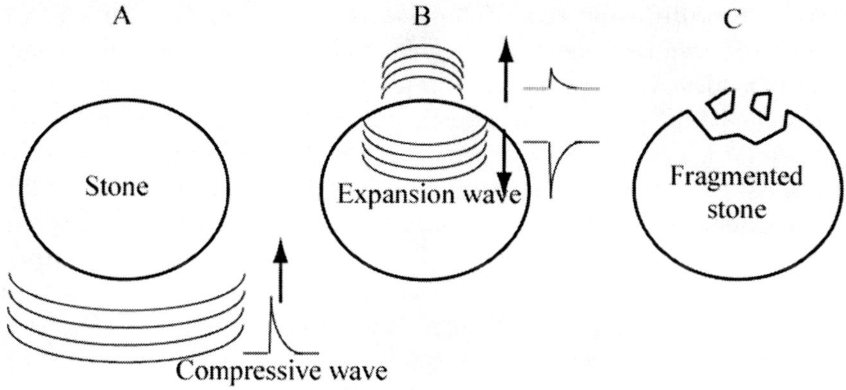

Figure 7. The compression wave reaches the stone (A). On the rear face, owing to tissue-stone impedance ratios, the reflected part of the wave turns into an expansion wave (B), which is responsible for the erosion of the stone (C).

3. High-intensity focused ultrasound for ultrasound surgery

The current trend is to move towards mini-invasive surgical techniques (which are more comfortable for patients), not only to reduce morbidity, but also to reduce costs. The therapeutic potential of high-intensity focused ultrasound has been known for a long time (Lynn 1942). The recent advances in imaging and electronics that have made accurate and controlled treatment possible show that HIFU now fits in perfectly with this approach. According to theoretical and experimental studies, the dominant mechanism for the induction of tissue damage is by thermal means. Cavitation and boiling resulting from high pressure and temperature conditions also play a role in the development of lesions. By focusing the ultrasound wave from outside the patient's body, coagulation necroses can be obtained at a depth of 10 cm without damaging the intermediate layers. Where the target is inaccessible from the outside, endocavitary or interstitial applicators may be brought closer to the target. Since the early 1990s, HIFU has found its place in clinical procedures for the treatment of benign and malignant tumors in several specialist medical fields.

3.1. HIFU GENERATORS

The use of high-frequency (several megahertz) and highly focused transducers allows well-delimited tissue damage to be obtained using the transducers' focal spot (measuring a few millimeters in diameter and a few centimeters in length). The ultrasound intensities used allow a temperature of over 50°C to be obtained within periods of a few seconds, limiting the effects of heat diffusion. This technique is particularly well suited to the treatment of localized tumors, whether malignant or benign. Focusing is generally achieved using concave transducers, but flat transducers can also be used in conjunction with acoustic lenses (Figure 8). Treatment of a whole tumor is effected by sequentially moving the ultrasound beam (using mechanical means) from one point to another, covering the whole tumor. More sophisticated applicators have now been developed for electronic tumor scanning. The transducer is divided into independently controlled elements. The phase differences between the elements allows the focal spot to be moved. We talk about 1D dynamic focusing when the focal point can be moved along the transducer network's acoustic axis only. For a 1.5D matrix, the focal point can be moved only in a plane sagittal to the matrix, as the columns symmetrical to the central column are interconnected. Finally, with a network of N×N independently powered elements, it is possible to achieve 2D focusing.

The most satisfactory treatments use focused extracorporeal transducers, which make totally non-invasive remote treatment of tumors possible. However, methods using endocavitary or interstitial applicators have been proposed in order to reach tumors situated too deep to be accessible to extracorporeal HIFU (when ultrasound waves encounter bones or gas pockets, they naturally undergo some degree of attenuation or deformation). Attenuation and phase aberrations cause a drop in pressure gain. The pressure may be increased at the transducer surface in the hope that enough energy will be supplied to the focal point; however, this compensation will be to the detriment of the intermediate tissue layers, which will also increase in temperature. The aim of using mini-invasive interstitial or endocavitary probes is to bring the ultrasound source as close as possible to the target using natural pathways in order to minimize the effects of attenuation and phase aberrations. It therefore becomes possible to work at higher frequencies in order to make heating more efficient. Compared to extracorporeal applicators, these probes are subject to greater constraints in terms of size and ergonomics.

3.2. TISSUE DESTRUCTION MECHANISMS INVOLVED DURING HIFU APPLICATION

Thermal ablation by HIFU or ultrasound surgery must not be confused with ultrasonic hyperthermia. Hyperthermia, used in cancer treatments, involves heating the tissue concerned to a temperature of 43°C over a period of an hour. This heating is applied at the same time as radiotherapy. The aim is to make the tumor cells sensitive to radiation. Because of the relatively long heating times, perfusion and thermal conduction make it difficult to obtain uniform heating effects. The application, through HIFU, of higher temperatures (> 50°C) over a period of a few seconds allows the thermal sink linked to diffusion to be minimized (Billard 1990). The principal mechanism for obtaining lesions is therefore by thermal means. However, the pressure generated at the transducers' focal point in order to achieve these temperatures can cause cavitation, which can in turn affect the development of coagulation necroses.

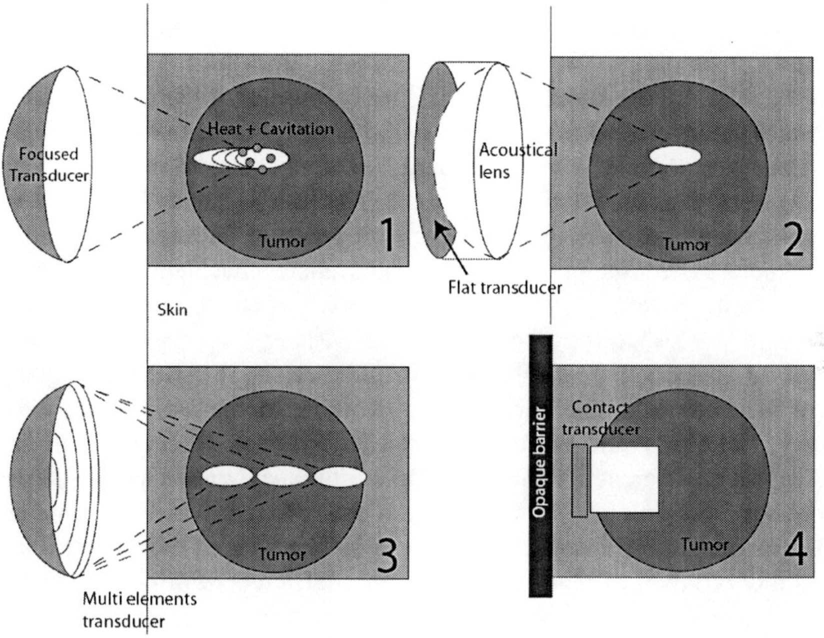

Figure 8. Different types of HIFU generator: single-element focused extracorporeal applicator (1), flat transducer with acoustic lens (2), Focused transducer with ring-shaped network of elements for 1D dynamic focusing (3), and flat interstitial transducer in direct contact with the tumor (4).

3.2.1. *Temperature*

An ultrasound wave which propagates in biological tissue will be partly absorbed. This absorption mechanism induces a rise in temperature as a function of (a) the thermal conductivity of the medium, and (b) the perfusion rate.

If the intensity of a plane wave is I_0 at the surface of the transducer ($x = 0$), the intensity at x is given by the exponential decay:

$$I(x) = I_0 e^{-\alpha x} \quad (1)$$

where α is the tissue's absorption coefficient. If we consider that the absorption coefficient is equal to the attenuation coefficient (μ), the local deposited heat per unit volume (q) is obtained by taking the gradient over x of the intensity:

$$q(x) = \mu I(x) \tag{2}$$

The absorption coefficient increases with frequency, hence the advantage of using a higher frequency for better heating. There is always a compromise to be made between the depth of treatment and the temperature increase induced. Similarly, for high pressures, non-linear propagation is accompanied by the generation of higher harmonics, which are absorbed to a greater extent and which thus add to the heating of the tissue.

The absorption of a wave's energy in a viscoelastic medium (such as tissue) occurs when fluctuations in this medium's density when the wave passes through are no longer in phase with pressure fluctuations. Relaxation phenomena reflect the presence of energy transfer mechanisms which occur over a finite period of time. The ultrasound wave transmits mechanical energy to the tissue through which it is passing; this tissue will then no longer be at energetic equilibrium. Since the return to equilibrium does not occur in phase with the variations in pressure at the entry point into the system, the ultrasound wave is attenuated. We then observe a peak in the absorption coefficient corresponding to a frequency known as the relaxation frequency. In general, several phenomena coexist which are linked to different relaxation times (or frequencies). The sum of these contributions gives the variation of the total absorption coefficient with frequency due to relaxation phenomena. The absorption phenomenon is described in detail by Hill et al. (2004).

The rise in temperature, resulting from ultrasound-wave absorption, can be obtained by solving the Bio-Heat Transfer Equation or BHTE (Pennes 1948). In a perfused medium, the BHTE takes the following form:

$$\rho_t c_t \frac{\partial T}{\partial t} = k_t \nabla^2 T + m_b c_b (T_a - T) + q \tag{3}$$

where T is the temperature, ρ_t the tissue density, c_t the specific heat of the tissue, k_t the thermal conductivity of the tissue, m_b the local perfusion rate and c_b the specific heat of blood. The term to the left of the « = » sign represents the variation in internal energy per unit volume. To the right of the « = » sign, the Laplacian expresses the exchanges by thermal conduction and the second term introduces the phenomenon of perfusion. These two terms can be ignored in the case of very brief applications of power.

3.2.2. *Cavitation*

Acoustic cavitation can be defined as the formation and activity of cavities filled with gas or vapor (bubbles) in a medium subjected to an ultrasonic field. Two types of cavitation are commonly described in medical literature: inertial (or transient) cavitation and non-inertial (or stable) cavitation. In the case of inertial cavitation, a cavity filled with gas will enlarge in the expansion phase of an acoustic cycle before violently shrinking to a fraction of its initial size. The inertia of the bubble governs deformation during the sudden compression phase. Very high temperatures, as well as light (sonoluminescence) can be generated locally. The collapse of the bubble is accompanied by the emission of a shock wave or the formation of free radicals. In non-inertial cavitation conditions, the bubbles exposed to the ultrasonic field exhibit oscillatory behavior instead. By rectified diffusion, they tend towards their resonance size. The oscillation of the bubbles may be accompanied by surface distortion or microstreaming phenomena.

The two types of cavitation occur during HIFU treatments. The generation of bubbles (cavitation bubbles, or through boiling in the case of excessive heating) at the focal point contributes significantly to the displacement of the coagulation necrosis towards the transducer, as shown theoretically by Chavrier (2000). Compared to the purely linear model, which has the development of the thermal lesion modelled on the ellipsoidal focal zone of the transducer, the experiments show cone-shaped lesions whose bases are directed towards the transducer (Meaney 2000). The thermal lens phenomena and the increase in the attenuation coefficient linked to the rise in temperature at the front of the focal point also contribute to the lesion's migration towards the transducer, but these are secondary factors.

3.3. APPLICATIONS OF HIFU

Many experimental applications of ultrasound surgery have been proposed in recent years. HIFU is used either to treat tumors (benign or malignant) or to control hemorrhages (Vaezy 2001). However, only a few applications (for which equipment is commercially available) are used routinely in clinics. These can be divided into two broad categories: extracorporeal approaches in gynecology and hepatic surgery; and endocavitary applications in urology.

3.3.1. *Extracorporeal applications of HIFU*

Extracorporeal transducers generally function at a lower frequency than endocavitary transducers, in order to limit the heating of tissue at the front of the focal point. They require quite a long focal distance to reach their target. These transducers – less limited in ergonomic terms – are larger than endocavitary transducers so as to increase the gain in the transducer's surface pressure at the focal length.

ExAblate® 2000 is manufactured by Insightec (Tirat Carmel, Israel). This device is approved in many countries for the non-invasive treatment of uterine fibromae. It is a new surgical device which combines HIFU with magnetic resonance imaging (MRI). The use of MRI makes it possible to monitor temperature and the extent of treatment during application.

Since 1997, the JC-model Tumour Therapy System (Chongqing Haifu™ Technology Co., China) has successfully treated over 8 000 patients suffering from various cancers in more than 20 centers throughout the Far East. This equipment, approved following conclusive clinical trials carried out at Churchill Hospital in Oxford, UK, and created by Ultrasound Therapeutics Ltd, a British company, has recently been granted CE certification. HIFU is generated by a flat transducer equipped with a lens. The focal point is positioned on the target by the operator using ultrasound imaging techniques. To ensure that the patient does not move at all during the procedure, the operation is carried out under general anesthetic. The trials at Churchill Hospital concerned the treatment of liver tumors. They showed that in 93% of cases of treated liver tumors, ablation of the targeted focal point was successful (Illing 2005). During more recent trials, kidney tumors have also been successfully treated. The associated literature explains the benefits of this procedure for patients suffering from pancreatic cancer, breast cancer and certain types of bone cancer (Wu 2004).

3.3.2. *Endocavitary applications of HIFU in urology*

Devices are now commercially available for the treatment of prostate tumors. Although the extracorporeal approach is feasible (Hacker 2005), the two devices available commercially use an endocavitary approach and HIFU is applied via the rectum. In both cases, treatment is carried out using ultrasound imaging together with a rectal probe.

The very first clinical application of HIFU was developed by EDAP Technomed SA (Vaulx-en-Velin, France). The Ablatherm is designed for the treatment of prostate cancer – the most common cancer in men, and the second most common fatal cancer (after lung cancer). The first prostate-cancer patient was treated in 1993. The HIFU treatment is relatively non-invasive and morbidity is very low. It is carried out under spinal anesthesia, so the time spent in hospital is very short. Patients who have experienced failure with radiotherapy can receive HIFU treatment (Gelet 2004). Equally, HIFU treatment is not a therapeutic impasse: the treatment can be repeated, and additional treatments can be carried out at the same time.

Very positive results have been obtained. A European multi-center study of 402 patients, each followed over an average of 13 months, showed that biopsies carried out were negative for 87.2% of patients and that PSA (Prostate-Specific Antigen) levels were normal for 81.4%.

After being initially limited to treating prostate adenoma (Sanghvi 1999), the Sonablate machines (Focus Surgery, Indianapolis, Indiana, USA) are now used to treat prostate cancer (Uchida 2002). With regard to the work on prostate adenoma, the results in terms of urinary flow, quality of life and IPSS (International Prostate Symptom Score) are very positive. HIFU treatment is safe and durable over time.

4. Conclusion

Ultrasound is now widely used in clinics for therapy. For example, extracorporeal lithotripsy by shock waves is now the first choice for the treatment of kidney stones. Ultrasound surgery by HIFU is slowly but surely moving out of the laboratory and into the clinic for the treatment of various tumors, both malignant and benign. New, revolutionary treatment methods using ultrasound abound in medical literature (Tachibana 2004). Because of cavitation, for instance, ultrasound can now increase the efficiency of thrombolytic agents in the case of acute myocardial infarction. Mention should also be made of advances in the following fields (among others):

- the development of new ultrasound catheters;
- sonodynamic therapy, i.e. the chemical activation of medication using ultrasound;
- gene therapy – transfection can be obtained by combining a gene, a microbubble and ultrasound;
- the local delivery of medication encapsulated in targeted transporting agents, which release their content when triggered by ultrasound.

The emergence of these new therapeutic ultrasound applications augurs well for even more varied uses of ultrasound in therapy.

ULTRASOUND IMAGE POST-PROCESSING – APPLICATION TO SEGMENTATION

OLIVIER BASSET[*] AND CHRISTIAN CACHARD
CREATIS CNRS 5515, Inserm U630, Université Claude Bernard, Lyon 1, 69621 Villeurbanne, France

Abstract. The post-processing of 2D or 3D ultrasound data is a very attractive research field to envisage an automatic analysis and/or quantitative measurements. For example, quantitative volume parameters recovery is a unique mean of making objective reproducible and operator independent diagnosis. Thus, it is important to perform a successful segmentation. Among the large variety of post processing devoted to ultrasound data, different segmentation approaches are discussed here and illustrated by some examples.

Keywords: Ultrasonic Imaging; Image Segmentation; Tissue Characterization; Parametric Imaging.

1. Introduction

Ultrasound images are of common use in medical imaging for the diagnosis of many diseases, in particular soft tissue diseases. The probe can be positioned freely by the clinician and the short acquisition time enables the possibility of getting real time data.

The quality of images used in conventional B-scan exams has been greatly improved in the recent years. Nevertheless, in a large majority of the applications, only a small part of the available information is used. Hence, the development of ultrasound techniques as a diagnosis aid tool is

[*] Olivier Basset, CREATIS CNRS 5515, Inserm U630, Université Claude Bernard, Lyon 1, 69621 Villeurbanne, France ; e-mail: Olivier.Basset@creatis.insa-lyon.fr.

Y. Lemoigne et al. (eds.), Physics for Medical Imaging Applications, 227–239.

potentially large. Several researches are carried out in different ways, but few have been implemented up to now on commercialised devices. Among the research field explored, the three-dimensional imaging, the tissue characterization, the data fusion, the contrast imaging, the harmonic imaging, new techniques of image formation, Doppler imaging, can be quoted. All these techniques tend either to improve the visual quality of 2D or 3D images, to provide quantitative information or to identify healthy or pathological tissues. These techniques are complementary to conventional B-scan imaging and improve it by taking advantage of available information (tissue characterisation, data fusion), of information accessible thanks to specific equipment (3D imaging) or after injection of a physical agent (contrast imaging).

However, the physical principle involved to provide ultrasound images yields to the speckle texture and makes noisy images, despite the recent improvements. The interpretation of images is often not obvious and the quantitative measurements difficult. In view to facilitate the image interpretation, it becomes useful to develop robust method of data post-processing to improve the visualisation or lead to accurate quantitative measurements.

The segmentation of these data is a very important step toward an automatic analysis and/or quantitative measurements. For example, quantitative volume parameters recovery is a unique mean of making objective reproducible and operator independent diagnosis. Thus, it is all the more important to perform a successful segmentation. Several attempts have been developed to segment ultrasound data from the echographic images: the cavities of the heart[1], liver[2] or prostate[3] tumours, for instance. As ultrasound data lead to very noisy images with low contrasted regions and blurred boundaries, the conventional edge-based segmentation methods such as Canny edge detection[4] or Prewitt operators[5] are not suited to find the boundaries between tissues.

To compensate for the low quality of ultrasound images, a segmentation tool based on 3D data and combining multiparametric information would be a good approach to emphasise the robustness of the post-processing. The advantages of a 3D approach and of parametric images given by tissue characterisation in a segmentation process are discussed hereafter.

2. 3D ultrasound imaging

The study of segmentation techniques devoted to 3D ultrasonic data presents several applications for medical imaging. The development of a 3D acquisition scanner addresses the problem of visualization of the 3D structures embedded in the volume. The segmentation allows extraction of the object of interest before a 3D representation is constructed. The establishment of a diagnosis is facilitated when an accurate volume measurement of an organ or of a lesion is available, for example, when tracking the evolution of a tumor. Moreover, the shape of a lesion (regular or uneven) can provide valuable information about the benign or malignant status of a specific lesion. Taking into account 3D data also improves the 2D segmentation of an image. The continuity of data in the third dimension allows a more robust detection of object boundaries. This is true even when the volume of data is built from the temporal evolution of a dynamic image in a fixed plane (3D M-mode imaging), and is particularly interesting for cardiac imaging.

3. Parametric imaging

The different tissues on an image (representing different organs or pathological and healthy tissues) are characterised by their acoustic impedance, which reflect differences in the structural organisation of the scatterers[6-9]. The conventional image used by the radiologist during an ultrasonic examination is the B-mode image computed from the envelope of the Radio Frequency (RF) signal. The envelope signal contains information related to amplitude, but the frequency and phase information, related to the attenuation and to the spatial distribution of individual scatterers, is lost irretrievably by the envelope extraction operation.

Several studies have proposed various methods for extracting information about the characteristics of the tissues in order to characterize the tissues and detect pathologies. Some measurements are closely related to the acoustical properties of the tissues[10] and are performed on RF signals or envelope images, some others, like textural parameters, are not[11] .

A multiparametric segmentation approach consists in calculating local features from the ultrasonic data that yield to images that are maps of the feature[12-13]. Then, to ensure the robustness of the method, the segmentation is not performed from the envelope image only, but from several images.

Some authors have proposed to perform the segmentation of ultrasonic B-scan data from a texture analysis of the various kinds of tissues[12-14]. The different constitution (size, distribution, type of reflectors) of the various tissues leads to different textural properties, which can help to distinguish the different structures. Textural features based on co-occurrence matrices have been widely used for the characterization of ultrasonic data. Originally proposed by Haralick[15], these features measure characteristics of the Gray level spatial dependencies (second order statistics). However, on ultrasonic images, the texture of the various tissues depends on the imaging device.

This drawback makes the choice of pertinent textural features that characterize the various tissues system dependent.

Acoustic parameters have also been extensively used for tissue characterization. Most of the applications concern liver and prostate tissues[3,16]. Contrary to textural features, acoustical features are calculated from the unprocessed RF signals. Information on the attenuation, scattering, elastographic characteristics of tissues, and the size, distribution and concentration of scatterers can be derived from the backscattered RF signal. These features are often associated with a classification procedure to identify the various tissues[17]. They can provide two kinds of information. On one hand, Integrated BackScatter coefficients (IBS)[18] and attenuation estimators[19] measure the echogenicity properties of the scatterers. On the other hand, parameters such as scatterers' density, scatterers' size or mean scatterer spacing estimators characterise the spatial organisation of the scatterers[20]. Some methods propose to derive acoustical parameters from the envelope image. They are based on the numerous studies that try to link the particularities of a tissue with a specific intensity distribution (Rayleigh, Rice, K-distribution, Nakagami) on the envelope image[21,27,28]. These latter parameters are representative of the scattering conditions in the tissue.

So, for each parameter, the computed map shows the variations of the parameter throughout the investigated area. Local measurements are thus needed for this mapping and therefore a trade-off between the size of the local measurements windows and the resolution of the parametric image has to be found. The window (either 1D, 2D or 3D) must be large enough to exhibit stationary statistics, representative of the local specificity of the tissue.

Most of the parametric measurement methods provide qualitative measurements. Because the purpose is image segmentation, accurate quantitative measurements are not required but reliable relative estimations that clearly emphasise the differences between various tissues.

4. Different segmentation approaches

The segmentation of acoustic data needs to be robust to speckle noise, low contrast, and attenuation. Several approaches have been proposed to segment ultrasonic images. Most of them are based on the pixel intensity and use a Bayesian framework to define an energy function that characterizes a homogeneous region or a contour. Ashton and Parker[22] have proposed a modification of an adaptive clustering algorithm to take into account the particularities of ultrasonic data. It is assumed that pixel intensities are given by a slowly varying class mean corrupted globally by uniform additive white Gaussian noise. Spatial smoothness constraints are incorporated in the algorithm by using a Markov Random Field (MRF) to model the region process. The Central Limit Theorem makes this model reasonably acceptable. It states that the distribution of the mean of a large number of independent random observations tends toward a Gaussian distribution centred to their collective mean. This is the case in a multiresolution implementation of the algorithm, for low-pass filtered and decimated ultrasonic images that are originally governed by non-Gaussian statistics. The segmentation process of this algorithm takes into account only the pixel Gray level at a site and its surroundings to classify each pixel.

Other schemes for segmentation of ultrasonic images have been proposed. Mulet-Parada and Noble[23] suggest an intensity-amplitude invariant approach using a phased-based feature detection method. Applied to an echocardiographic image sequence, the algorithm takes advantage of the temporal inconsistency of speckle to detect the acoustic boundaries.

A multiresolution texture segmentation approach was proposed by Muzzolini et al.[14]. Their approach generalized the conventional Simulated Annealing (SA) methods to a multiresolution framework and minimized an energy function, which is dependent on the resolution and the size of the texture blocks of the image. SA remains a computationally expensive method of minimization. The use of Besag's Iterated Conditional Mode (ICM) [24], which corresponds to instantaneous freezing in SA for energy minimization, would improve the applicability of this algorithm. However, ICM requires accurate initial parameters estimation, which is difficult under non-Gaussian statistics.

Active contours or ``snakes" can be used to segment objects automatically. They are used in ultrasound data to segment intravascular data[30], prostate or breast tumors. The basic idea is the evolution of a curve, subject to constraints from the input data. The curve should evolve until its boundary segments the object of interest. The model converges from an initial shape to its equilibrium iteratively, by a weighted sum of the internal and external forces. Internal forces are based on the local curvature of the surface and provides for a smooth contour and external forces drive the expansion of the contour and are extracted from the volumetric image data by applying an appropriate edge filter[31].

5. Segmentation examples

This section presents some segmentation results from multiparametric data. The implemented method is based on a Bayesian approach using a multiresolution Markov random field. The several parameters are computed from the image or from the Radio-Frequency (RF) signals. The multiparametric approach is implemented to obtain a more robust segmentation and the multiresolution approach is interesting to improve simultaneously the robustness of the segmentation and the speed of the algorithm.

The segmentation method, based on the Ashton and Parker algorithm[26] and adapted for multi-parametric data[12-25] is briefly presented. We also detail how the parametric images can be taken into account in the segmentation process.

We assume that the observed data Y is a random field defined on a 2D rectangular grid S. Y_s denotes the value of Y at the site $s \in S$. A segmentation of the image into regions will be noted by X, where $X_s = i$ means that the pixel at s belongs to region i. The number of different regions type is k. The conditional density function of Y given X, is assumed to exist and to be strictly positive and is denoted by $f(y/x)$. The probability $P(X = x)$ is written as $P(x)$. The image may be segmented by estimating the pixel classification X given the observed image Y using the Maximum A Posteriori estimation of X.

$$\begin{aligned} \hat{x}_{map} &= \arg\max_x \{P(X = x / Y = y)\} \\ &= \arg\min_x \{-\ln f(y/x) - \ln P(x)\} \end{aligned} \tag{1}$$

Hence, once the distributions of $f(y/x)$ and $P(x)$ are defined, the problem of segmenting an image will be reduced to a problem of minimising an energy function.

We use a Markov Random Field to model the region process X, due to its restriction to local interaction. So, for a given neighbourhood system, the *prior* density $P(x)$ can be written as a Gibbs density which has the following form[24]:

$$P(x) = \frac{1}{Z} \exp\left\{- \sum_{all\ cliques\ C} V_c(x)\right\}. \tag{2}$$

Here Z is a normalising constant called the partition function. $V_c(x)$ are the clique potentials. A clique is a set of points that are neighbours of each other. In our work, we used a 6-points neighbourhood in the 3D space and consider only the two-point clique potentials, which are defined as follows:

$$V_c(x) = \begin{Bmatrix} -\beta, & \text{if } x_s = x_q & \text{and } s,q \in c \\ +\beta, & \text{if } x_s \neq x_q & \text{and } s,q \in c \end{Bmatrix}, \ \beta > 0. \tag{3}$$

The intensity of a region i at the site s is assumed to be a Gaussian distribution with mean μ_s^i and variance σ_s^{i2}. The local class mean μ_s^i is a slowly varying function of s. σ_s^{i2} Is estimated independently for each class, and is proportional to μ_s^i. Under these assumptions, $\ln f(y/x)$ may be written as:

$$\ln f(y/x) \propto \sum_s \left(\ln(\sigma_s^{x_s}) + \frac{1}{2(\sigma_s^{x_s})^2} \left(y_s - \mu_s^{x_s}\right)^2 \right). \tag{4}$$

Substituting $P(x)$ from Eq. (2) and $\ln f(y/x)$ from Eq. (4) into Eq. (1) leads to the following energy function:

$$U(x/y) = \sum_s \left(\ln(\sigma_s^{x_s}) + \frac{1}{2(\sigma_s^{x_s})^2} \left(y_s - \mu_s^{x_s}\right)^2 \right) + \sum_{all\ cliques\ C} V_c(x) \tag{5}$$

This function has two components. One constrains the region intensity to be close to the data, the other is a regularisation term that imposes a smoothness constraint.

To improve the robustness of the algorithm, we consider the possibility of modifying the energy function by adding other constraints based on acoustical properties representative of each region. Let $\{Y_1,...,Y_n\}$ be a set of acoustic features calculated on each site of the volume of data. The features used must be chosen as relevant measurements of the characteristics of each region. They must be adapted to the objectives of the segmentation and to the particularities of the data processed. The parametric images are modelled in the same way than grey scale images in Eq. (4). A parametric image is assumed to be a collection of uniform or slowly varying intensities. The sharp transitions in grey level may only occur at region boundary. The parametric measurement of feature Y_j at a pixel location s is denoted $(y_j)_s$. We assume that the values given by a feature Y_j on a class i is modelled by a normal distribution of mean $(m_j)_s^{x_s}$ and variance $(\sigma_j^{x_s})^2$. Then the complete energy function has the following form:

$$\begin{aligned} U(x/y, y_1...y_n) = & \sum_s \left(\ln(\sigma_s^{x_s}) + \frac{1}{2(\sigma_s^{x_s})^2} \left(y_s - \mu_s^{x_s}\right)^2 \right) \\ & + \sum_{all\ cliques\ C} V_c(x) + \sum_j \sum_s \left(\ln(\sigma_j^{x_s}) + \frac{1}{2.(\sigma_j^{x_s})^2} \left((y_j)_s - (m_j)_s^{x_s}\right)^2 \right). \end{aligned} \tag{6}$$

Finding the global minimum of this function requires a lot of computation. Besag has proposed the Iterated Conditional Modes algorithm[24] as an alternative to Simulated Annealing. It modifies the label of sites in x in view to maximise the conditional density function at each site. The algorithm converges to a solution, which is reasonably acceptable under a good initialisation.

The initialization is performed using the K-Means clustering algorithm. K-Means is a least-squares partitioning method that divide a collection of pixels into K classes. The algorithm iterates over the two following steps: 1) it computes the mean of each cluster and 2) it computes the distance of each pixel from each cluster mean and assigns each pixel to the cluster it is nearest to. The initial assignment of points to clusters is done randomly.

Two segmentation examples are presented hereafter. The first one corresponds to simulated data obtained with the field software[29]. The medium is composed of two objects as shown on Fig. 1, with different characteristics (different density of scatterers and different echogenicity). From these data, parameters maps have been computed using 32x32 pixels window size, giving IBS, density and Nakagami parameters images.

Segmentation has been performed with the following parameters: Number of classes: 2; Number of scale decompositions: 2; (β regularisation coefficient at the highest resolution: 0.6; $\Delta\beta$ (increment of β at each resolution): 0.2). Fig. 2 shows an example of segmentation results.

Figure 1. Geometrical shape of the phantom.

The results illustrated by Fig. 2 show clearly that the segmentation depends on the data involved in the segmentation process and using several images that represent different properties of the tissues improves the delineation of the object. When four images are used, the resulting segmentation is close to the ideal result shown in Fig. 1. The complementarity and/or the redundancy of the different data reinforce the robustness of the segmentation.

Figure 3 shows a segmentation result performed on echocardiographic data of the left ventricle of a dog's heart. It can be seen that the accuracy and smoothness of the resulting contours corresponding to the inside and outside wall depends on the parametric images involved in the segmentation.

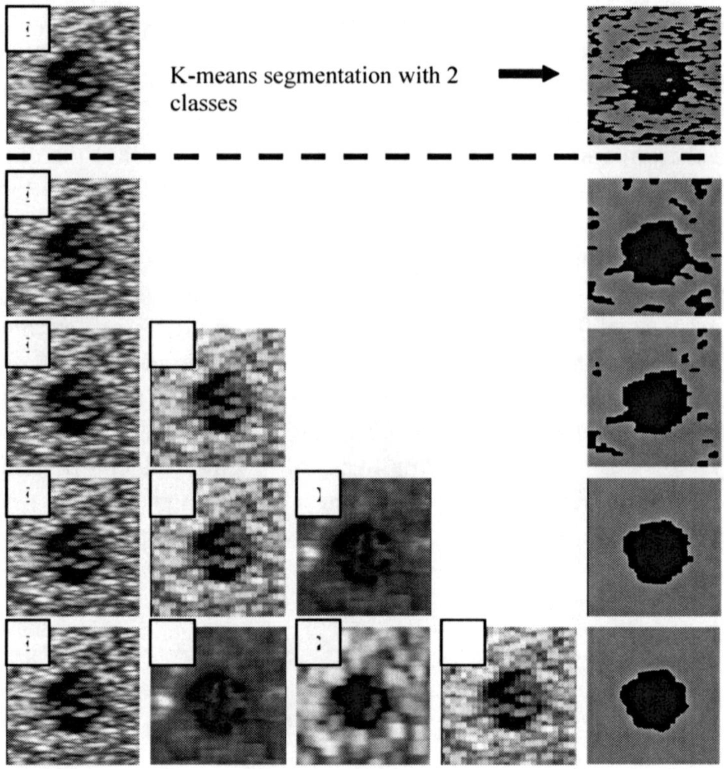

Figure 2. Segmentation results of a numerical phantom. The right column shows the segmentation results involving the parametric images on the corresponding row. The first row shows a segmentation with the k-means algorithm. E : Envelope, D : estimation, I : *IBS,* Ω : Nakagami parameter.

6. Discussion

Figures 2 and 3, illustrating the presented segmentation approach show that involving several different data in the segmentation process, complementary to the envelope image leads to a more accurate segmentation, less sensitive to the speckle noise. Nevertheless, several limitations of the proposed methods can be pointed out.

The window size used to calculate the local information in parametric images is rather large. Consequently, the multi-parametric segmentation approach doesn't allow obtaining better resolution on segmented images than on envelope images and then is not adapted to improve the detection of very small lesions. Although the presented results exhibit generally improvements of the segmentation using several parameters, one can remark that the choice of the parameters involved in the segmentation is of major importance.

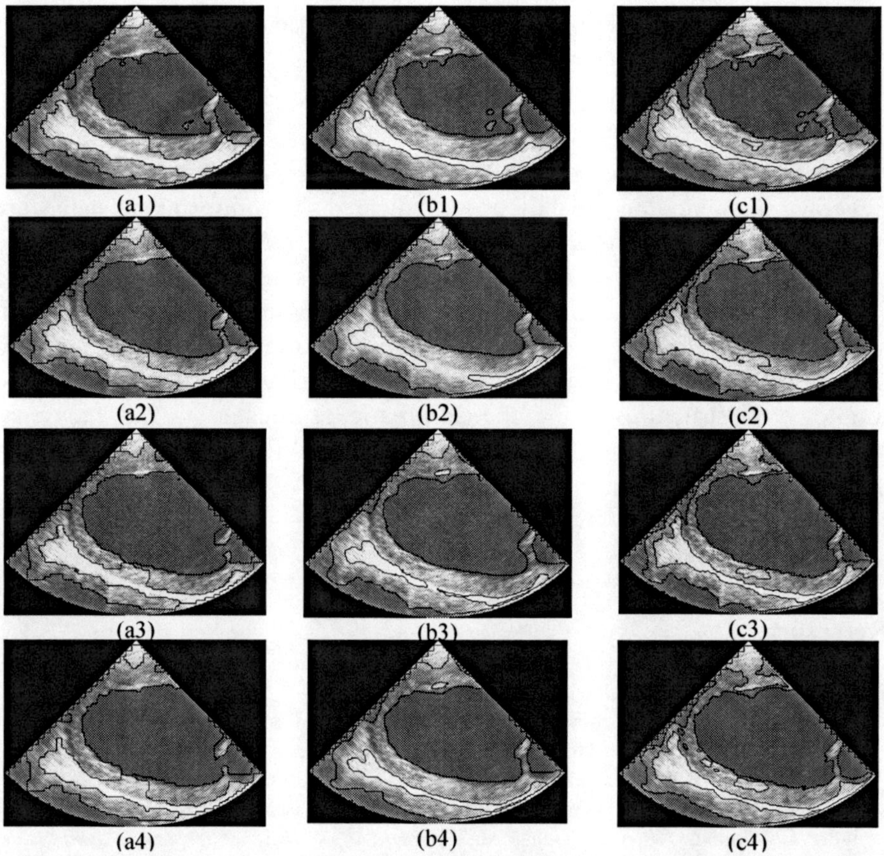

Figure 3. The region contours detected are superimposed on the original envelope slices at 4 instants of the cardiac cycle. (a, 1-4) Segmentation without tissue characterization (b, 1-4) with texture characterization; (c, 1-4) with texture and acoustical characterization.

Another drawback of such an approach lies in the supervised aspect. To overcome this limitation "mean shift" technique can be used[32-33]. A multidimensional filtering based on the mean shift technique can be performed in the joint spatial-range domain. This technique consists in assigning each pixel with a local mode of the underlying distribution. The local modes are identified by an iteratively moving kernel in both spatial and range domains. The final unsupervised segmentation is obtained from the resulting multidimensional filtered image, by fusing regions associated with nearby modes.

Contour based segmentation approaches are an alternative to region-based approaches. They leads to closed contours but the limitation with this approach is that the topology of the region to be segmented must be known in advance and the accuracy and speed of convergence to the desired final contour depend on a reasonable initial contour. The application of deformable model to 2D or 3D data is often restricted to data of particular nature, such as intravascular ultrasound imaging.

7. Conclusion

Several segmentation techniques dedicated to ultrasound data are mentioned in this paper. The multiparametric technique, based on the combination of different information extracted from the tissue characterisation performed on 3D data is a promising approach for robust segmentation. However, the noisy nature of the data, the non sharp boundaries between multiple structures, the various shapes, the variable topology of structures make the automatic segmentation and the post-processing of ultrasound data in general, a very difficult task, and an open problem.

References

1. I. Dydenko, D. Friboulet, J.-M. Gorce, J. D'hooge, B. Bijnens, and I. E. Magnin, "Towards ultrasound cardiac image segmentation based on the radiofrequency signal," *Medical Image Analysis*, vol. 7, pp. 353–367, 2003.
2. H. Yoshida, B. Keserci, D. D. Casalino, A. Coskun, O. Ozturk, and A. Savranlar, "Segmentation of Liver Tumor in Ultrasound Images based on Scale-Space Analysis of the Continuous Wavelet Transform," *presented at IEEE Ultrasonics Symposium, Sendai, Japan,* 1998.
3. F. L. Lizzi, E. J. Feleppa, M. Astor, and A. Kalish, "Statistics of Ultrasonic Spectral Parameters for Prostate and Liver Examination," *IEEE Transaction on Ultrasonics, Ferroelectrics and Frequency Control*, vol. 44, pp. 935–942, 1997.
4. J. Canny, "A computationnal approach to edge detection," *IEEE Trans. Pattern Anal. Machine Intell.*, vol. 8, pp. 679–698, 1986.
5. J. M. S. Prewitt, "Object enhancement and extraction," in *Picture processing and psychopichtorics*, B. S. L. a. A. Rosenfeld, Ed. New York: Academic Press, 1970.
6. U. Haberkorn, I. Zuna, A. Lorentz, H. Zerban, G. Layer, G. V. Kaick, and U. Rath, "Echographic Tissue Characterisation in Diffuse Parenchymal Liver Disease: Correlation of Image Structure with Histology," *Ultrasonic Imaging*, vol. 12, pp. 155–170, 1990.
7. L. Landini and L. Verrazzani, "Spectral characterization of tissues microstructure by ultrasounds: a stochastic approach," *IEEE Transactions on Ultrasonics. Ferroelectrics. and Frequency Control.*, vol. 37, pp. 448–456, 1990.

8. P. D. Edmonds, C. L. Mortensen, J. R. Hill, S. K. Holland, J. F. Jensen, P. Schattner, and A. D. Valdes, "Ultrasound Tissue Characterization of Breast Biopsy Specimens," *Ultrasonic Imaging*, vol. 13, pp. 162–185, 1991.
9. T. J. Hall, M. F. Insana, L. A. Harrison, and G. G. Cox, "Ultrasonic Measurement of Glomerular Diameters in Normal Adult Humans," *Ultrasound in Med. & Biol.*, vol. 22, p. 987, 1996.
10. M. J. T. M. Cloostermans, H. Mol, J. A. Verhoef, and J. M. Thijssen, "In vitro estimation of acoustic parameters of the liver and correlations with histology," *Ultrasound in Medicine and Biology*, vol. 12, pp. 39–51, 1986.
11. F. M. J. Valckx and J. M. Thijssen, "Characterization of Echographic Image Texture by Cooccurrence Matrix Parameters," *Ultrasound in Medicine and Biology*, vol. 23, pp. 559–571, 1997.
12. D. Boukerroui, O. Basset, A. Baskurt, and G. Gimenez, "A Multiparametric and Multiresolution Segmentation Algorithm of 3–D Ultrasound Data," *IEEE Trans. on Ultrasonics, Ferroelectrics, and Frequency Control*, vol. 48, pp. 64-77, 2001.
13. D. Boukerroui, O. Basset, A. Baskurt, and A. Noble, "Segmentation of ultrasound images – Multiresolution 2D and 3D algorithm based on global and local statistics," *Pattern Recognition Letters*, vol. 24, pp. 779–790, 2003.
14. R. Muzzolini, Y.-H. Yang, and R. Pierson, "Multiresolution texture segmentation with application to diagnostic ultrasound images," *IEEE Trans. Medical Imaging*, vol. 12, no. 1, pp. 108–123, 1993.
15. R. M. Haralik, K. Shanmugam, and I. Dinstein, "Textural features for image classification," *IEEE Trans. System, Man and Cybernetics*, vol. 3, no. 6, pp. 610–621, 1973.
16. G. Schmitz, H. Ermert, and T. Senge, "Tissue-characterization of the prostate using radio frequency ultrasonic signals," *IEEE Trans. on Ultrasonics, Ferroelectrics, and Frequency Control*, vol. 46, no. 1, pp. 126–138, 1998.
17. G. Georgiou and F. Cohen, "Unsupervised segmentation of RF echo into regions with different scattering characteristics," *IEEE Trans. on Ultrasonics, Ferroelectrics and Frequency Control*, vol. 45, pp. 779–787, 1998.
18. D. Nicholas, "Evaluation of backscattering coefficients for excised tissues : results, interpretation and associated measurements," *Ultrasound in Medicine and Biology*, vol. 8, pp. 17–28, 1982.
19. H. A. H. Jongen, J. M. Thijssen, M. V. D. Aarssen, and W. A. Verhoef, "A general model for the absorption of ultrasound by biological tissues," *J. Acoustic. Soc. Am.*, vol. 79, pp. 535–540, 1986.
20. T. Varghese and K. D. Donohue, "Estimating Mean Scatterer Spacing with the Frequency-Smoothed Spectral Autocorrelation Function," *IEEE Trans. on Ultrasonics, Ferroelectrics, and Frequency Control*, vol. 42, pp. 451–463, 1995.
21. G. E. Sleefe and P. P. Lele, "Tissue characterization based on scatterer number density estimation," *IEEE Trans. on Ultrasonics, Ferroelectrics, and Frequency Control*, vol. 35, pp. 749–757, 1988.
22. E. A. Ashton and K. J. Parker, "Multiple resolution bayesian segmentation of ultrasound Images," *Ultrasonic Imaging*, vol. 17, pp. 291–304, 1995.
23. M. Mulet–Parada and J.A. Noble "2D+T acoustic boundary detection in echocardiography." *Medical Image Analysis,* vol 4, pp 21–30, 2000.
24. J. Besag, "On the statistical analysis of dirty pictures (with Discussion)," *Journal Royal Statistical Society B*, vol. 48, pp. 259–302, 1986.
25. F. Davignon, J.F. Deprez and O. Basset "Parametric imaging for the segmentation of ultrasound data " Ultrasonics, 43, pp. 789–801, 2005.

26. B. J. Besag, "Spatial interaction and the statistical analysis of lattice systems," *J. Royal Stat. Soc. B*, vol. 26, pp. 192–236, 1974.
27. F. Ossant, F. Patat, M. Lebertre, M.-L. Teriierooiterai, and L. Pourcelot, "Effective density estimators based on the K-distribution: Interest of low and fractional order moments," *Ultrasonic Imaging*, vol. 20, pp. 243–259, 1998.
28. P. M. Shankar, "A general statistical model for ultrasonic scattering from tissues," *IEEE Trans. Ultrason., Ferroelect., Freq. Contr*, vol. 47, pp. 727–736, 2000.
29. J. A. Jensen and P. Munk, "Computer phantoms for simulation ultrasound B-mode and CFL Images," *presented at 23rd Acoust. Imag. Symp.*, Boston, MA, 1997.
30. E. Brusseau, CL de Korte, F. Mastik, J. Schaar AF. van der Steen, "Fully automatic luminal contour segmentation in intracoronary ultrasound imaging--a statistical approach." *IEEE Trans Med Imaging.*;23(5):554–66, 2004.
31. RF. Chang, WJ. Wu, WK. Moon, WM. Chen, W. Lee, DR. Chen, "Segmentation of breast tumor in three-dimensional ultrasound images using three-dimensional discrete active contour model" *Ultrasound Med Biol.* 29(11):1571–81 Nov; 2003.
32. D. Comaniciu and P. Meer, "Mean shift: A robust approach toward feature space analysis," *IEEE Transactions on Pattern Analysis and Machine Intelligence*, vol. 24, pp. 603–619, 2002.
33. T. Grenier, C. Revol-Muller, F. Davignon, O. Basset, G. Gimenez "Multiparametric smoothing based on mean shift procedure for ultrasound data segmentation" *Eusipco,* 2005.

INTRAVASCULAR IMAGING

NICO DE JONG*, NICOLAAS BOM, JOHANNES SCHAAR, DAVID GOERTZ, MARTIJN FRIJLINK, ANTON FW VAN DER STEEN
Thoraxcentre; Erasmus Medical Center Rotterdam, P.O. Box 1738, NL-3000 DR Rotterdam, The Netherlands

Abstracts. IVUS is used for diagnostics, therapy guidance and scientific purposes. It is the only clinical available technique that can assess plaque burden and free lumen diameter at high accuracy. Contrast angiography, which was the golden standard before IVUS, can only give a shadow projection of the lumen. Especially with the advent of 3D IVUS using pull backs it became an important tool for monitoring treatment and follow up of interventions like balloon angioplasty and placing of stents (wire prostheses that are used to prevent the arterial wall from recoiling). 3D IVUS in combination with biplane angiography allows assessment of true 3D reconstructions of arteries, pre and post treatment. Using computational fluid dynamics the velocity profile and thus the shear stress at the vascular wall can be calculated. This can be related to biological markers, which gives insight in formation of atherosclerosis, restenosis and remodelling.

Keywords: IVUS catheter; echotransducer; synthetic apertures system; elastography; palpography; contrast agents.

1. Introduction

Currently, IVUS is, beside OCT the only clinically available technique, capable of providing real time cross-sectional images *in vivo.* It delivers information that is not available from X-ray angiography. Since the number of available interventional techniques for treatment of atherosclerotic luminal narrowing increases, specific diagnostic information that can aid in the selection becomes increasingly important. For this reason, IVUS is more and more routinely used for guiding interventional procedures and for studying the mechanisms for restenosis[1].

* Nico de Jong, Erasmus University Rotterdam P.O. Box 1738, NL-3000 DR Rotterdam; e-mail: n.dejong@erasmusmc.nl

Y. Lemoigne et al. (eds.), Physics for Medical Imaging Applications, 241–251.

2. Construction of an IVUS catheter

2.1. ROTATING ELEMENT

In Figure 1, the rotating tip system is schematically illustrated. The driving motor is connected to the proximal end of the catheter. The shaft (1) must be very flexible and contains the electric wires for the transducer.

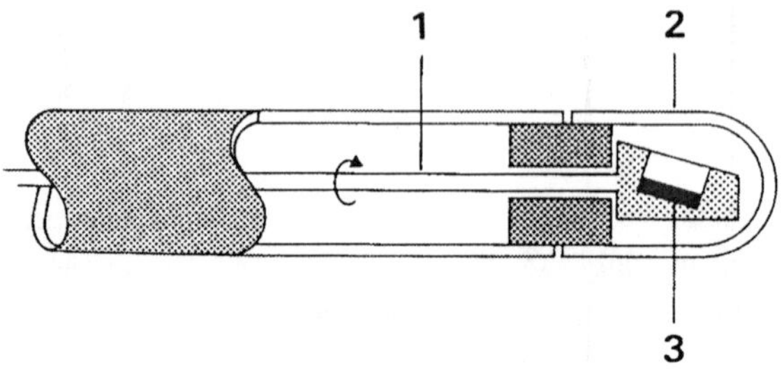

Figure 1. Flex-shaft rotating element principle. This contains a rotating shaft (1); acoustically transparent dome (2); and echo element (3) in a catheter tip. The flex shaft is powered by a proximal motor outside the catheter. In some practical application.

The shaft needs to be flexible; yet it must drive the tip in a predictable way. This is not always possible when the catheter must follow a tortuous path. As a result, the beam deflection on the display may not correspond with the acoustic beam deflection. This causes errors in the image. Capacitive and optical feed-back techniques have been suggested to cope with this problem. The element (3) is positioned in such a way that no transmission pulse effect appears on the display, since the echo travel time to the dome (2) is of sufficient duration. It allows imaging very close to the catheter outer wall since no 'dead zone' is present. The dome must be acoustically transparent. Results obtained in-vitro are often reported without such a dome, thus increasing image quality and avoiding the liquid filling process necessary for acoustic coupling.

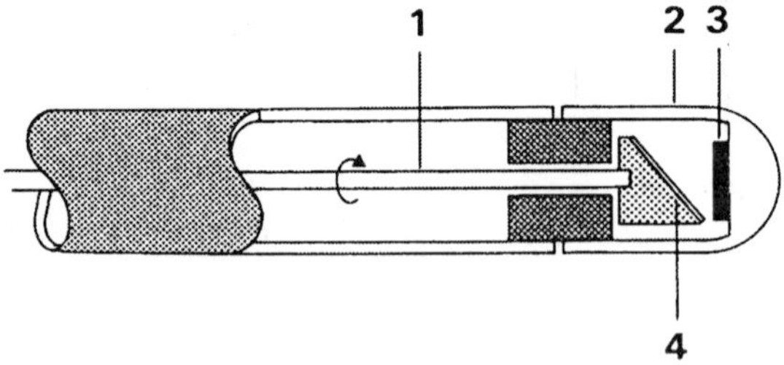

Figure 2. The rotating mirror principle. This contains the flex shaft (1); the transparent dome (2); the acoustic element (3) and the rotating mirror (4).

2.2. ROTATING MIRROR

The rotating mirror technique is similar to the previously described method. Schematically this method is shown in Figure 2. The flexible shaft (1); the transparent dome (2); the echotransducer (3) are complemented with a mirror. The mirror creates an even shorter dead zone outside the catheter due to the longer acoustic pathway inside the catheter. The non-moving transducer avoids the necessity of rotating electric wires. Acoustic lenses and focussing shapes of the mirror have been described.

2.3. ELECTRONICALLY SWITCHED PHASED ARRAY

The catheter shown in Figure 3 contains many small acoustic elements (2) which are positioned cylindrically around the catheter tip. The number of elements may be any practical number such as 16, 32, 64 or 128. The tip may contain an electronic component to reduce the number of electric wires.

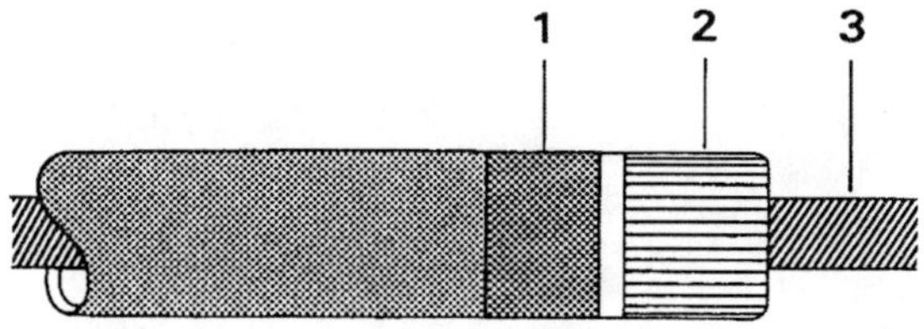

Figure 3. Electronically switched phased array catheter tip with integrated circuitry for reduction of the number of wires (1); the elements (2) and a guide wire (3).

The construction allows for the introduction of a central guidewire (3). A first prototype, using the same principle but meant for intracardiac imaging was already described in 1972[2] . By introduction of time delays subgroups of elements may together form a "single larger echo transducer". This process can be repeated with any other subgroup. This allows aperture variation and electronic focussing methods. On the other hand, acoustic element geometry is not optimal and a near-field dead zone may exist. In a second principle, all elements are used individually for transmission and reception. Beam forming is created with computer algorithms using a Synthetic Aperture System (SAS).

A summary of today's IVUS methods is illustrated in Figure 4. Mostly, the intravascular procedure is applied for decision-making when angiographic data are less conclusive. For obvious reasons, there is a strong urge to combine "see and do" in interventional procedures. This leads to catheters in which, for instance, an angioplasty balloon is combined with ultrasonic imaging in or close to the balloon. Another combination may provide guidance during stent implantation (Figure 4d). Since intravascular imaging provides accurate geometrical information within the cross-section, combination with other interventional procedures is likely to expand in future.

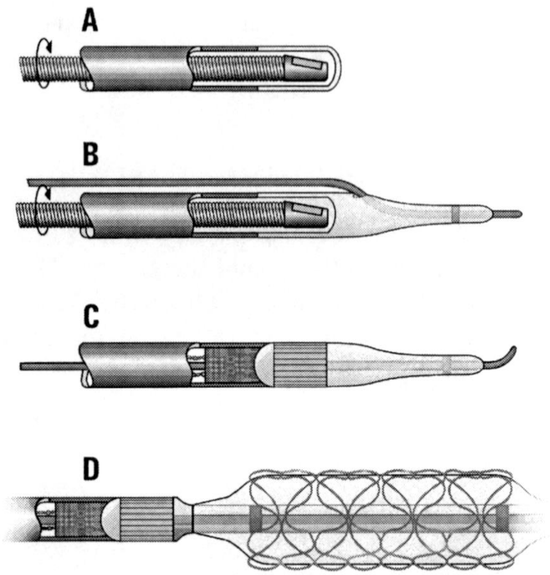

Figure 4. Overview of presently most frequently used IVUS catheters. A) Flex shaft rotating element system. B) Identical to A with added guide wire capabilities. C) Phased array system with guide wire and multiplexing electronics. D) Catheter combining image.

3. IVUS in the clinic

Intravascular coronary ultrasound (IVUS) provides real-time high-resolution images of the vessel wall and lumen[3]. The size of IVUS catheters is between 2.9 to 3.5 French. Depending on the distance from the catheter the axial resolution is about 150 microns, the lateral 300 microns. The images appear real time at a frequency of up to 30 frames/sec. Features of the vessel can be detected based on the echogenicity and the thickness of the material. Small structures can be visualized, however only those sized over 160 microns can be estimated accurately. The normal thickness of the media is about 125–350 μm. IVUS provides some insight into the composition of coronary plaques, among them the vulnerable plaque. Rupture of vulnerable plaques is the main cause of acute coronary syndrome and myocardial infarction. Identification of vulnerable plaque is therefore essential to enable the development of treatment modalities to stabilize such plaque. Because myocardial infarction and its consequences are so important, we must investigate options to identify those areas that will be responsible for future events. A wide variety exists in the stability of coronary atherosclerotic plaques. A plaque may be stable for years, however abrupt disruption of his structure is the main cause of acute coronary syndrome.

The susceptibility of plaques to rupture is known to be related to their composition, stress distribution and degree of inflammation. Increasingly it is becoming recognized that, while not fully understood, two factors play a major role in plaque progression and vulnerability: the presence of increased plaque invasion by microvessels (vasa vasorum) and the expression of specific molecules by cells within the plaques. Both issues can be assessed with an ultrasound contrast agent (An ultrasound contrast agent contains small bubbles (~ 3 μm), which are proven to be blood tracers) using special signatures of these bubbles under ultrasound insonification.

The vulnerable plaque contains certain features that could be diagnosed by various specialized methods. The ideal technique would provide morphological, mechanical and chemical information, however at present, no diagnostic modality providing such all-embracing assessment is available. The main modalities are:

- Elastography and Palpography
- Thermography
- Contrast agents microvessel imaging
- Harmonics
- Modulography

Only the first 3 will be discussed in this manuscript.

3.1. ELASTOGRAPHY AND PALPOGRAPHY

In 1991, a new technique was introduced to measure the mechanical properties of tissue using ultrasound: elastography[4]. The underlying concept is that upon uniform loading, the local relative amount of deformation (strain) of a tissue is related to the local mechanical properties of that tissue. If we apply this concept to determine the local properties of arterial tissue, blood pressure acts as a stressor. At a given pressure difference, soft plaque components will deform more than hard components. Measurement of local plaque deformation in the radial direction can be obtained with ultrasound. For intravascular purposes, a derivate of elastography called palpography may be a suitable tool[5]. In this approach, one strain value per angle is determined and plotted as a colour-coded contour at the lumen vessel boundary. Since radial strain is obtained, the technique may have the potential to detect regions with elevated stress: increased circumferential stress results in an increased radial deformation of the plaque components. In vitro studies with histological confirmation have shown that there are differences of strain normalized to pressure between fibrous, fibro-fatty and fatty components of the plaque of coronary as well as femoral arteries[6].[i] This difference was mainly evident between fibrous and fatty tissue. The plaque types could not be differentiated by echo-intensity differences on the IVUS echogram.

The principal of elastography is as follows: Two IVUS images are made with a short time interval. Due to cardiac pressure there will be a small difference in intraluminal pressure. In case of little difference of pressure the pictures will be almost identical. However, thorough analysis of the ultrasound RF data tells a different story. Proper data processing will allow visualisation of the pressure caused differences. This may be called an "elastogram" where a colour coding is used for identification of the "softness". A principle is illustrated in figure 5.

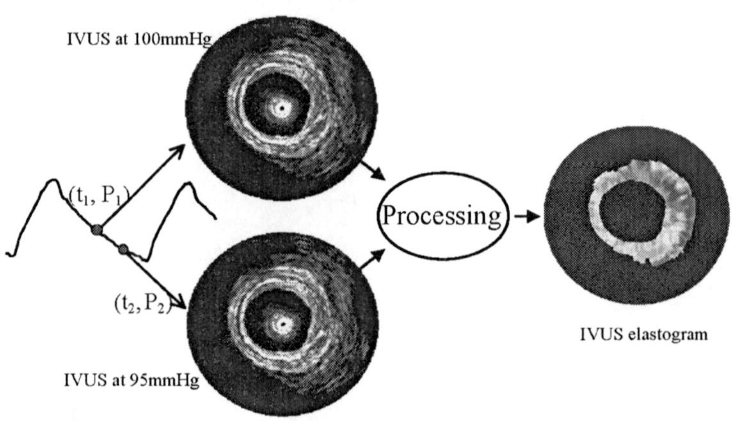

Figure 5. IVUS elastography: two IVUS images are acquired at two slightly different intraluminal pressures. A deformation image is obtained by data processing. This image displays soft regions in green.

"Palpography" is a further development that is simulating putting your finger inside the vessel and palpating it to feel how hard it is. It is a new name for elastography in only a thin internal layer of the artery. Palpography reveals information that is not seen on IVUS. To differentiate between hard and soft tissue may be important for the detection of an instable plaque that is prone to rupture. Since palpography is based on clinically available IVUS catheters, the technique can be easily introduced into the catheterization laboratory. By acquiring data at the end of the filling phase, when catheter motion is minimal, the quality and reliability of the palpogram is increased. The clinical value of this technique is currently under investigation.

It is feasible to apply intravascular palpography during interventional catheterization procedures. In a recent study, data were acquired in patients (n=12) during PTCA procedures with echo apparatus equipped with radiofrequency output. The systemic pressure was used to strain the tissue. Significantly higher strain values were found in non-calcified plaques than in calcified plaques[7]. Another in vivo validation study in atherosclerotic Yucatan pigs showed that fatty plaques have an increased mean strain value. High-strain spots were also associated with the presence of macrophages, a further feature of vulnerable plaques[8].

3.2. THERMOGRAPHY

Inflammation produces a temperature rise in the affected tissue. Since atherosclerosis is accompanied by inflammation the hypothesis was created that a temperature rise could be measured at the surface of a plaque. As vulnerable plaque is a very active metabolic area, the hypothesis was extended that vulnerable plaques may have an even higher temperature rise.

Casscells et al. reported that carotid plaques taken at endarterectomy from 48 patients have temperature heterogeneity[9]. The temperature difference between different areas was up to 2.2°C, and correlated with cell density ($R2= -0.47$, $p=0.0001$). There was a negative correlation between temperature difference and cap thickness ($R2= -0.34$, $p=0.0001$). The same group reported approximately the same in vitro findings in atherosclerotic rabbits[10]. A correlation between temperature rise and macrophage infiltration has also been suggested in an in-vivo rabbit trial[11]. Stefanadis et al. performed studies in humans. Patients with stable angina, unstable angina, and acute myocardial infarction were studied. The thermistor of the thermography catheter has a temperature accuracy of 0.05°C, a time constant of 300 ms, and a spatial resolution of 0.5 mm. The thermistor of the catheter was driven against the vessel wall by the force of blood flow, without the help of a mechanical device like a balloon. Temperature was constant within the arteries of the control subjects, whereas most atherosclerotic plaques showed higher temperatures compared with healthy vessel wall. Temperature differences between atherosclerotic plaque and healthy vessel wall increased progressively from stable angina to patients with acute myocardial infarction with a maximum difference of 1.5 +/– 0.7 degrees celcius[12]. Furthermore patients with a high temperature gradient have a significantly worse outcome than patients with a low gradient[13].

However, this data has yet to be confirmed prospectively in other centres and the influence of parameters such as coronary blood flow or catheter design has to be studied in the future.

3.3. CONTRAST IMAGING

Unfortunately, in smaller vessels and capillaries blood detection is not possible due to low signal strengths from blood, tissue motion effects, and limited resolution (~ 0.5–1 mm). Ultrasound contrast agents (UCA, consisting of small (encapsulated) gas bubbles), will increase the reflection of ultrasound by the blood pool, after intravenously administer, and by that make it possible to provide perfusion images and imaging the vascularity in plaques.

When a gas bubble in liquid is subjected to pressure variations induced by an acoustic wave it acts as a forced, damped, oscillator, with a resonant frequency that is inversely related to its diameter. If the incident acoustic wave is of sufficient amplitude and of a frequency near to the resonant frequency of the bubble, the induced radial oscillations will become more pronounced and track the incident pressure wave in a nonlinear manner. As a result, re-radiated acoustic energy will be coupled into other frequency bands. The coupling of acoustic energy into the second harmonic frequency region occurs most efficiently when a bubble is insonated at or near its resonant frequency. Subharmonics (at half the insonation frequency) can be efficiently generated at twice the natural resonant frequency.

Optical observation of the vibrating bubble with a fast frame camera and acoustic measurements with high frequency ultrasound reveal these nonlinear vibrations. The fast-frame camera 'Brandaris 128', developed at the Erasmus medical centre in collaboration with Twente University, can be used to visualize micro bubble agent behavior at high ultrasound frequencies. Designed to investigate micro bubble dynamics at conventional ultrasound frequencies, it has a unique capability to image bubble dynamics at a frame rate of up to 25 Mega frames-per-second for 128 consecutive frames. For bubble sizes of 4 micrometer figure 6 shows both radial oscillations and surface modes. In this figure only 5 frames from the 128 available frames are plotted, with an interframe time of 100 ns.

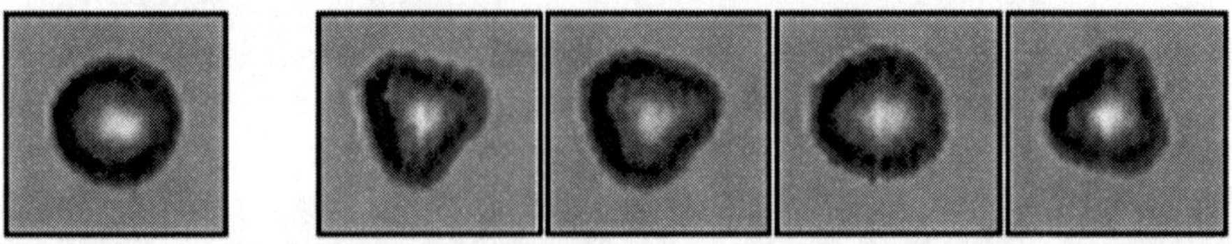

Figure 6. Mode vibrations for a bubble with a diameter of 4 μm. Here a mode n = 3 is observed, however a range of modes is observed for a transmit frequency of 3.5 Mhz.

3.4. IN VIVO EXPERIMENTS

In ultrasound contrast imaging, these nonlinear emissions are the foundation of specific micro bubble detection schemes, which are essential for detecting small vessel *in vivo.* Such *in vivo* studies were conducted in atherosclerotic rabbit abdominal aortas. Atherosclerosis was initiated using endothelial cell injury procedures followed by a high cholesterol diet[14]. Experiments were performed 10–11 weeks after initiation of atherosclerosis. Example *in vivo* results are shown in figure 7 for fundamental and harmonic imaging modes. In fundamental mode (top) the injected contrast can hardly be discriminated from the tissue. In the special contrast mode (harmonic imaging) results are substantially different. Post-injection, agent was first visualized in the main lumen. By approximately 5 seconds post injection a ring at the boundary of the main lumen could be seen, which can be attributed to the presence of a small amount of more slowly moving agent adjacent to the aortic wall. This effect enabled the lumen boundary to be easily distinguished. There is then an eccentric circumferential region devoid of enhancement, with a thickness larger in the 2 O'clock direction than in the 8 O'clock direction. Outside this hypoechogenic region, numerous locations of enhancement are observed that are associated with the presence of agent. While towards 10 O'clock the enhancement is associated with the vena cava, the other locations are consistent with the detection of microvessels outside the main vessel lumen[15].

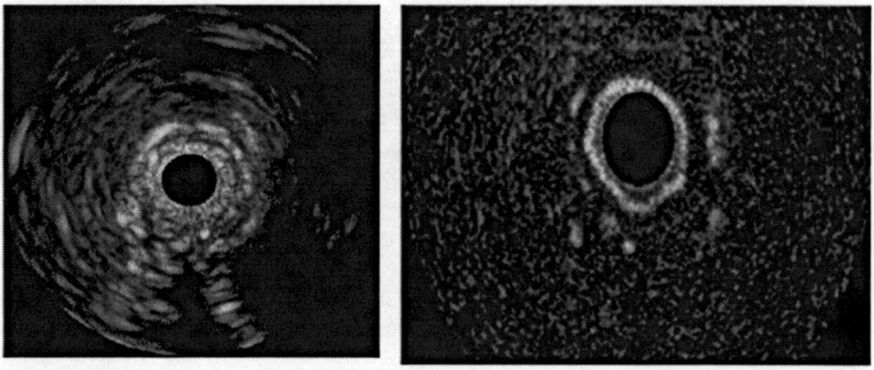

Figure 7. In vivo results in an atherosclerotic rabbit aorta using decanted Definity TM. Left normal imaging. Right: 10 second post-injection harmonic mode shows significant adventitial enhancement, consistent with the detection of adventitial microvessels.

4. Conclusion

In the last couple of years the possibilities to access a sufficient amount of RF data at adequate speed have improved significantly. This has given the possibility to study IVUS data in a totally different way. In the traditional way the emphasis was always on imaging itself while RF-analysis also allows tissue characterization, elasticity imaging, flow estimation and enhanced boundary detection.

References

1. Mintz G. S., Popma J. J., Kent K. M., Satler L. F., Wong S. C., Hong M. K.,. Kovach J.A, and Leon M. B., "Arterial remodeling after coronary angioplasty. A serial intravascular ultrasound study," *Circulation*, vol. 94, pp. 35-43, 1996.
2. Bom N, Lancée CT, Van Egmond FC, An ultrasonic intracardiac scanner. *Ultrasonics* 10: 72 1972.
3. Bom N, Li W, van der Steen AF, Lancee CT, Cespedes EI, Slager CJ, de Korte CL. Intravascular imaging. *Ultrasonics*; 36: 625-8 1998.
4. Ophir J, Cespedes I, Ponnekanti H, et al. Elastography: a quantitative method for imaging the elasticity of biological tissues. *Ultrason Imaging*; 13: 111-34 1991.
5. Doyley MM, Mastik F, de Korte CL, et al. Advancing intravascular ultrasonic palpation toward clinical applications. *Ultrasound Med Biol*; 27: 1471-80 2001.
6. De Korte CL, Pasterkamp G, van der Steen AF, et al. Characterization of plaque components with intravascular ultrasound elastography in human femoral and coronary arteries in vitro. *Circulation*; 102: 617-23 2000.
7. De Korte CL, Carlier SG, Mastik F, et al. Morphological and mechanical information of coronary arteries obtained with intravascular elastography; feasibility study in vivo. *Eur Heart J*; 23: 405-13 2002.
8. De Korte CL, Sierevogel MJ, Mastik F, Strijder C, et al. Identification of atherosclerotic plaque components with intravascular ultrasound elastography in vivo: a Yucatan pig study. *Circulation*; 105: 1627-30 2002.
9. Casscells W, Hathorn B, David M, et al. Thermal detection of cellular infiltrates in living atherosclerotic plaques: possible implications for plaque rupture and thrombosis. Lancet; 347: 1447-51 1996.
10. Casscells W, David M, Bearman G et al. Thermography. In: The Vulnerable Atherosclerotic Plaque. Edited by V. Fuster, Futura Publishing Company,: 231-42 1999.
11. Verheye S, De Meyer GR, Van Langenhove G et al. In vivo temperature heterogeneity of atherosclerotic plaques is determined by plaque composition. *Circulation* 2002; 105: 1596-601.
12. Stefanadis C, Diamantopoulos L, Vlachopoulos C et al. Thermal heterogeneity within human atherosclerotic coronary arteries detected in vivo: A new method of detection by application of a special thermography catheter. *Circulation*; 99:1965-71 1999.
13. Stefanadis C, Toutouzas K, Tsiamis E, et al. Increased local temperature in human coronary atherosclerotic plaques: an independent predictor of clinical outcome in

patients undergoing a percutaneous coronary intervention. J Am Coll Cardiol;37: 1277-83 2001.

14. Schaar J. A., de Korte C. L., Mastik F., van Damme L. C., Krams R., Serruys P. W., and van der Steen A. F. W., Three-dimensional palpography of human coronary arteries. *Herz*, 30(2):125-133, 2005.
15. Goertz D. E., Frijlink M. E., Tempel D., Van Damme L. C. A., Krams R., Schaar J. A., Ten Cate F. J., Serruys P. W., De Jong N. and Van der Steen A. F. W. Invest. Radiol. *"Contrast Harmonic Intravascular Ultrasound: a Feasibility Study for Vasa Vasorum Imaging "* (*in press 2006).*

EARLY, RECENT AND FUTURE APPLICATIONS OF ECHOCARDIOGRAPHY*

NICO DE JONG*, NICOLAAS BOM, ANTON FW VAN DER STEEN, JOS RTC ROELANDT*
Thoraxcentre; Erasmus Medical Center Rotterdam, P.O. Box 1738, NL-3000 DR Rotterdam, The Netherlands

Abstract. In Cardiology the first use on ultrasound was described in 1953 by Inge Edler and Helmuth Herz in Lund, Sweden. They used the M-mode technique which is, in brief, the registration of moving echoes as recorded in a single ultrasound beam. Two dimensional (2D) real time echography appeared in the late sixties with mechanically and electronically swept spoundbeams. The linear- and phased array approaches are examples of electronic real time systems which today represent the majority of apparatus. Use of frequency analyses of the echoes allowed indication of blood velocity over the 2D image. This is called Colour Doppler where a colour coding separates velocity towards the transducer from blood velocity away from the transducer. This is very powerful since now 2D geometry becomes available together with the blood flow information. Echo contrast is an old technique to enhance the echo image at the location of the contrast fluid. This fluid contains small gas containing bubbles. Present use aims at better identification of cardiac chambers contours as well as myocardial perfusion. Semi invasive is the Trans Oesophageal Isographic (TEE) approach. Here a transducer is mounted at the tip of a long tube. It allows contact with the heart with improved image quality. The IntraVascular Ultrasound System (IVUS) uses echography at the tip of a catheter. Following this, techniques for real time 3D will be presented and a brief description of portable echo equipment is given.

Keywords: echogram; transducers; M-mode; colour doppler; transesophageal echo; contrast agents; echo catheters; speckle pattern; elastography; palpography; three-dimensional imaging.

---

* Nico de Jong, Erasmus University Rotterdam, P.O. Box 1738, NL-3000 DR Rotterdam, The Netherlands; e-mail: n.dejong@erasmusmc.nl.

* in part the material has been published in "From hearing to seeing" by N. Bom and "Echocardiography in the 3rd millenium" by AFW van der Steen in a special issue of the Thoraxcentre Journal volume 15, issue 1, June 2003 with emphasis on echocardiography research activities in Rotterdam.

Y. Lemoigne et al. (eds.), Physics for Medical Imaging Applications, 253–264.

1. Introduction

It is clear that a great number of scientists, in former times as well as more recently, contributed to raise diagnostic ultrasound to its present level. This lecture will focus on achievements in cardiology over the last 30 years.

Around 1952, the first reports on echo images appeared. Wild and Reid[1] and Howry and Bliss[2] operated with a single narrow sound beam which scanned in a plane through the organ. On the basis of this work, the compound-scan method was developed. Unfortunately this method was so time-consuming, that it was absolutely unsuitable for acquisition of cross sectional images of a moving organ such as the heart.

2. M-mode

With a single sound beam aimed in a fixed direction through the heart it was possible to record the various echoes in this soundbeam as a function of time. This was the so-called M-mode (M for Motion) registration, introduced by Edler and Hertz around 1954[3]. An illustration of their early work is given in figure 1. Transducer sensitivity was poor since quartz crystals were used as transducer. The highly reflecting calcified mitral leaflet produced visible echoes. A normal, less reflecting mitral leaflet was not visible.

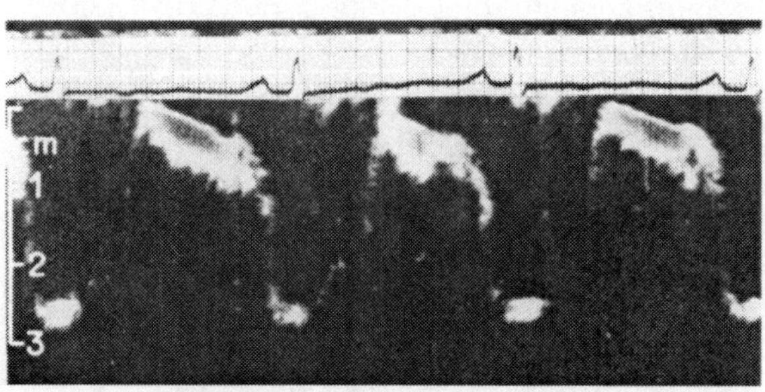

Figure 1. Early echogram of mitral stenosis as recorded by Edler and Hertz in December 1953.

This improved drastically when Edler and Hertz started to use barium titanate as transducer material after 1956. Early M-modes were photographed off the oscilloscope screen and displayed a limited number of heartbeats. Later on fiber optic paper recorders were introduced. M-mode registrations eventually developed into long registrations on paper. All the recorded "wiggles" represented echo motion of, for instance, a valve leaflet in the single sound beam as a function of time. Further orientation being absent, it proved difficult to understand the origin of all the echoes. It appeared that only in the hands of very skilled clinicians the M-mode could be useful. At the same time it became clear that, for imaging a fast moving organ such as the heart, there was an urgent need to create a method capable of scanning a sound beam rapidly through an entire cardiac cross-section. If this were possible, a two-dimensional image of a cross-section of the moving heart would be the result. In cooperation with Hertz, in 1967 first sequential cardiac cross sections were echographically documented with a frame rate of 7 frames per second in Lund by Åsberg[4]. This mechanical mirror system was not fast enough and difficult to handle. Here and there ideas on two dimensional (2D) real time imaging started, nevertheless it was apparent that the early M-mode work by Edler and Hertz was the basis of the breakthrough of Echocardiography.

3. The linear array

A very simple real-time 2D approach, based on a multiplicity of transducers positioned in a row, was proposed in the Thoraxcentre, Rotterdam in 1969. Fast electronic switching should yield high frame rate. A first prototype of this system, later called the Multiscan, was built, patented in 1971 and subsequently described in literature[5]. The first transducer contained 20 small round transducers in a row. Demonstrations in a water tank proved its capability to instantaneously image the cross- sectional plane through a moving cylindrical gauze. One of the very first cardiac cross-sections non-invasively obtained in a patient is illustrated in figure 2. Results in the very first hundred patients were reported by Jos Roelandt[6].

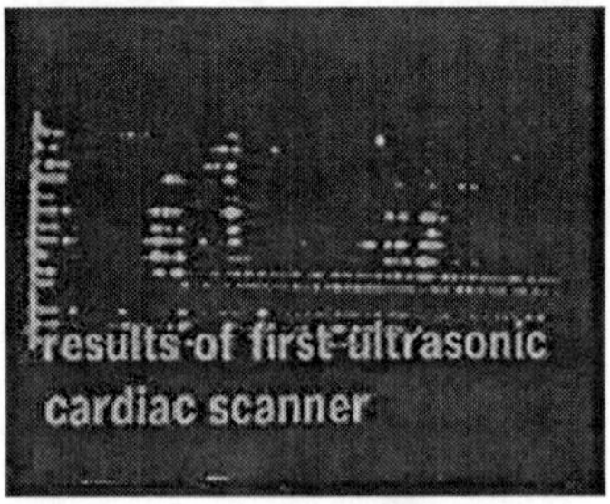

Figure 2. Still frame showing cross sectional image from the moving heart as obtained with the linear array electronic scanner.

As remarked earlier, orientation for proper M-mode registration was not so easy. It was stated by Edler and Hertz that the introduction of real-time 2D echography greatly facilitated orientation within the heart and recognition of cardiac structures. This in turn boosted M-mode as well.

4. Other fast imaging methods

In the period 1968–1974 the need to rapidly obtain diagnostic ultrasonic images became apparent and various approaches were followed. Immediate feedback while moving the probe to select the diagnostic plane through the organ was one of the reasons. Jan Somer[7] presented a phased array for neurological studies in 1968. He suggested a multiple element transducer with small "footprint" where, with varying individual element time delays, the sound beam could be steered. This resulted in a sectorial image format and a transducer, which fitted much better between the ribs than the linear array transducer. The phased array method eventually became the method of choice in cardiology. The linear array technique proved later on to be optimal in abdominal and obstetrical applications. Jim Griffith[8] described a mechanical method that was based on a pivoting single transducer. Today most diagnostic echo apparatuses all over the world are equipped with both electronic scanning methods: the phased array and the linear array, or combinations thereof. For external use a frequency between 2 and 5 MHz is used. This allows a depth penetration of approximately 16 cm.

5. Colour doppler

Change in horn-pitch, which is experienced when a fast automobile is passing, is a well-known phenomenon. An observer notes an increase in sound frequency when the car is coming towards him and a decrease when the distance is getting larger. This is the Doppler effect, which is also used in diagnostic ultrasound. The frequency of the transmitted sound pulse and that of the received echo will not be the same if the distance between the transducer and the reflecting structure changes. It then becomes possible to detect the change in frequency of echoes originating from moving blood particles and thus to derive blood velocity. This was first described by Satomura[9] in 1957. In Lund Continuous Wave (CW) Doppler recordings were made in an isolated heart model. Edler and Lindström presented clinical results in 1969[10]. Péronneau described Pulsed Doppler were it became possible to select blood velocity measurement at a given depth in 1969[11] and so did Baker[12]. In the Netherlands, much Doppler work has been reported by Arnold Hoeks[13]. He proposed Doppler measurements at multiple selected depths. This developed into two-dimensional Colour

Doppler imaging. In clinical practice, colour coding in red means blood velocity towards the transducer; blue means blood velocity away from the transducer.

6. Trans esophageal echo

The heart is in close proximity of the oesophagus, and very little acoustically disturbing tissue is in between. To obtain high quality images, it was therefore suggested to position the transducer on the tip of a long tube like the ones used for gastric diagnosis. Various versions of the TransEsophageal Echo (TEE) transducer have been developed in Rotterdam in cooperation with Oldelft industry. The first TEE transducers were limited to a single imaging plane. Later Jacques Souquet[14] brought forward the idea to incorporate the capability of rotating the imaging plane. The Rotterdam effort included a special rotation technique, which allowed further miniaturization. Today a very small TEE probe is available for diagnosis in babies.

7. Ultrasound contrast agents

Ultrasound contrast agents are fluids containing little spheres with a diameter in the order of 3 micrometer (3/1000 of a millimetre). The bubbles contain gas and have a very thin shell. Through their specific acoustic properties these bubbles have a high reflectivity and thus appear clearly visible in the echo image. The bubbles are part of the "contrast fluid" which is injected and then transported through the blood circulation.

Early contrast studies were used to identify structures and to confirm congenital defects. Leakedge of contrast through the septum indicated a septal defect which could be made visible. Thereafter contrast applications were aimed at enhancing Doppler signals or improvement of left ventricular wall identification.

Newer and more future applications aim at visualization of myocardial perfusion. Studying the blood circulation of the muscle at each location gives insight in the presence and size of heart infarcts. Those parts of the heart that do not get blood will loose their functionality. A method to study the perfusion of the heart muscle is to make use of ultrasound contrast. Where the bubbles are visible, there is a good perfusion of the muscle. By now we have been investigating ultrasound contrast agents in Rotterdam for over 20 years[15].

A crucial part of the research is focusing on the behaviour of ultrasound contrast bubbles when they are irradiated by ultrasound. For this goal an apparatus, called Brandaris, was developed which can produce 25 million (!) pictures per second. The principle is based on a microscope, underneath the microscope an ultrasound contrast agent irradiated with ultrasound. The image of the microscope is directed towards a fast rotating prism, which projects the image on 128 CCD-cameras. Because of the rotating light beam we have baptised it "Brandaris", after the Netherlands most famous lighthouse. The ultrasound will make the bubble vibrate. This vibration can be filmed using Brandaris. The bubble shell may contain drugs. When the bubble collapses[16], the drug will be transported to and in many cases through the surface; if this surface happens to be a cell, we have a "nano-syringe".

8. Echo catheters

Already in 1966, Peter Well described a long thin tube with an echo element at its tip for imaging in veins. By mechanical rotation of the acoustic element or a mirror, an echographic cross-section of an artery or vein could be produced. In Rotterdam, we developed a phased array catheter with 32 elements positioned in a cylindrical geometry on a catheter in 1972[17]. The outer diameter of this catheter was 3 mm. Rotation of the sound beam was carried out by way of electronic switching and the use of time delays. Production of the echo tip, and in particular the sawing of the 32 elements, was a major effort. For that purpose, we constructed a wire saw where a long 30-micron diameter tungsten wire served as saw. The catheter functioned well, but seemed, given the success of the non-invasive linear array, too invasive at the time.

Serious requests for intravascular echo catheters came only after the start of large-scale interventional cardiology with the introduction of Dotter methods around 1985. It became important to see plaque geometry in a coronary artery. We dusted off our phased array catheter experience, concluded that large-scale production of this catheter would require too much funding, and started research on the subject of a mechanically rotated echo catheter. In such a catheter, the element is rotated with a thin internal flexible shaft driven by a distal motor. Examples of the catheter in 1972 and the catheter in 1992 are illustrated in figure 3.

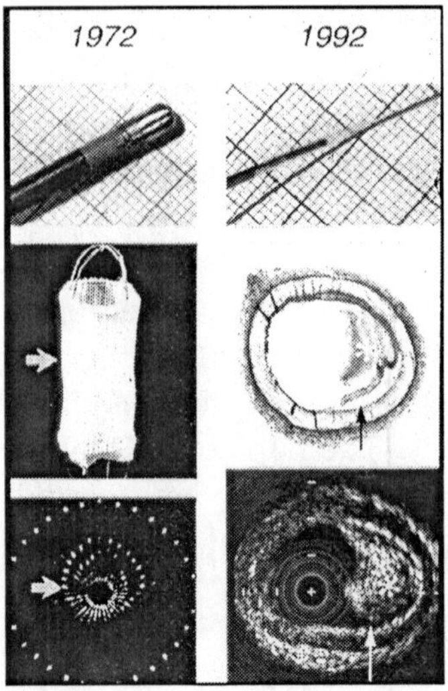

Figure 3. Comparison of echo catheters as made in Rotterdam in 1972 (a phased array technique) and in 1992 (a flex shaft mechanically rotated transducer technique).

Like with all echo systems, also here we note an enormous improvement of image quality over the years. A clinical example is shown in Figure 4.

Echo catheters are expensive. Images obtained with flex shaft technology may be subject to artefacts due to rotation. Phased array catheters are complex and relatively rigid when it comes to change of parameters. We therefore undertook an endeavour to mount a micro motor in the tip of an echo catheter. The outer diameter of the motor was 1 mm. Part of the collector was constructed as flex-print. The prototypes worked well and even images obtained in patients became available. The project was supported by industry, but unfortunately the micro motor echo catheter was never commercialised.

9. Blood flow

Standard use of intravascular echography is imaging of the selected cross-section. However many additional parameters can be obtained from the echo data. One example is the derivation of blood as it passes the catheter tip with its rotating soundbeam. The principle can be described as follows. As function of depth the echo signals are documented in a specific beam. These signals are caused by scattering of passing blood particles or arterial tissue. Signals recorded in the next following soundbeam, which covers almost the same area, can be compared for each depth position. If the echo (Radiofrequency signal) RF speckle pattern does not change, the echo must have been caused by a non-moving structure. Fast moving blood however will cause a different set of speckles to create the echo signal. High decorrelation means high blood velocity. This can be measured at all positions in the arterial cross section. With colour coding the velocity can be indicated in the image[18]. Integration over the entire cross-section yields blood volume as function of time.

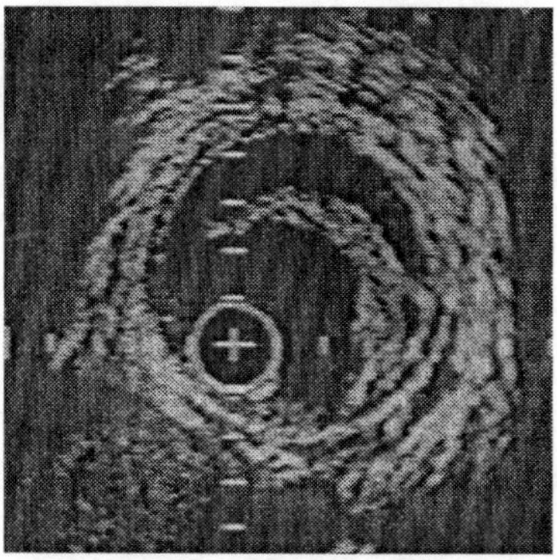

Figure 4. Intravascular ultrasound cross section of the superficial femoral artery revealing a dissection following balloonangioplasty (Courtesy E. Gussenhoven. From Intravascular Ultrasound 1991; Kluwer Academic Publishers).

10. Elastography

The electrographic technique is using the elastic properties of the tissues as recorded by ultrasound[19]. Two IVUS images are made with a short time interval. Due to cardiac pressure there will be a small difference in intraluminal pressure. In case of little difference of pressure the pictures will be almost identical. However, thorough analysis of the ultrasound RF data tells a different story. Proper data processing will allow visualisation of the pressure caused differences. This may be called an "elastogram" where a colour coding is used for identification of the "softness". A principle is illustrated in figure 5.

"Palpography" is a further development that is simulating putting your finger inside the vessel and palpating it to feel how hard it is. It is a new name for elastography in only a thin internal layer of the artery[20]. It is the essential area to look for vulnerable plaque, and, with palpography, data processing can be carried out fast. One of the strong aspects of the technique is that we can characterise the full inside of a vessel with pull back of the catheter. Over a hundred patients have been diagnosed so far using this technique.

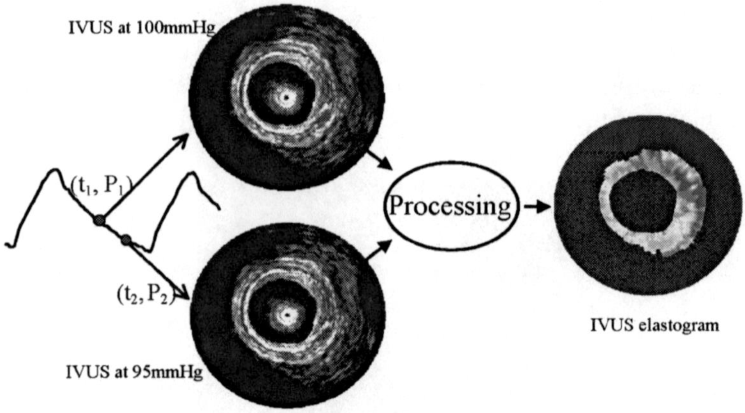

Figure 5. IVUS elastography: two IVUS images are acquired at two slightly different intraluminal pressures. A deformation image is obtained by data processing. This image displays soft regions in green.

11. Three-dimensional imaging

So far, we discussed two-dimensional imaging of the heart. However, organs have three dimensions. It is difficult to obtain a three-dimensional (3D) echographic view of the moving heart. Accumulation of echo data synchronized with the electrocardiogram in a multiplicity of echo planes was our first approach to obtain such a view. Data were accumulated over a series of cardiac beats. Acquisition took many minutes. An example of a rendered 3D view of the heart is shown in figure 6.

At the Thoraxcentre in Rotterdam, Jos Roelandt carried out much work on 3D imaging of the heart[21] and pushed this field substantially forward with his ideas.

It is apparent that, in order to construct such a 3D view with instantaneous display, all echo data must be accumulated very rapidly. Von Ramm[22] described an electronic method for that purpose and we in Rotterdam developed a motor driven rotating phased array technique. At present, a prototype is being tested in the clinic.

12. Portable echo apparatus

At present, there is a tendency to make all apparatuses small and portable. A well-known example is the mobile telephone. Well, the echo apparatus is no exception. Since approximately 1996, a variety of portable echo equipment has become commercially available. Where the stethoscope is exclusively made for sound (and thus should have been named stethophone), the portable echo devises are sometimes, but then correctly, called echo-stethoscope.

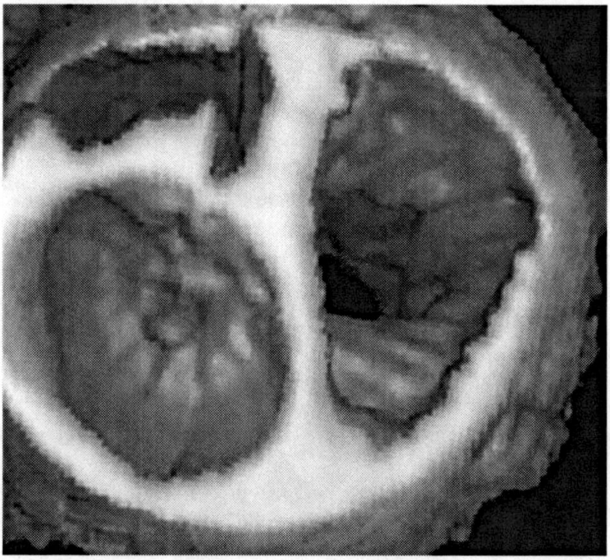

Figure 6. Rendered 3D image of the heart.

We developed a portable device, which we called "Minivisor" in cooperation with Organon Teknika around 1976. This apparatus is illustrated in figure 7. The Minivisor was battery-powered and functioned for 2 hours before the battery had to be recharged. The size of the image was that of a large stamp. In total 1000 Minivisors have been built, than production was stopped. Apparently we were 20 years too early.

13. Conclusion

Based on the initial work in 1953 of Inge Edler and Helmuth Hertz, and in particular their M-mode approach, echocardiography has started. To day echocardiography has matured into one of the best non-invasive diagnostic tools in cardiology. Many later developments such as Colour Doppler; Real-Time 2D imaging; the TEE approach and for instance the portable apparatus have further pushed use of echocardiography. Also invasive tools such as Intravascular Echo as well as Contrast have gained importance. Image quality has drastically improved as well. In all we must be vary thankful for the initial creativity of the fathers of echocardioghraphy: Inge Edler and Helmuth Hertz.

References

1. Wild JJ, Reid JM (1955) Ultrasonic rectal endoscope for tumour location. Am. Inst Ultrasonics Med 4: 59
2. Howry DH, Bliss WR (1952) The ultrasonic visualization of soft tissue structures and disease processes. J Lab Clin Med 40:812-813.
3. Edler I, Hertz C (1954) The use of the ultrasonic reflectoscope for the continuous recording of movements of heart walls. Kungl Fysiogr Sällsk Lund Förhandl: 24(5):40-58.
4. Åsberg A.: Ultrasonic cinematography of the living heart. Ultrasonics April: 113-117, 1967.
5. Bom N, Lancée CT, Honkoop J, Hugenholtz PG (1971) Ultrasonic viewer for cross-sectional analysis of moving cardiac structures. Biomed-Engng 6: 500-503 and 508.
6. Roelandt J, Kloster F, Ten Cate FJ, Bom N, Lance CT, Hugenholtz PG: Multiscan echocardiography – Description of the system and initial results in 100 patients, Hart Bulletin 4: 51-56, 1973.
7. Somer JC (1968). Electronic sector scanning for ultrasonic diagnosis. Progress Report Institute of Medical Physics TNO: August: 37-43.
8. Griffith JM, Henry WL (1973) A real-time system for two-dimensional echocardiography. 26th ECEMB 15: 422.
9. Satomura S (1957). The ultrasonic Doppler method for the cardiac functional test. J Acoust Soc Jpn 13:138-143.
10. Lindström K, Edler I. Ultrasonic Doppler. Vienna conference proceedings 1969.
11. Péronneau P. Deloche A, Bui-Mong-Hung, and Hinglais J: Débitmétrie sanguine par ultrasons. Développements et applications expérimentals. European Surgical research 1:147, 1969.
12. Baker DW. Pulsed ultrasonic Doppler blood-flow sensing. IEEE Trans. Sonics Ultrason. 17: 17, 1970.
13. Hoeks APG (1982) On the development of a multi-gate pulsed Doppler system with serial data-processing. Proefschrift Rijksuniversiteit Limburg, Maastricht 1982.
14. Souquet J, Hanrath P, Zitelli L et al. (1982). Transesophageal phased array for imaging the heart. IEEE Trans Biomed Engng BME-29:707-712.
15. Jong N de, Cate FJ ten, Lancee CT, Roelandt JRTC, Bom N. Principles and recent developments in ultrasound contrast agents. Ultrasonics;29:324-30 1991.
16. Postema M, Bouakaz A, Chin CT, Jong N de. Optically observed microbubble coalescence and collapse. Munchen: IEEE Ultrason SYMP, 2002.
17. Bom N, Lancée CT, Van Egmond FC (1972) An ultrasonic intracardiac scanner. Ultrasonics 10: 72-76.
18. Li W, Steen AFW van der, Lancee CT, Céspedes El, Bom N. Blood flow imaging and volume flow quantitation with intravascular ultrasound. Ultras. Med Biol 1998;24: 203-14.
19. Steen AFW van der, Korte CL de, Cespedes El. Intravascular Ultrasound Elastography. Ultraschal Med 1998;19:196-201.

20. Doyley MM, Mastik F, Korte CL de, Carlier SG, Cespecies EI, et al. Advancing intravascular ultrasonic palpation toward clinical applications. Ultrasound Med Biol 2001;27:1471-80.
21. Roelandt JRTC (ed.) (2000) Three-dimensional Echocardiography of the Heart and Coronary Arteries. Alphen aan de Rijn, Van Zuiden Publications: p. 171.
22. Von Ramm OT, Smith SW (1990). Real time volumetric ultrasound imaging system. J Digit Imaging 3: 261-266.

Part IV

Nuclear Medicine Techniques

ISOTOPES IN MEDICINE

GERD-J. BEYER*
Prof. Dr.rer.nat.habil.i.R., ESI, bat ; Salève, Site d'Archamps, 74166-Archamps, France[*]

Abstract. In this paper the different techniques of radionuclide application in medicine will be outlined in some details. The corresponding chemical requirements such as radionuclide purity, pharmaceutical requirements, carrier influence and others will be underlined. An overview will be given on the different production modes of radionuclides based on reactors, small or medium cyclotrons, high-energy particle accelerators and short outlook on future aspects of medical isotope application will be given.

Keywords: radionuclides; application in medicine; medical imaging; pharmaceuticals; radiotherapy; production of radionuclides; reactor; cyclotron; high energy accelerator.

1. Introduction

Immediately after discovery of the natural and later artificial radioactivity the radioisotopes have been used for very different kind of medical applications. Today about 80 % of the artificially produced radioactive isotopes are used in medicine. The application of radioisotopes is very different and ranging from in vitro diagnosis, over in vivo diagnosis to the very different approaches for therapy. An overview (Beyer 2000; Beyer 2002; Beyer & Ruth 2003) of the different fields of medical application is given in Fig.1.

2. Applications in medical imaging

The very different fields of application require dedicated nuclear decay properties of the isotopes as a well as defined chemical quality. In vitro application (RIA = radio immune assay) requires long-lived isotopes with biogenic behavior. Nuclear medical imaging techniques (gamma scintigraphy, Single Photon Emission Tomography = SPECT) require

[*] Gerd-J. Beyer, ESI, 74166-Archamps, France ; e-mail : gerd.beyer@cern.ch, gerd.beyer@hcuge.ch.

Y. Lemoigne et al. (eds.), Physics for Medical Imaging Applications, 267–271.

radionuclides, which decay with single photon emission, preferable with photon energy of 100–200 keV. Closest to this requirement comes ^{99m}Tc (6h half life), which is consequently the most commonly used radioisotope in nuclear medicine diagnosis.

BOLD: most important nuclides with high clinical and commercial potential

Arial: reactor produced nuclides

ISOTOPES IN MEDICINE				
DIAGNOSIS		THERAPY		
in vitro	in vivo	internal		external
		systemic	sources	Tele-Co
^{14}C ^{3}H ^{125}I others	**SPECT** **^{99}Mo/^{99m}Tc** *^{201}Tl* *^{123}I* *^{111}In* *^{67}Ga* *^{81}Rb/^{81m}Kr* others **PET** ^{18}F, ^{11}C ^{15}O, ^{13}N ^{86}Y, ^{124}I ^{68}Ge/^{68}Ga ^{82}Sr/^{82}Rb others	**ß$^{-}$-emitters** **^{131}I, ^{90}Y** ^{153}Sm, ^{166}Ho, ^{177}Lu ^{186}Re ^{188}W/^{188}Re others **α-emitter** *^{211}At* ^{225}Ac-^{213}Bi ^{223}Ra ^{224}Ra *^{149}Tb* others	^{192}Ir, ^{182}Ta ^{137}Cs ^{103}Pd, ^{125}I ^{90}Sr or ^{90}Y ^{137}Cs others	^{60}Co Gamma Knife ^{137}Cs Blood Cell Irradiation

Figure 1. Overview of the different fields of radioactive isotope application in medicine [Beyer 2002a].

Enormous efforts in research and investments were required to bring this technique to the high reliable standard available today (^{99}Mo/^{99m}Tc generator and the related Kit's). Other suitable radionuclides for SPECT are ^{123}I, ^{201}Tl, ^{67}Ga or ^{111}In, which needs to be produced using accelerators. Positron Emission Tomography (PET) is the first and only functional imaging technique, which can quantitatively based on an objective physical principle determine tracer concentration with high special resolution in vivo. Since the needed basic positron emitters (^{11}C, ^{13}N, ^{15}O and ^{18}F) are of short half-life (20, 10, 2 and 110 min respectively) a dedicated cyclotron has to be installed on site the hospital. Today PET technique has gained significant clinical relevance, especially in oncology. The technique is very fast expanding not only in the so-called Western hemisphere.

3. Applications in radiotherapy

Different concepts of cancer therapy require very different nuclear properties of the nuclides. The question is always how to bring sufficient radiation dose to the cancer tissue or cancer cells without damage of the surrounding healthy tissue. The different principles are illustrated in Tab. 1.

TABLE 1. Parameters and requirements for the radionuclides to be used in different approaches of cancer therapy.

	Gamma Knife	β-Knife	α-Knife	Auger Knife
Isotope	^{60}Co $E_{\gamma} > 1$ MeV	^{131}I, 90Y ^{153}Sm ^{166}Ho Others E_{β} 1-3 MeV	$^{212,213}Bi$, ^{211}At, ^{149}Tb, $^{223,224}Ra$ E_{α} 4 – 8 MeV	^{125}I ^{165}Er E_e few eV
Range	Full body penetration	About 1 cm	30 – 80 μm	1 μm
Application	Head Cancer	RIT Radio immune-therapy	Leukaemia	Future
	Tissue surgery	Tissue surgery	Cell surgery	Molecular surgery

Classical radiation therapy makes use of long-lived isotopes with showing high energetic gamma radiation. Most popular is ^{60}Co. A special design of this technology is realized in the so called “gamma knife”: the radiation of around 200 thin ^{60}Co sources of very high specific activity is focussed through corresponding collimators from outside to a tumor (mainly brain) suggesting, that then gamma radiation works like a scalpel or knife. In another approach the isotopes are enclosed in capsules and the corresponding seats are inserted into the tumor tissue (brachy therapy). The isotopes of choice for this technique are ^{125}I and ^{103}Pd, when prostate cancer is the tissue for treatment. Instead of small seats we may use “intelligent” molecules” (100 years ago Ehrlich: named them “magic bullets”) and use them as vehicles to bring radionuclides into the tumor tissue. In analogy to the gamma knife we may name this approach “beta knife” and understand it as a kind of tissue surgery, since the range of the beta radiation is of the order of 1 cm. If we wish to target single cells in circulation then we need a radiation with a shorter range and higher LET (linear energy transfer). Consequently we can speak on “alpha knife” as approach for cell surgery. We may also assume targeting the DNA directly, in this case the Auger electrons have the corresponding short range and the approach can be named “Auger knife” performing a molecular surgery. The related “intelligent” molecules are bio-specific tracer molecules that bind specifically to certain receptor sides that exist exclusively in tumor cells. The types of molecules are generally monoclonal antibodies or the related fragments, other proteins, different types of peptides or olegonucleotides. The field is very fast developing and one can expect high isotope product selling along the therapy line in the very near future.

4. Production of radio isotopes

As seen from Tab.1 there are different approaches to access the required isotopes. Most common and known is the reactor based isotope production. Economically most important products are ^{99}Mo, ^{131}I and ^{125}I, and than the isotopes needed for therapy ^{177}Lu, ^{166}Ho, ^{153}Sm and others. The ^{192}Ir and ^{60}Co needed for medical and industrial application play also a significant role in the field of reactor based isotope production. A reactor used for medical isotope production is usually a water tank reactor, operated at 50–60°C. Reactors allow irradiating many targets of relatively large volume at a time simultaneously. Only targets in the form of the element or as oxides can withstand the high radiation dose in a long activation process at high neutron flux intensities.

Since neutron deficient isotopes are hardly available in reactor based production routes several of the suitable single photon emitting isotopes need to be produced with an accelerators, mainly with a cyclotron. Most important in this concern are ^{201}Tl (cardiology), ^{123}I (labelled compounds), ^{111}In (labeling of blood cells and peptides) and ^{81}Rb (^{81}Rb/^{81}mKr-generators for lung ventilation studies). In difference to the reactor at a cyclotron one can usually produce only one isotope at a time with a relative specific and selective nuclear reaction. Heat deposition of the accelerated particle beam inside the targets determines the practical limits for this production routes. The preferred energy of the particle beam (preferable protons) is 30 MeV. This energy allows to run the ^{203}Pb(p, 3n)^{201}Pb process, which is the reaction for large-scale ^{201}Tl-production (which is the daughter of the primary generated ^{201}Pb). In addition the universal (p, 2n)-process leads to all the other mentioned isotopes.

For generating the short-lived positron emitters for PET a lower energy particle beam (protons of 10–18 MeV) is sufficient. Consequently dedicated small cyclotrons have been developed, that are now widely distributed in hospitals (PET-centres). The production of short-live radiopharmaceuticals is an interesting and challenging task that requires qualification in organic -, radio- and pharmaceutical chemistry.

As said before, the reactor provides access to the neutron rich isotopes (needed for therapy), the accelerator production routes provides the access to the neutron deficient isotopes, high energy proton induced reactions are unspecific and provides access to the whole chart of nuclides via spallation, fission and fragmentation reactions. This universal production route in combination with advanced radiochemical and physical separation techniques (on-line or off-line isotope separation techniques as developed at ISOLDE at CERN for example) can become a universal large-scale production route for medical isotopes in the future. Multi-MW proton driver accelerators are under construction as new basic instrumentation for fundamental research in particle physics. Those accelerators provides potential for modern isotope production programs covering the needs for routine diagnosis, for therapy as well as the research in a parasitic operation mode, which means under very economical conditions.

References

Beyer G.J. // Isotopes in Medicine – Requirements – Production and Application // European School on Medical Physics, ESI, Archamps (France), 31.Oct.-3.Dec.2002, Proceedings Week 2: Medical Imaging, *http://www.cur-archamps.fr/esi.*

Beyer G.J. // Radioactive Ion Beams for Biomedical Research and Application //Hyperfine Interactions **129** (2000) 529-553.

Beyer G.J. & Ruth T.J.// The role of electromagnetic separators in the production of radiotracers for bio-medical research and nuclear medical applications// NIM B **204** (2003) 694-700.

ELECTRONICS SIGNAL PROCESSING FOR MEDICAL IMAGING

RENATO TURCHETTA*
CCLRC - Rutherford Appleton Laboratory, Instrument. Dept., Chilton, Didcot, Oxon OX11 0QX, UK

Abstract. This paper describes the way the signal coming from a radiation detector is conditioned and processed to produce images useful for medical applications. First of all, the small signal produce by the radiation is processed by analogue electronics specifically designed to produce a good signal-over-noise ratio. The optimised analogue signal produced at this stage can then be processed and transformed into digital information that is eventually stored in a computer, where it can be further processed as required. After an introduction to the general requirements of the processing electronics, we will review the basic building blocks that process the 'tiny' analogue signal coming from a radiation detector. We will in particular analyse how it is possible to optimise the signal-over-noise ratio of the electronics. Some exercises, developed in the tutorial, will help to understand this fundamental part. The blocks needed to process the analogue signal and transform it into a digital code will be described. The description of electronics systems used for medical imaging systems will conclude the lecture.

Keywords: electronics; imaging; processing.

1. Introduction

The role of any electronic signal processing architecture is to condition the radiation detectors so that it can ultimately be transferred to a computer.

In general, radiation, made of particles and/or photons, impinging on a detector will generate a small signal that will then be fed to the signal processing architecture. Different types of detectors exist. In a scintillator, light is generated by the radiation and subsequently transformed into an electric signal by a photomultiplier. In gas or solid-state detectors, electron – ion or electron – hole pairs respectively are created. The number N of

---

*Renato Turchetta, Rutherford Appleton Laboratory, Instrument. Dept., Bd 25a, room 79, Chilton, Didcot, Oxon, OX11 0QX, UK; e-mail: R.Turchetta@rl.ac.uk.

Y. Lemoigne et al. (eds.), Physics for Medical Imaging Applications, 273–276.

generated pairs is proportional to the energy ΔE lost by the radiation in the detector and the proportionality constant W depends on the detector material and has no dependency on the type of particle and little dependency on temperature. The proportionality constant W is in the order of tens of eV for gaseous detectors and of eV for solid-state detectors. As an example, let us consider a 20 keV photon, typical energy of mammography, fully absorbed in a detector. In silicon, where W is 3.6 eV / pair, it would generate about 5500 electron – hole pairs, while in a gaseous detectors it would generate about ten times fewer electron – ion pairs, although internal amplification could increase the amount of charge collected in the detector. Whatever material you use, the charge signal generated by a single 20 keV photon is fairly small, corresponding for example to less than 1 fC for the silicon detector. From this simple example, it is clear that low noise is one the main requirements for the electronic signal processing chain.

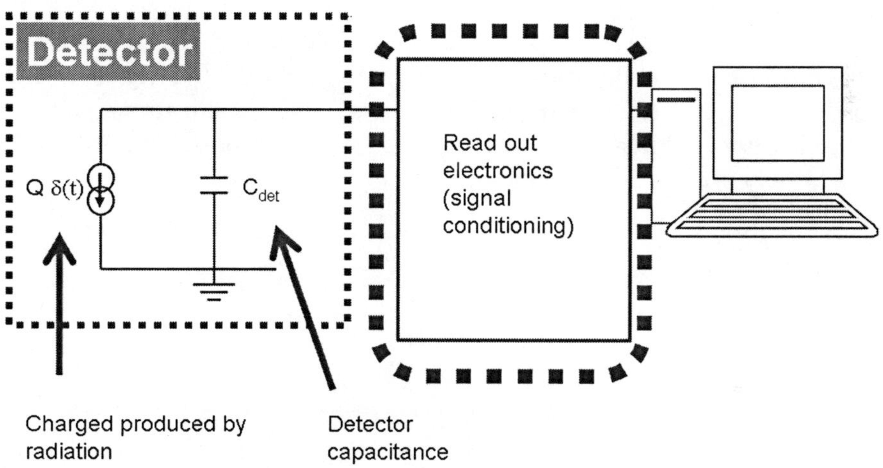

Figure 1. Basic model of a signal processing chain.

Figure 1 shows a typical detector system: on the left, the detector, in the centre the signal conditioning block or readout electronics and on the right a computer. In the figure the detector is identified by its simplest electrical model: a radiation-induced current source, with a short, delta-like pulse and a capacitance. Although more detailed detector models can be built, this simple form is already very useful for the electrical simulation of the readout electronics. An excellent introduction to detectors can be found in[1].

2. Electronics for medical imaging

Two ways of operating a detector exist: either integrating or in single photon detection. The distinction between the two depends on the relative speed of the electronics and of the rate of arrival of radiation. If T_{rad} is the average time between two radiation-generated pulses and τ is the time the electronics take to process a single pulse, then the electronics will work in integrating mode if $T_{rad} \ll \tau$ or in single particle detection mode if $\tau \ll T_{rad}$. Depending on the application, the same system can integrate or process single photons.

In figure 2, simplified schematics of an integrating electronics are shown. On the left, photons generate a given amount of charge and the total charge Q_{tot} is integrated over the detector capacitor C_{det}. The electronics can be a simple unity gain amplifier (buffer or follower). The output voltage of the buffer is then converted into a digital value by an analogue-to-digital converter (ADC).

Because of their readout time, charge-couple device (CCD) and CMOS image sensors normally operate in integrating mode [2, 3]. They are widely used for intraoral dental imaging and find applications also in mammography. Other integrating devices are amorphous silicon flat panel, now widely used in digital radiography.

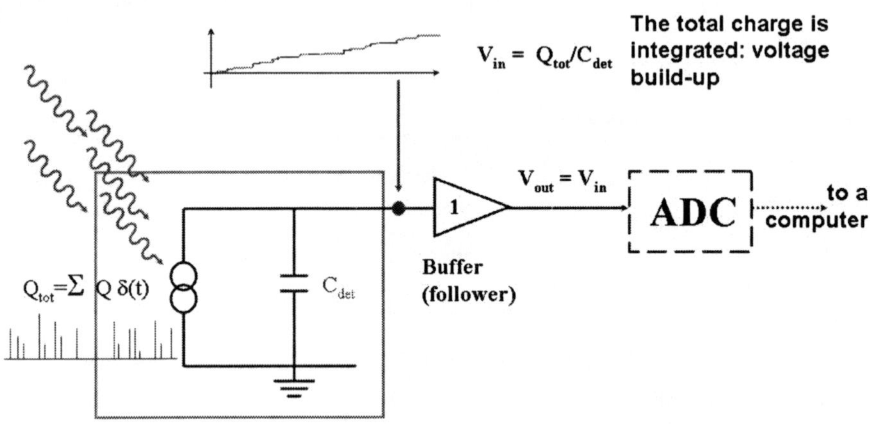

Figure 2. Integrating mode. Simplified schematics.

In figure 3, simplified schematics for a single photon detecting system is shown. The first stage is normally a low-noise amplifier. In the figure is mounted in the charge sensitive configuration, but current or voltage sensing configurations are also used. The preamplified voltage is then fed into a block, indicated with its transfer function $H(\omega)$, which filters the noise and shapes the signal. It is then often called a ‘shaper’. In the diagram, a unity gain buffer completes the circuitry, but often a variable gain amplifier is used to adapt the voltage range to the following circuitry, which could be an ADC. As the input signals are small, an important parameter of the preamplifier – shaper set is the noise performance. Noise optimization is thoroughly described in [4, 5], while introduction to microelectronic architecture, including ADCs can be found in [6].

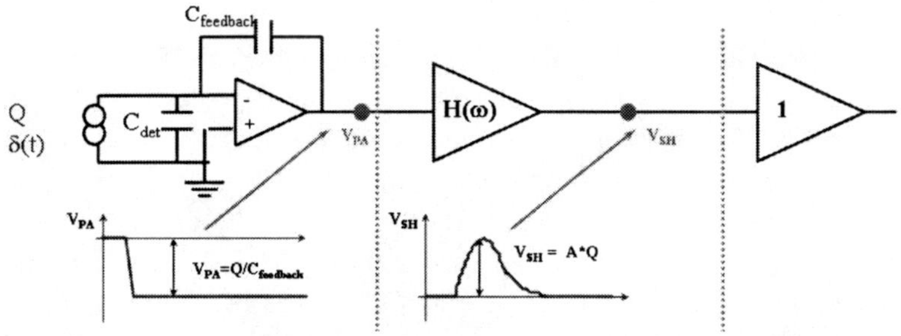

Figure 3. Single particle detection mode. Simplified schematics.

References

1. G. F. Knoll, Radiation Detection and Measurement, John Wiley & Sons, New York, 3rd Edition, 2000.
2. M W. Davidson from National High Magnetic Field Laboratory for the image of solid-state imaging devices from the Microscopy Primer web-site:**http://www.microscopy.fsu.edu.**
3. M. Sze: *Semiconductor sensors*, McGraw-Hill.
4. E. Gatti, P. F. Manfredi: *Processing the signals from solid-state detectors in elementary particle physics*, La Rivista del Nuovo Cimento, vol. 9, serie 3, 1986 (~100 pages volume in English).
5. Z. Y. Chang, W. Sansen: *Low noise, wide-band amplifiers in bipolar and CMOS technology*, Kluwer Academic Publishers.
6. A. S. Sedra, K. C. Smith: *Microelectronics Circuits*, Oxford University Press.

EMISSION TOMOGRAPHY

ROBERT J. OTT*
Physics Department, Institute of Cancer Research and Royal Marsden Hospital, Downs Road, Sutton, Surrey, SM2 5PT, UK

Abstract. Positron (PET) and Single photon tomography (SPECT) are both methods used to acquire data in nuclear medicine from the detection of gamma photons. While SPECT is normally performed by rotating a device such as one or more gamma cameras around a patient, conventional PET devices are normally composed of rings of individual detectors, such as small BGO crystals. The total number of such detectors can be very high, and requires clever electronic design to reduce the total number of components. In both PET and SPECT the data recorded must be reconstructed to form sets of transaxial slices, after various types of correction to remove sources of degradation such as scatter etc. Filtered backprojection as originally employed in both PET and SPECT is now being widely superseded by iterative techniques such as OSEM, which are increasing being operated in true 3D. Considerable progress has been achieved recently in instrumentation design, and in image quality, an example of which is illustrated in several PET designs for scanning small animals. New detector materials have been proposed in particular for PET such as LSO+NaI multi-layer crystals, solid state detectors such as HgI, detectors linked to arrays of various crystals with decoding schemes to identify both position total energy and timing. Some related information about the development of direct capture devices (solid state and scintillating) in so-called digital radiology will be included. For PET, fast timing is critical, and systems have even been developed to perform time of flight measurement of coincident gammas, but problems remain. A particular topic of interest has been the development of devices capable of both PET and SPECT, for example using multiple gamma cameras in coincidence, with some limited success. Additional corrections such as depth of interaction are then required. The chain of processing is described in such systems, going from detection (and collimation) electronics and algorithms involved in pre-processing, reconstruction (2d and 3D) and typical processing algorithms to produce clinically valid physiological data. Quantitation has been a long term aim in nuclear medicine, and some of the critical design parameters and limitations (such as positron range) are indicated. The differences and relationships between physiological and anatomical imaging will be stressed.

* RJ Ott, Physics Department, Institute of Cancer Research and Royal Marsden Hospital, Downs Road, Sutton, Surrey, SM2 5PT, United Kingdom; e-mail:bob@ICR.ac.uk.

Y. Lemoigne et al. (eds.), Physics for Medical Imaging Applications, 277–296.

1. Part I – Single photon emission computed tomography

1.1. INTRODUCTION

Nuclear Medicine Imaging is a technique that involves the use of radioactive tracers to image the function of the body. The most common form of imaging is 2D planar scintigraphy. Here a patient is injected with a ^{99m}Tc labeled tracer, for example, and an image is acquired with the patient placed against a gamma camera. This type of image provides a single time-point (snap-shot) of function. Dynamic scintigraphy is the production of a time-series of 2D planar images which allow a tracer to be tracked through the tissue under study providing dynamic functional information. Single photon emission computed tomography (SPECT) is a form of 3D imaging produced by taking a set of snap-shot images around the patient. An alternative to SPECT is the use of positron emitting tracers such as ^{18}F together with a positron camera again to provide 3D images. This is known as positron emission tomography (PET). A schematic diagram of a gamma camera for single photon imaging is shown in figure 1 and whole body planar images of the skeleton from the front (anterior) and back (posterior) of the patient shown in figure 2.

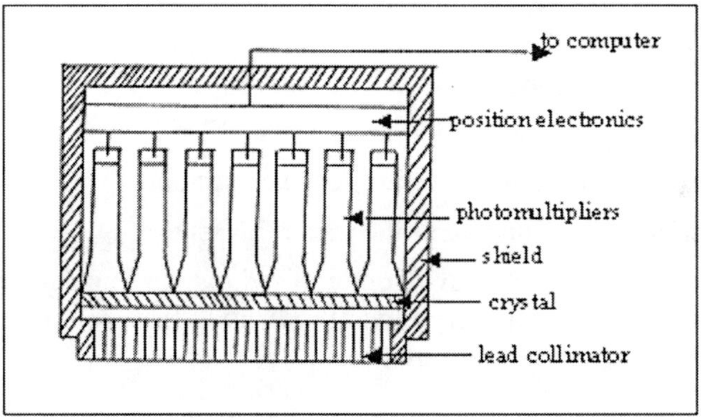

Figure 1. Schematic diagram of a Gamma camera.

Gamma rays entering the lead collimator along the axes of the holes will reach the scintillating crystal. The photons that deposit energy in the crystal produce a flash of light which is detected by the photomultipliers and turned into electronic pulses. The energy deposited in the crystal is proportional to the total amount of light produced. The position where the photon enters the crystal is determined by the difference in the pulses produced by the photomultipliers using the position electronics. Positional and energy information is used to form an image in a computer system.

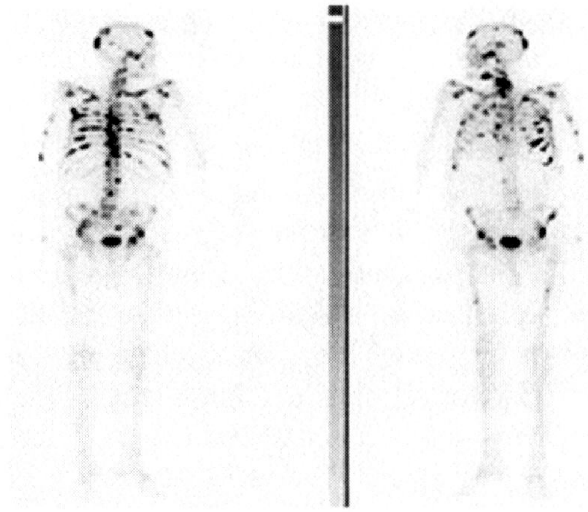

Figure 2. Whole-body skeletal images with ^{99m}Tc MDP. The images were formed by scanning the patient longitudinally between the heads of a double-headed gamma camera.

1.2. EMISSION COMPUTED TOMOGRAPHY (ECT)

The aim of emission computed tomography is to measure and display the concentration of a gamma ray-emitting radioisotope within individual slices of the body – the production of a 3D distribution as compared to the 2D distribution produced by planar scintigraphy. The advantages of ECT over planar imaging are:

1. improved image contrast between different tissues
2. better localisation of the different tissues in 3D
3. improved detection rates of lesions such as tumours
4. quantification of the uptake of the tracer in tissue.

Figure 3 shows cross-sectional X-ray and radioisotope images though the brain of a patient with a primary tumour indicating that the tumour has high blood flow in comparison to normal brain.

2. Types of emission computed tomography

As indicated above there are two major types of emission computed tomography – single photon emission computed tomography (SPECT) performed with tracers such as ^{99m}Tc and positron emission tomography (PET) with positron emitter labeled tracers. The former is the most common as the cost of SPECT studies is quite modest and the imaging systems are readily affordable. SPECT can be carried out using a rotating gamma camera or a special purpose ring PET scanner.

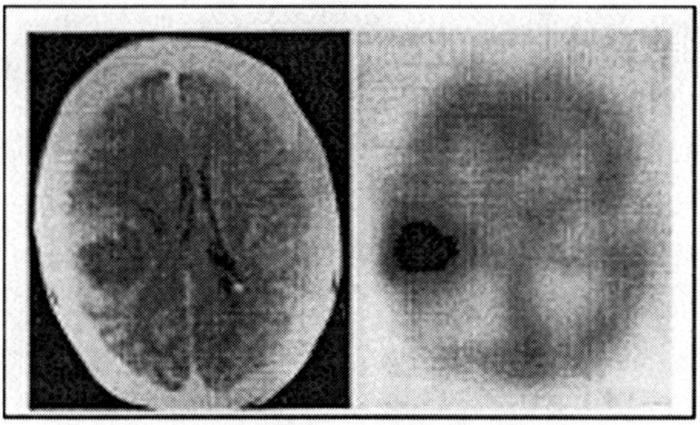

Figure 3. A CT scan (left) and a SPECT brain scan (right) using a 99mTc labelled blood flow tracer showing high perfusion in the tumour.

2.1. SPECT WITH ROTATING GAMMA CAMERA(S)

The process of performing a SPECT scan is similar to all radioisotope scanning. The patient is injected with a tracer and usually waits between 1–3 hours for some of the tracer become trapped in the target tissues/organs and the rest to be excreted from the blood pool. The patient is then placed on a couch and imaged with camera(s) mounted on a rotable gantry, figure 4. Multiple 2D projection images are taken at 64/128 angles at 180°/360° around the patient and stored on a computer system. 3D images are subsequently reconstructed using analytical or iterative methods. Corrections for photon attenuation and scatter in the patient are important for quantification of tracer uptake.

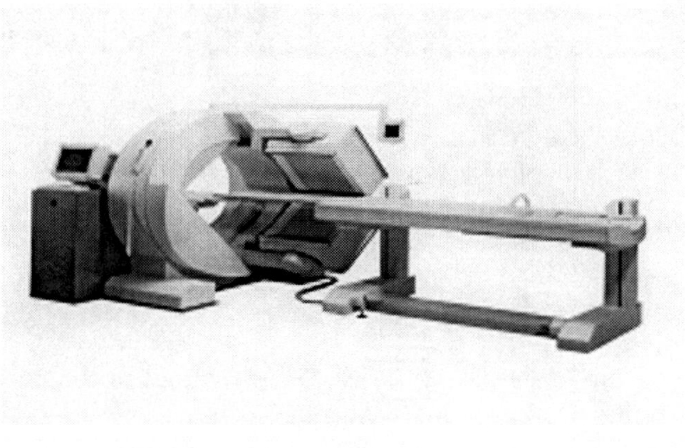

Figure 4. Single-headed (left) and double-headed (right) SPECT systems.

Figure 5 shows a schematic diagram of the acquisition by the gamma camera in a single transaxial slice though the patient. Figure 6 shows how photon trajectories perpendicular to the face of the camera are acquired and shows that the number of photons detected at any position along a projection is an integral along the projection line, known as a line of response or LOR. The analytical reconstruction method is similar to that used in X-ray CT.

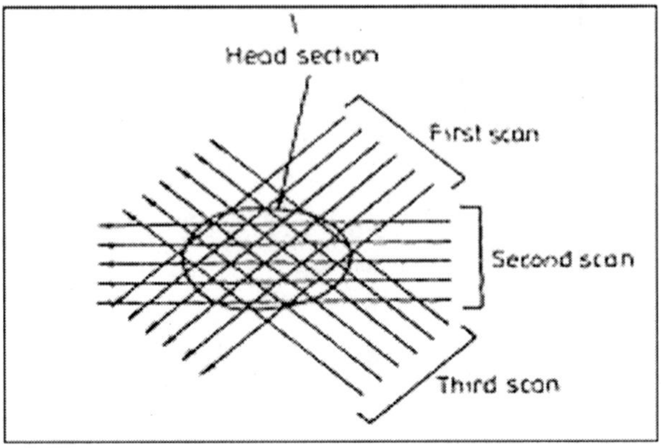

Figure 5. Multi-view acquisition for SPECT showing how the parallel LORs are acquired at each angular position around the patient.

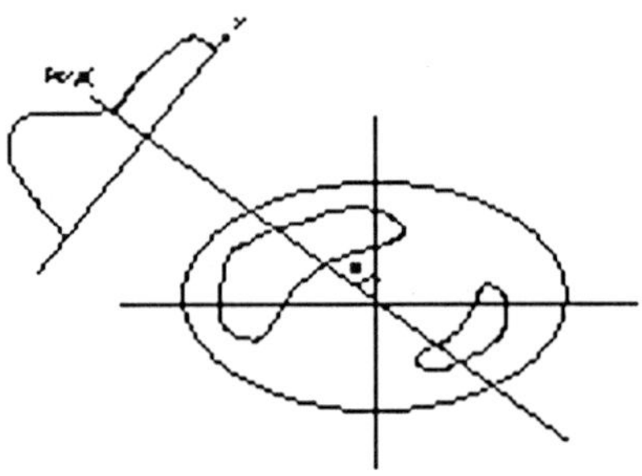

Figure 6. Basic Method of ECT image reconstruction. Here P(x,θ) is the number of counts or LORs at position x in the projection at angle θ.

The raw projection data can be displayed as a sinogram, figure 7.

2.2. IMAGE RECONSTRUCTION USING ANALYTICAL METHODS

Figure 8 shows the basic principles of SPECT image reconstruction using the back projection method.

The acquisition process (figure 8a) is known as forward projection and to reconstruct an image the projection data can be simply back-projected into image space (figure 8b). However due to the lack of information about the position along the LOR that each photon originated from, the image produced is blurred by the 'star' artefact shown.

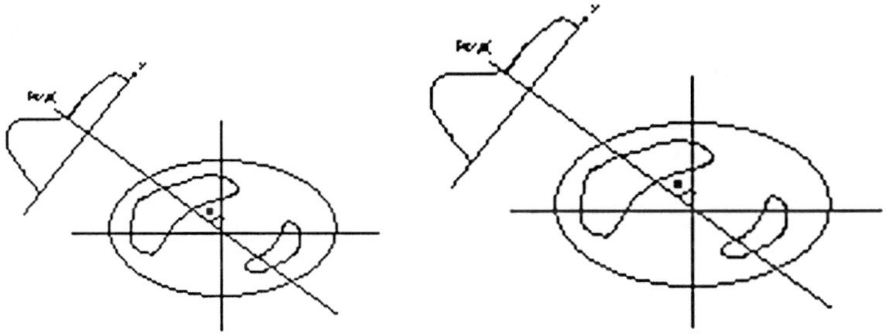

Figure 7. A sinogram. For each cross section through the patient, data is stored by the distance x across the projection and the angle of the projection θ around the patient. The plot is called a sinogram as any single point in the object describes a sine wave trajectory in the plot.

The solution to this problem in simple terms is to mathematically filter each projection as shown in figure 8c so that each element of the project contains a negative component. When the filtered projections are back-projected that negative components cancel out the star artefact. This method of image reconstruction is known as filtered back-projection.

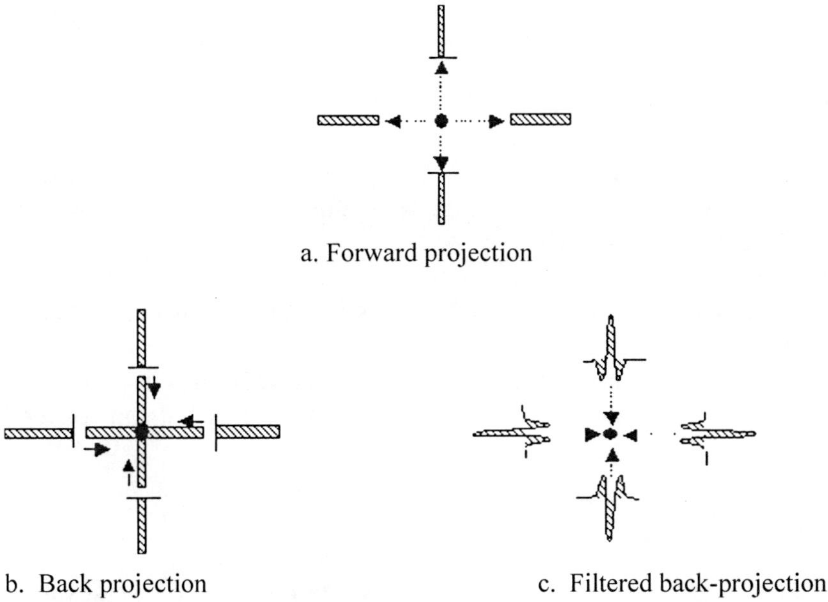

Figure 8. The processes of projection, back-projection and filtered back-projection in SPECT. Only four projections of a single point source are shown for simplicity.

3D sections from SPECT can be displayed, figure 9, in any direction but most commonly in the transaxial (across the body), coronal (along the body from viewed from the front) and sagittal (along the body viewed from the side. The head, feet, front and back of the image are known as superior, inferior, anterior and posterior views.

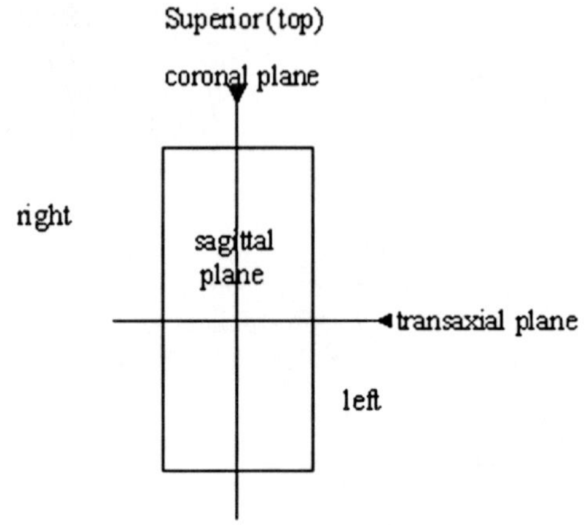

Figure 9. SPECT data as displayed in the 3 most common planes used for diagnosis as viewed from the front (anterior) of the patient.

3. Attenuation correction for SPECT

Photons of energy between 100keV and 200 keV are highly attenuated in tissues with the majority of the effect being via the Compton scatter process. Hence, in order to quantify SPECT images it is essential to correct for this attenuation process. Simplistically this can be done using the method illustrated in figure 10.

The correction factor (CF) estimated above assumes that both the attenuation and that the isotope distribution is also uniform which is clearly not true in most cases.

Transmission imaging to provide values for μ and (t1+t2) is carried out with either scanning point or line sources or, more recently with an attached x-ray system.

3.1. SCATTER CORRECTION

The simplest way to perform a scatter correction on SPECT images is to acquire projection images with 3 energy windows as shown in figure 11. A proportion of the two scatter projections acquired using windows W2 and W3 are subtracted from the projection acquired using window W1. This fraction is determined experimentally. The method assumes that the scatter in the photopeak under W1 is linearly related to the scatter in the two adjacent windows. This assumption is strictly not true but the method works because of the modest energy resolution of gamma cameras.

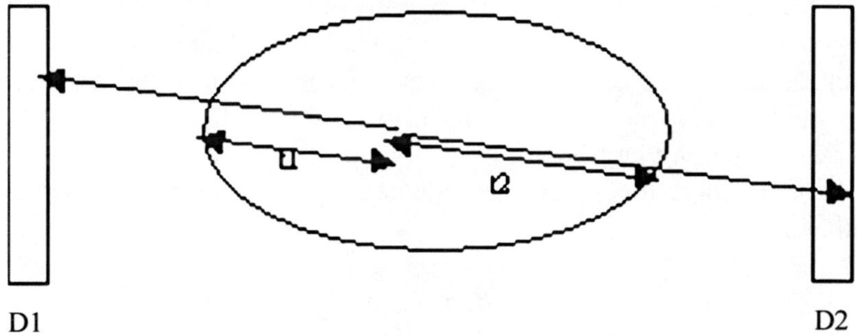

Figure 10. Attenuation correction in SPECT. The counts detected in D1 ~ exp(-μt1), in D2 ~ exp(-μt2) and D1*D2 ~ exp(-μ (t1+t2)). The combined image from the two camera heads is dependent on object thickness (t1+t2) and μ, the attenuation coefficient and the correction factor (CF) for each pixel is given by CF=1/2μTexp((1/2μT)/sinh(1/2 μT).

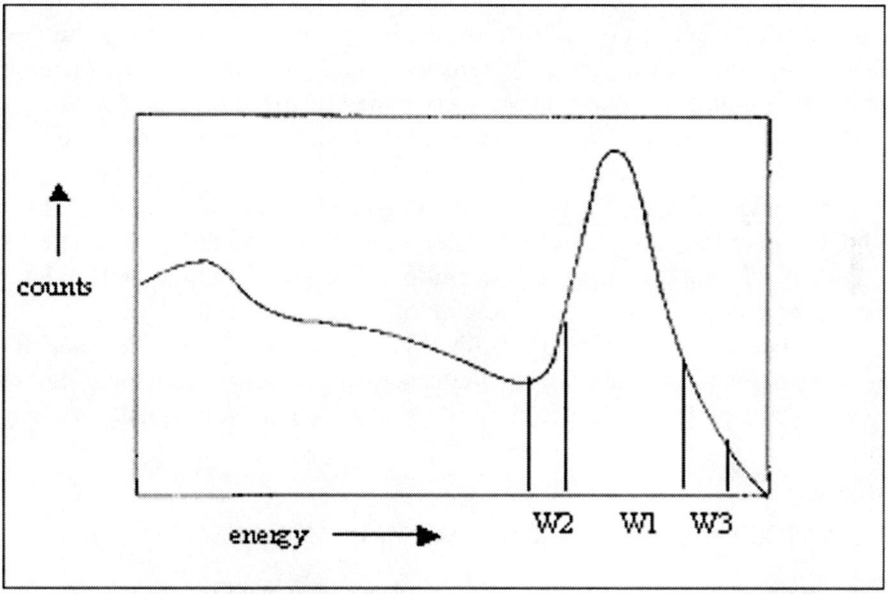

Figure 11. A typical photon spectrum acquired with a gamma camera showing the three energy windows used for scatter correction.

Following the subtraction process images are reconstructed using FBP or an iterative method.

3.2. OPTIMISATION OF SPECT ACQUISITION PARAMETERS

In order to maximum the quality of SPECT images it is necessary to ensure the most appropriate parameters are used for data acquisition. Due to the imperfections in the camera/gantry system it is essential that the centre of rotation of the camera is checked – any shift in this by more than a fraction of a pixel as the camera rotates will produce 'ring artefacts in the images. Non-uniformity of the camera/collimator system must also be corrected to avoid circular artefacts – this is carried out by imaging a uniformly filled cylindrical phantom that ha a diameter equal to the camera sensitive area. A 30 M count flood image is recommended to allow a correction with sufficient accuracy.

The geometrical properties of collimator used are very important as using the wrong collimator can seriously affect image quality. The thickness of the collimator and its septa must match the energy of the photons being imaged. The diameter and lengths of the collimator holes are important as the longer and narrower the hole is the better the spatial resolution at the expense of sensitivity, and vice versa. The energy windows should match the energy of the photons being imaged and the width of these windows is a compromise between sensitivity and scatter contamination, figure 11. The minimum radius of rotation of the camera around the patient is clearly determined by the patient diameter. The best spatial resolution is achieved by ensuring that the camera is as close as possible to the patient and this can be obtained by using elliptical orbits.

The number of angles or views acquired depends somewhat on the time being allowed for acquisition. 128 views for a 360° rotation are recommended. The acquisition matrix size needs to be chosen up-front – a 64×64 matrix for each projection with a 6 mm pixel size this is adequate for a system with an18 mm spatial resolution. A128×128 matrix with a 3 mm pixel size is needed for a spatial resolution better than 10 mm.

It is possible to zoom the acquisition if the object is smaller than the FOV of the camera, maximizing the sensitive area used. The time/view is typically set at 15 – 30s per view depending upon the acquisition count rates.

3.3. OPTIMISATION OF SPECT RECONSTRUCTION PARAMETERS

Prior to reconstruction it is necessary to chose the slice thickness required – this must be ~1 pixel for the detection of small lesions to minimize the so-called 'partial volume' effect where small objects are poorly resolved.

The number of slices chosen for reconstruction is determined by the need to minimize the time and the need to cover organ of interest. Images can be reconstructed using a range of algorithms including the filtered back projection (FBP) method outlined above and one of a number of iterative methods. In FBP it is necessary to chose a reconstruction filter applied in real or frequency space which will minimise image noise whilst maximising the resolution that is achievable in the image. Various filters are available (e.g. Ramp-Hanning, Ramp-Butterworth, Shepp-Logan, Figure 12) and the choice is almost a personal preference depending on the final image quality required. For example minimal high frequency filtering leads to noisy images although edges are well-resolved. Substantial high frequency filtering produces smooth images but with loss of detail and the production of mottle artefacts.

Smoothing can also be applied before or after reconstruction and in 2D or 3D.

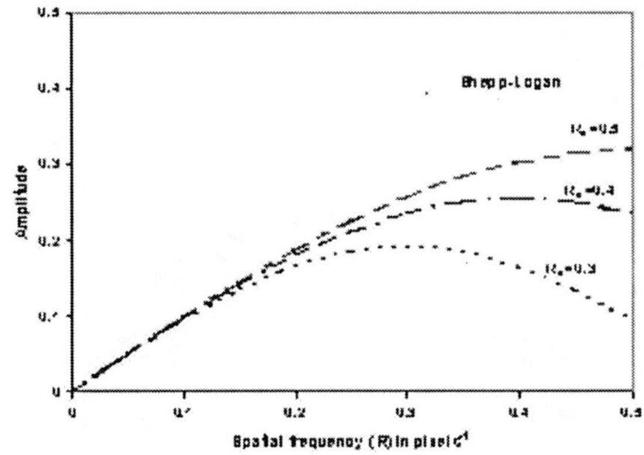

Figure 12. Examples of the Shepp-Logan filter used in image reconstruction – the values of R_m represent different parameters used to change the shape of the filter.

3.4. SUMMARY

SPECT provides 3D radioisotope distributions with improved 3D contrast compared to planar scintigraphy. However many corrections to the data are needed for image quantification and most SPECT studies are semi-quantitative at best.

4. Part II - positron emission tomography

Positron emission tomography (PET) is based on the use of positron emitting radionuclide which produces two back-to-back 511 keV gamma rays following positron annihilation. Figure 13 shows the process whereby a proton converts into a positron plus a neutrino (not shown).

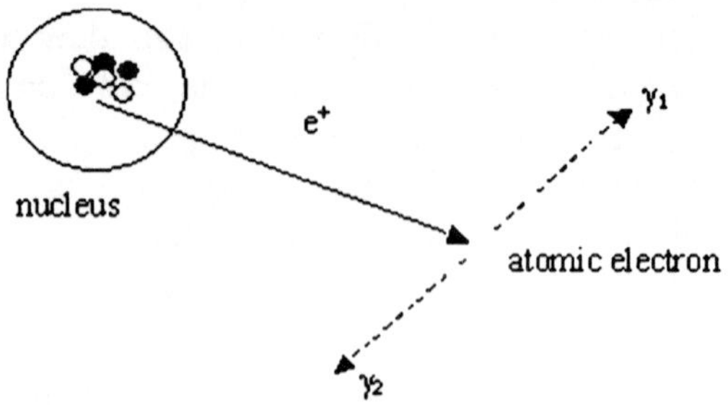

Figure 13. The positron annihilation process. A proton rich nucleus decays via positron emission and the positron annihilates with an atomic electron after travelling a short distance in the medium. The annihilation process produces two back-to-back photons of energy 511 keV.

Commonly used radionuclides in PET include ^{18}F which has a half life of 110 m, ^{11}C (half-life 20 m), ^{13}N (half-life 10 m) and ^{15}O (half-life 2 m). These nuclei are used to label a range of biologically active tracers such as amino acids, proteins etc. The two annihilation photons produced in the radioactive decays are detected by a positron camera as shown in Figure 14.

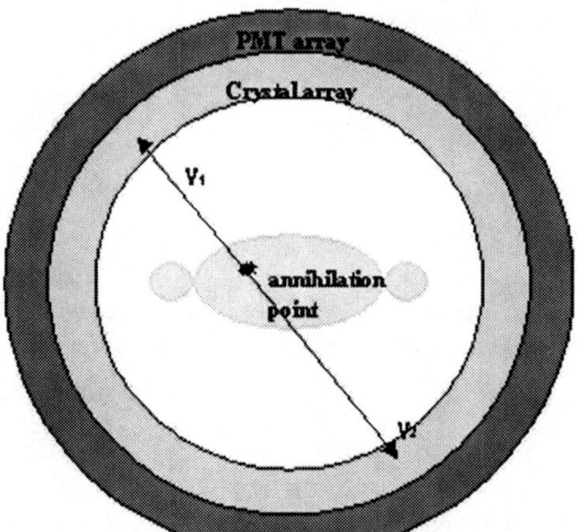

Figure 14. Illustration of annihilation photon detection in a ring positron camera.

The important part of this process is that by detecting the two photons in coincidence there is no need for a collimator to define the LOR as there is in SPECT. This makes the sensitivity of PET several orders of magnitude greater than SPECT for a given level of radioactivity imaged. The detector requirements for PET include high detection efficiency for 511 keV gamma rays, an intrinsic spatial resolution ~5 mm, high temporal resolution (~10 ns) to ensure a high level of good coincidences, an energy resolution (15-25%) to reject detected scattered photons and a high count rate performance (>100 kcps).

The spatial resolution in PET is determined by individual detector element size which is ~5mm, the range of positron in tissue which can be >1mm for some radionuclides and the acollinearity of photons (up to ±0.5°) which happens when the positron-electron pair annihilate before the positron has come to rest. Typically the intrinsic spatial resolution for whole body PET imaging is >5mm.

The sensitivity of PET is determined by stopping power (~ρ and Z) and the thickness of crystal (2-3cm), and total camera sensitive area (typically ~15cm axially, ~60cm diameter):

$$\text{Sensitivity} = \varepsilon^2 \Phi$$

where ε is efficiency of detectors and Φ is the fractional detector solid angle.

The count-rate performance of a positron camera is determined by the dead-time of the detectors which in turn is limited by the optical decay time of crystal and the time taken for the electronics to readout an event (>100ns).

The timing resolution which determines the coincidence speed is limited by light output and optical decay time of the crystal. The more light is emitted rapidly the higher the coincidence rate achievable.

The energy resolution of the PET detector is determined by light output and efficiency of light collection by the photodetectors.

4.1. PROPERTIES OF SCINTILLATORS USED FOR PET

Table 1 lists some of the properties of scintillating crystals that have been use for PET imaging. PET scanner configurations include multiple rings of crystal and photomultiplier, partial rings, multiple planar detectors and double-headed cameras. The majority of commercial positron cameras are based on the multi-ring design. Because photomultipliers are usually >10mm in diameter, coupling to crystals with a 5mm by 5mm cross section has been done using block formats similar to that shown in figure 15. In the example shown a segmented block of BGO has been coupled to four square photomultipliers. In this case the position of the gamma ray entering the block is determined by the size of the signals produced in the four photomultipliers.

4.2. RANDOMS AND SCATTERED EVENTS

Several types of coincidence events are possible during a PET acquisition. In addition to so called true events (Figure 14) where the two photons from the same nuclear annihilation are detected, accidental coincidences (random) and scatter events are detected, figure 16. Both these events produce false, detected LORs.

TABLE 1. Some properties of scintillating crystals used for PET imaging.

Material	NaI(Tl)	BaF2	BGO	LSO	GSO
Density (gm/cc)	3.67	4.89	7.13	7.4	6.71
Effective Z	51	54	75	66	58
1/e stopping length (mm)	25.6	22.7	11.2	11.4	13.9
Light yield	100	5,16	15	75	20
Decay time (ns)	230	0.6,620	300	40	60

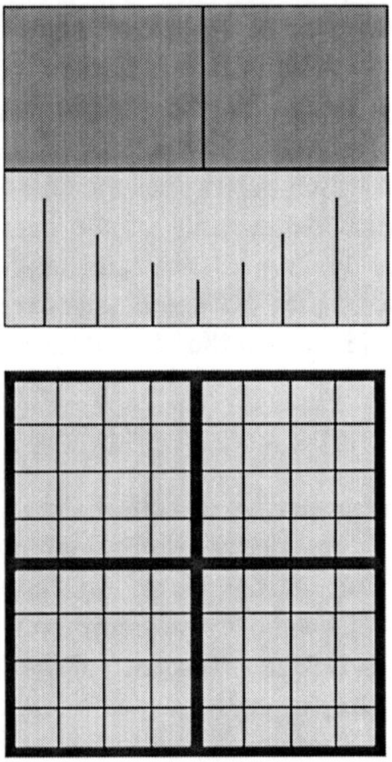

Figure 15. A PET block detector. Top shows how a segmented block of a phosphor is coupled to square photomultipliers. Bottom shows a front view of the phosphor block.

Random events are generated by the detection of two photons from two different nuclear decays and can be a very large fraction of the total event rate if the level of radioactivity being imaged is high. The random rate (R_r) is given by:

$$R_r = 2\tau S^2$$

where τ is detector resolving time and S is the detector singles count rates

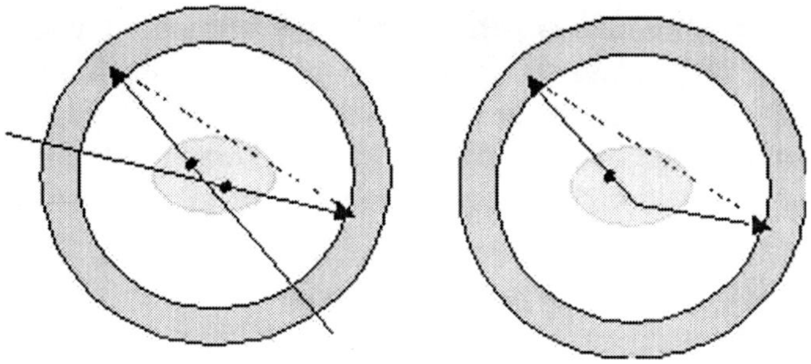

Figure 16. Randoms (left) and scatter (right) events detected in PET.

The true event rate (R_t) is given by:

$$R_t = kS$$

where k is a constant dependent upon system being used.

The ratio of true to random event rate is therefore given by:

$$R_t / R_r = k / 2\tau S$$

From this we can see that the best true to random ratio is achieved when the resolving time is small.

Scatter events are those in which one or both of the annihilation photons are scattered in the object being imaged. The fraction of scattered events can be high in PET when a large object is being imaged. Although the attenuation coefficient in tissue at 511keV is less than that at 140 keV, the probability of scatter is high as two photons must be detected. Most attenuation is due to Compton scattering as there is negligible photo-electric absorption at 511 keV. In addition photon energy resolution at this energy is poorer than at lower energies so that discrimination against scatter is not as effective in PET as it is in SPECT. Until recently all positron cameras included lead or tungsten shields between rings of detectors (figure 18) to reduce this scatter although this reduces the sensitivity of a positron camera by ~5. In the absence of scatter, attenuation correction in PET can be achieved fairly simply. If the path lengths in the object traversed by the two gamma rays are t_1 and t_2 (figure 10) and μ is the attenuation coefficient of the object material then:

$$\text{Attenuation of photon 1} \sim \exp(-\mu t_1)$$

$$\text{Attenuation of photon 2} \sim \exp(-\mu t_2)$$

$$\text{Combined attenuation} \quad \sim \exp(-\mu(t_1+t_2))$$

We can see that attenuation correction can be carried out with a knowledge of the thickness of the object traversed (t_1+t_2) and its attenuation coefficient.

Attenuation correction is performed by performing a transmission scan of the object to provide an attenuation map. As for SPECT imaging, transmission scanning is performed either by using a rotating line or point source or a CT scanner attached to the PET scanner. The transmission scan data are also used to reconstruct an anatomical image of the patient.

4.3. DEPTH OF INTERACTION CORRECTION

Due to the finite size of the detector elements used in PET the spatial resolution achieved deteriorates off axis, as shown in figure 17.

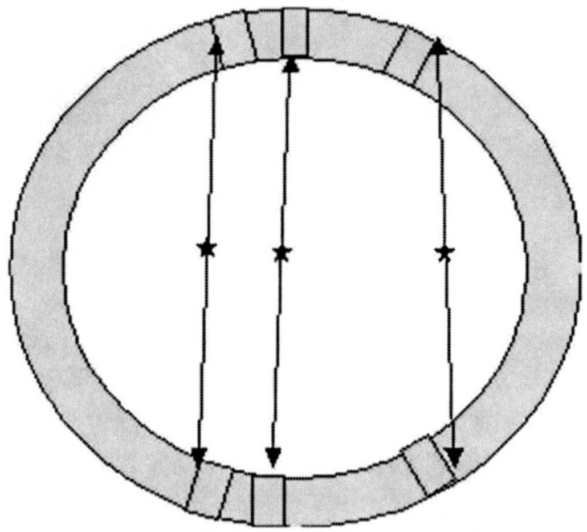

Figure 17. The depth of interaction effect on spatial resolution for off-axis points. As the emission point moves off the central axis the LOR shown are detected along the side of the crystals leading to a reduction in spatial resolution.

This depth-of-interaction effect means that a spatial resolution of 5mm on axis can fall to 8mm off axis. Several methods have been studied to reduce this effect including the use of two different crystal materials. None of these methods have been incorporated into commercial devices due to excessive cost.

4.4. SENSITIVITY IN 2D AND 3D PET

As indicated above the sensitivity of PET is dependent upon the fractional solid angle of the positron camera seen by the object being imaged. This solid angle depends upon the design of the camera and two common architectures are shown in figure 18.

The camera in figure 18a is known as a 2D PET system and has a severely limited sensitivity in the axial direction. The sensitivity is substantially increased in figure 18b by the removal of the septa. This is a 3D PET system having a solid angle of 5-10 times that of a 2D system. 3D PET provides much higher count rates than 2D PET but at the expense of the increased acceptance of scattered photons. The approximate absolute efficiencies are 0.5% for 2D PET and 3% for 3D PET and scatter fractions are typically 5-10% and 30-40% respectively.

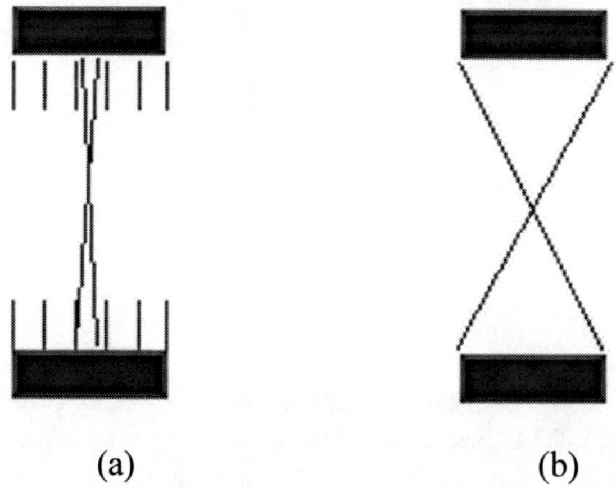

Figure 18. Two common architectures for positron cameras along the axial direction – a multi-ring camera with lead septa separating each ring (a) and the same detector with no inter-ring septa (b).

4.5. PET IMAGE RECONSTRUCTION

2D PET images are reconstructed using the same analytical or iterative techniques as described in SPECT after resorting the projection data into sinograms, figure 6.

3D PET image reconstruction is more complex (Bendriem and Townsend) and is usually approached by rebinning the 3D data into 2D projections and then using the methods described above. Rebinning can be carried out by several methods that are beyond the scope of this text. A method of fully 3D image reconstruction has been developed which uses the parallel projections from the 2D LORs to reconstruct a low statistics image. These data are then used to fill in the missing data that occurs for oblique LORs in 3D PET. This 'missing data' effect is illustrated in Figure 19.

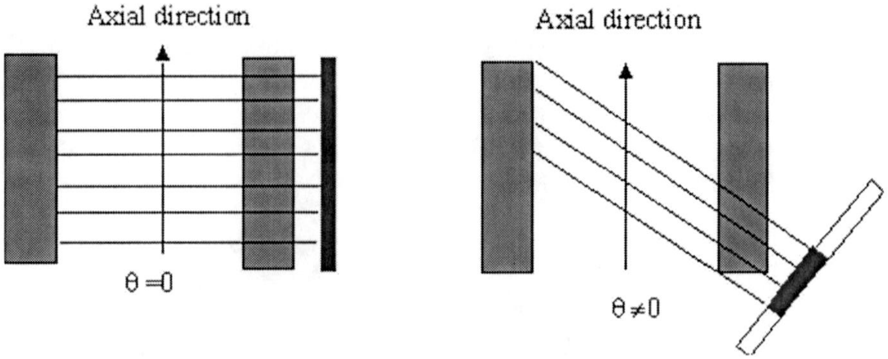

Figure 19. 2D PET (left) has axial LOR angles of θ=0. For 3D PET (right) most LORs have θ≠0 leading to missing parts of the oblique projections which must be estimated from the 2D reconstructions.

4.6. PET DATA CORRECTION

As with SPECT it is necessary to apply corrections in order to produce quantitative images. The first requirement in PET is to normalise the raw data to correct for individual detector efficiency variations. This is done by modifying the sinogram after imaging a uniform cylinder to correct each LOR.

Following this corrections must be made for randoms, photon scatter, attenuation and detector dead-time. The randoms fraction in the image is estimated using a delayed coincidence channel which allows the acquisition of randoms events only. The scatter distribution in an image is corrected for by convolving the image data with a scatter response function or by using a model (refs). The scatter distributions produced by these methods are subtracted from the sinogram. Attenuation correction is applied after generation of the transmission map as described above. Dead-time corrected using a dead-time model.

Finally calibration of the image is carried out usually by imaging a phantom with a known concentration of activity to convert counts per voxel into activity per voxel.

5. Whole-body PET imaging

The majority of clinical studies carried out in oncology are whole-body PET images. These kind of images are ideally suited to establishing the spread of caner prior to treatment and can subsequently be used to monitor the response to treatment. Figure 20 shows a whole-body PET image carried out using ^{18}F – fluorodeoxyglucose (FDG) illustrating the high uptake of this tracer in growing tumour.

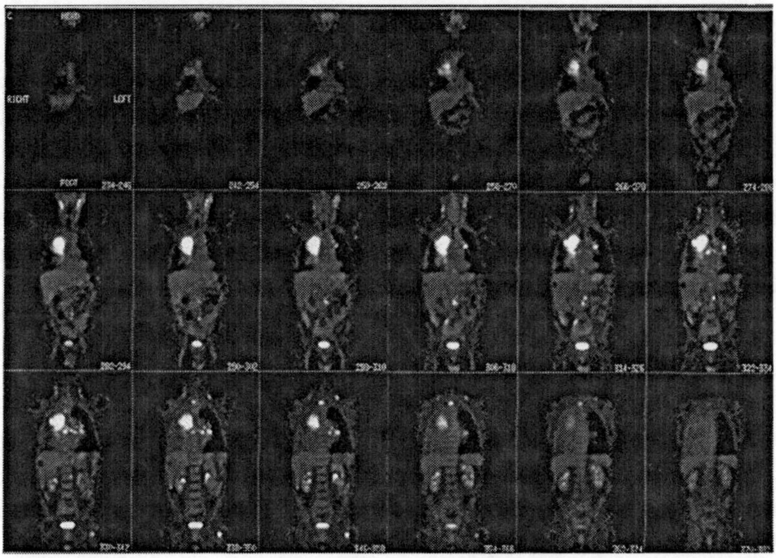

Figure 20. Multiple coronal whole-body sections of a patient with lung cancer showing high uptake of FDG in the bladder, the kidneys and in multiple lesions in the neck, chest and upper left leg. The images are shown from anterior to posterior.

With the recent development of PET/CT systems the power of PET has been increased by the addition of anatomy from the CT scan. Figure 21 shows an FDG PET scan superimposed on the CT scan of a patient with lymphatic disease, illustrating how the addition of anatomy to the PET image localises the disease accurately.

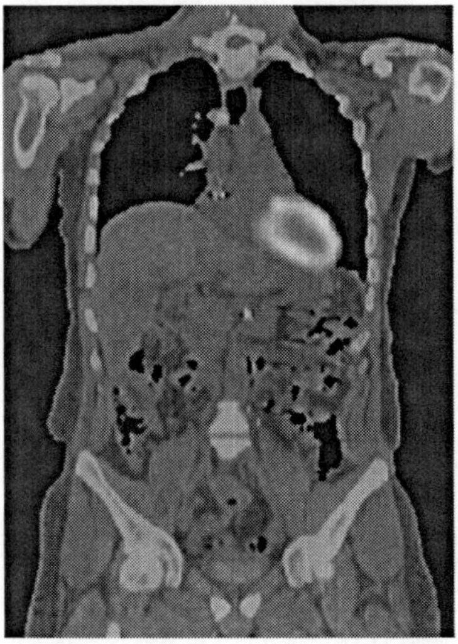

Figure 21. A PET/CT scan of a patient with a lymph node disease. The image shows high uptake of FDG in the lymph node, the heart and the bladder. (See color section).

5.1. GAMMA-CAMERA PET IMAGING

Modifications to a standard dual-head gamma camera system has allowed the production of PET images. The main changes have been the increase in the thickness of the NaI(Tl) crystal from about 8-10mm to 16-25mm, the removal of the collimator and the addition of coincidence circuitry. However NaI(Tl) is not ideally suited to the detection of 511 keV gamma rays and these devices are subject to high levels of scatter, modest timing resolution and low sensitivity (reference).

6. MWPC PETRRA system

A double-headed positron camera (PETRRA) based on coupling barium fluoride crystals to a multi-wire proportional chamber filled with a photosensitive gas has been developed (reference). The prototype system has a fast timing resolution, a high spatial resolution and modest sensitivity. Images taken with this system are encouraging (figures 22 and 23) and indicate that a low cost positron camera with higher sensitivity than the prototype.

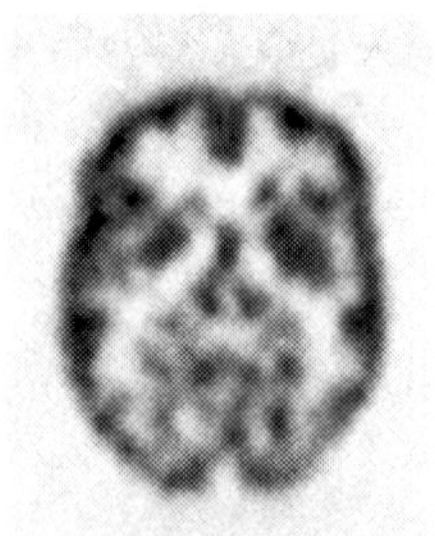

Figure 22. Image of a Hoffman brain phantom taken with the PETRRA positron camera.

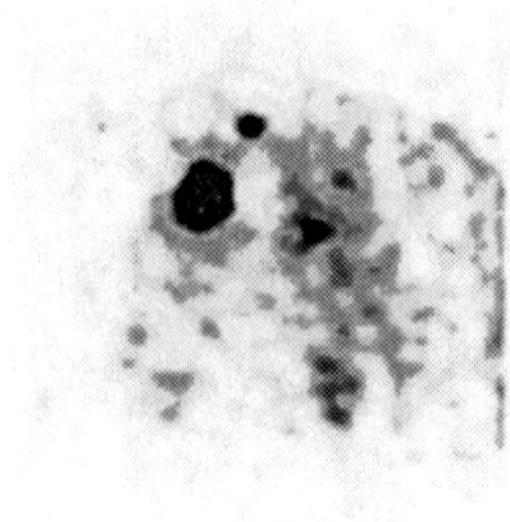

Figure 23. A half-body PETRRA FDG image of the patient shown in Figure 19. (See color section)

6.1. SUMMARY

PET is a high sensitivity, high resolution functional imaging technique which can provide quantitative images of tissue function. The tracers used are 'biological' in nature allowing tissue physiology to be measured. As yet it is still an expensive technique compared to SPECT but has become an essential tool for staging cancer, monitoring treatment and has a growing number of applications in cardiology and neurology.

References

1. The Physics of Medical Imaging. Ed. S Webb. IOP Publishing, ISBN 0-85274-349-1.
2. The Theory and Practice of 3D PET. Bendriem and Townsend, Kluwer Academic Publications,ISBN 0-7923-5108-8.
3. Basic Science of Nuclear Medicine, Parker, Smith & Taylor (Churchill/Livingstone).
4. Textbook of Nuclear Medicine. Vol. I - Basic Science. Vol II - Clinical Applications Harbert & Da Rocha, (Lea & Febiger).
5. Practical Nuclear Medicine. Sharp, Gemmell & Smith.(Oxford University Press).
6. Physics in Nuclear Medicine. Cherry, Sorenson and Phelps. (Saunders).
7. Positron Emission Tomography : Basic Science and Clinical Practice. Valk, Bailey, Townsend and Maisey. (Springer).
8. Positron Emission Tomography: Basic Sciences. Bailey, Townsend, Valk and Maisey. (Springer).
9. Computers in Nuclear Medicine: A Practical Approach. Kai H. Lee. (Society of Nuclear Medicine).

DATA ACQUISITION AND PROCESSING IN RADIOISOTOPE IMAGING

ROBERT J. OTT*
Physics Department, Institute of Cancer Research and Royal Marsden Hospital, Downs Road, Sutton, Surrey, SM2 5PT, UK

Abstract. The use of computers in radioisotope imaging is now well established for data acquisition, data correction, image reconstruction, image display and manipulation, data storage, system control and multimodal imaging and registration. This paper reviews these areas highlighting some of the applications and demands on computing facilities.

Keywords: radioisotope imaging; data acquisition; data correction; image reconstruction; image display; data storage.

1. The computer-based following topics are discussed in the article:

(a) Data acquisition

(b) Data correction

(c) Image reconstruction

(d) Image display and manipulation

(e) Data storage

(f) System control

(g) Multi-modality imaging/registration

---

* RJ Ott, Physics Department, Institute of Cancer Research and Royal Marsden Hospital, Downs Road, Sutton, Surrey, SM2 5PT, United Kingdom; e-mail:bob@ICR.ac.uk.

Y. Lemoigne et al. (eds.), Physics for Medical Imaging Applications, 297–309.

1.1. A) DATA ACQUISITION

All modern gamma cameras and positron cameras are interfaced to one or more digital computer(s) for data acquisition and processing purposes, figure 1.

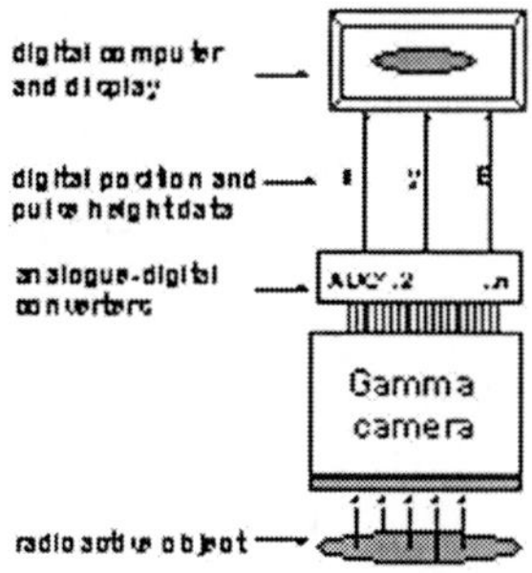

Figure 1. A schematic showing the coupling of a gamma camera to a digital computer system.

This allows a range of acquisition processing formats to be used and provides a greater number of studies to be carried out. Types of acquisition used include:

- 2D planar single photon imaging
- 2D and 3D whole body scanning
- 2D and 3D dynamic imaging
- Multiplanar tomography (SPECT & PET)

Within these are several modes of acquisition which provide flexibility of acquisition in terms of speed and size. List mode acquisition, in which raw data from the cameras is stored event-by-event, provides fast acquisition speeds at the expense of data volume.

Frame mode is most commonly used for 2D imaging allowing single or multiple snapshots of tracer distribution in the most compact way. Versions of this are include multiple projection/sinogram storage for SPECT and PET and multi-isotope mode for single photon imaging.

The different acquisition types produce different data sizes. Planar 2D imaging is generally acquired in frame mode with matrix sizes of 32^2, 64^2, 128^2, 256^2. Dynamic 2D imaging is acquired into multiple time frames and separate phases, usually into smaller matrix sizes as the quantity of data acquired per frame is smaller that in planar imaging.

SPECT is acquired into 64^2, 128^2 2D frames at 64/128 angular views around the patient.

Whole body scans require asymmetric frame sizes to optimise the acquisition from a patient. Multiple frames are needed if several energy windows are used for multi-isotope imaging or for single isotope multi-energy imaging, for scatter correction, for example.

Image sizes for typical single photon imaging are shown in Table 1.

TABLE 1. Typical single photon image sizes. All these numbers assume byte mode acquisition and single energy window.

Image type	Max. frame size	Number of frames	Mbytes
2D static	256^2	1	0.065
2D dynamic	128^2	100	1.6
2D whole body	512×1024	1	0.5
SPECT	128^2	128	2.0
List mode	$10^6 \times 4$ byte events	-	4.0

Dynamic data acquisition can be carried out using external pulses to trigger the time sequence. An example of this is the use of the R-wave produced by the heart, figure 2, as it enters diastolic mode.

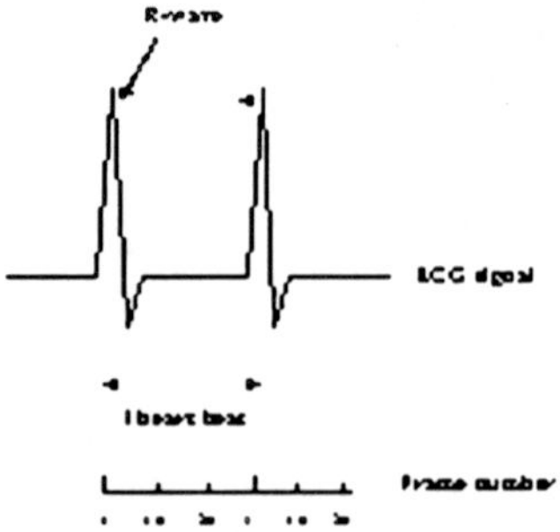

Figure 2. The R-wave produced by the heart can be used to correctly phase the acquisition of dynamic frames from multiple heart beats.

1.2. B) DATA CORRECTION

Computers are now used for whole range of data correction processes in planar and dynamic imaging. For example energy and positional information from a gamma camera is entirely dependent upon the production of analogue pulses from the photomultipliers (PMTs). These devices are notoriously unstable and corrections for changes in PMT gain is essential if high quality images are to be produced. PMT gain drift occurs continuously and can lead to substantial errors in positional and energy measurements. Compensation for gain drift is often performed by computer controlled adjustment of PMT voltages on millisecond time-scale by either monitoring PMT pulse heights. Once the PMT gains are stabilised it is possible to remove any residual spatial non-linearities by using a correction matrix. Additionally gamma cameras are usually tuned to optimally detect photons 140keV from ^{99m}Tc and this means that the response at other photon energies is non-uniform. This can be overcome by storing energy dependent spatial shifts in the computer system. Finally once all these corrections have been applied it is still necessary to correct for any non-uniformity caused by variations in the crystal or collimator and this is done by storing a flood correction obtained from imaging a uniform planar phantom. Figure 3 shows a uniformity corrected gamma camera image.

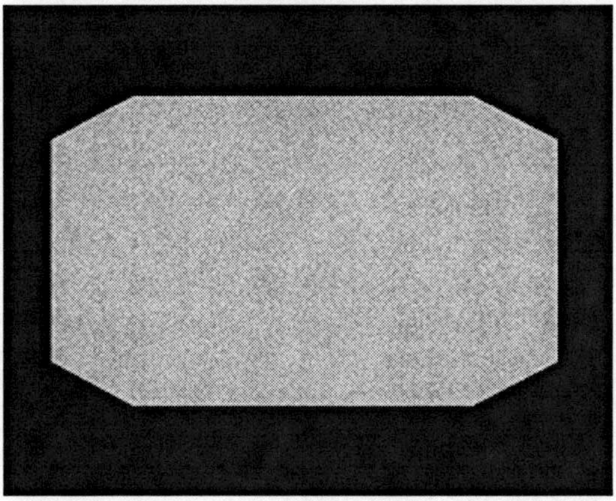

Figure 3. A uniformity corrected gamma camera image – typically residual non-uniformity is <5%.

Additional corrections are required for SPECT imaging. Particularly important is the need to ensure that the geometrical and electronic centres of rotation are the same. This is achieved by imaging a point source placed off the rotation axis. Figure 4 shows the results of a centre of rotation study. The plots of XC and YC show the X and Y positions of the source as a function of camera angle around the axis - XC should follow a sine wave and YC a flat straight line. The XCO and XCR plots show the differences between the actual source positions and the expected positions. Any variations in these greater than or equal to a pixel will cause rotational artefacts in the SPECT images and need to be corrected for by storing a correction matrix in the computer system.

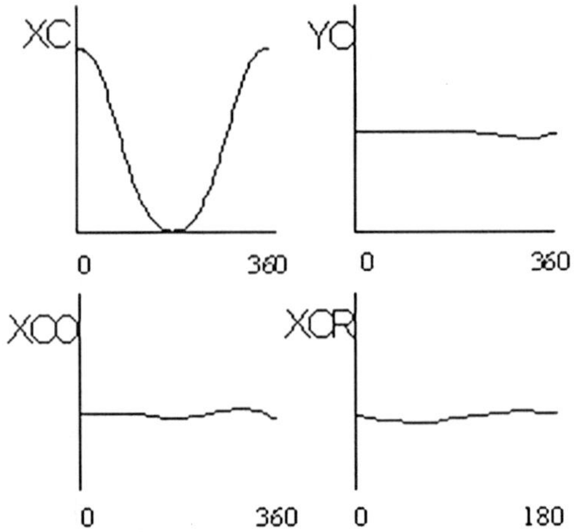

Figure 4. Results of a centre of rotation study – the top two plots show the position (in pixels) of the point source across the camera surface in the X and Y directions. In the bottom two plots, XCO=XC – sine wave and XCR = (XC(θ) and XC(θ + 180))/2 – both should show deviations of less than 1 pixel for a 'good' study.

As indicated in the chapter of tomography corrections for attenuation and scatter are needed if the images are to be quantitative. What ever methods are chosen to estimate the corrections they are usually carried out by modifying the sinograms prior to image reconstruction. The effect of applying attenuation correction to PET images is shown in figure 5. On the left is a whole-body PET image before attenuation correction illustrating an apparent high uptake of the tracer in skin which disappears in the corrected image on the right.

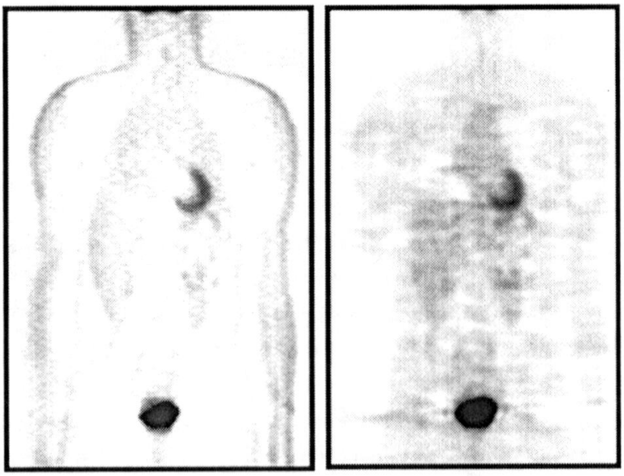

Figure 5. Whole-body PET images before (left) and after (right) attenuation correction.

For data acquired at high rates it will also be necessary to correct for dead-time especially if a short-lived isotope is being used. This is particularly important for dynamic imaging where the time-activity curves may be seriously affected by dead-time. Dead-time is usually corrected by measuring camera count rates and applying a computer-based mathematical model to the data.

Finally for PET imaging corrections for randoms are essential especially if quantification is required.

1.3. C) IMAGE RECONSTRUCTION

Tomographic image reconstruction for SPECT and PET can only be performed using a powerful computer system. The fastest and most conventional method is filtered back-projection (FBP) and a fully 3D image can usually be reconstructed in a few minutes using this technique. The method has been described briefly in a previous chapter.

More recently with the advent of faster and more powerful computer systems it has been possible to use iterative reconstruction methods. These methods are time consuming as they often required the data to be reconstructed many times but they produce images that appear better qualitatively than those reconstructed by FBP. Figure 6 shows schematically the process of iterative image reconstruction.

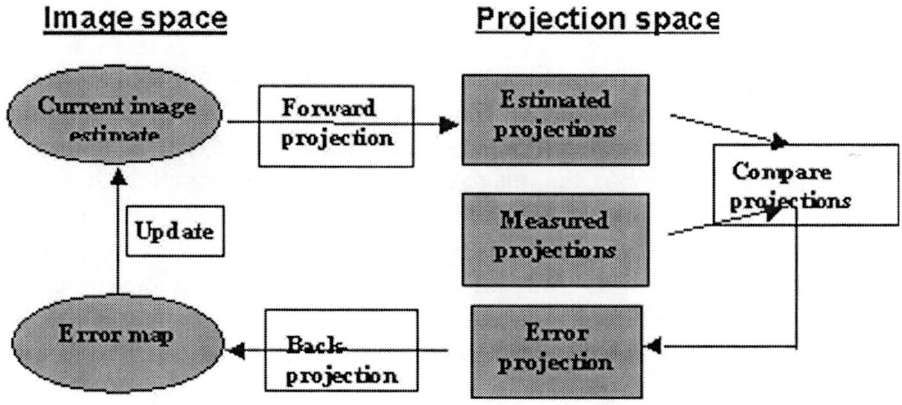

Figure 6. Schematic of the iterative reconstruction process. Here a guess at the image is used as a starting point for multiple forward and back-projection procedures which terminate when the projections produced by the guess match the real projections within a pre-defined uncertainty.

Figure 7 shows an example of the same data reconstructed with FBP and an iterative reconstruction method. NB picture numbering is wrong.

1.4. D) IMAGE DISPLAY AND MANIPULATION

Once an image is reconstructed it can be processed and manipulated using the computer system. Processes that can be performed include imaging smoothing and enhancement, rotation, magnification, reduction, image arithmetic such as subtraction and addition, region of interest analysis, numerical analysis and curve fitting.

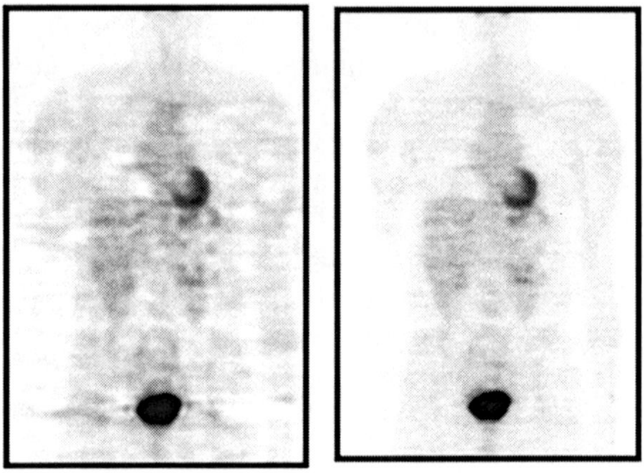

Figure 7. Whole-body PET images reconstructed with FBP (left) and an iterative method.

Image display and manipulation techniques include the ability to display images in various array sizes, to change the gray or colour display levels, overlay images on top of each other, display a series of dynamic frames in time sequence. Modifying the image display levels, for example, allows the clinician to enhance low or high activity features, enhancing the power of the images for clinical diagnosis.

Filtering is a powerful way of reducing noise or enhancing specific features. Image processing computers have a range of filters that can be used in image space directly on the image or in Fourier space if the former takes too long. The most simple image space filters are the 2D 9-point smooth, and the edge enhancement filter, figure 7. These filters are convolved with the 2D images to either smooth out noise or amplify the edges of an image.

TABLE 2. A 9-point smooth filter (left) and an edge enhancement filter (right) that replaces each element in an image with the average of the point and the 8 surrounding points after multiplication by the filter.

1	*2*	*1*	*0*	*-1*	*0*
2	**4**	**2**	**-1**	**4**	**-1**
1	**2**	**1**	**0**	**-1**	**0**

For large images faster filtering can be performed by transforming the image using the Fast Fourier Transform and multiplying by a Fourier space filter such as Hanning or Butterworth filters. The formulae for these filters are:

Hanning Filter

$$A(f) = 0.5[1 = \cos(\pi|f|F)] \qquad |f| \leq F$$

$$= 0 \qquad |f| > F$$

Butterworth Filter

$$A(f) = [1 + (|f|/Q)^{p}]^{-0.5}$$

where A is the filter amplitude (usually between 0 and 1), f is the spatial frequency in the image, F is the cut-off frequency (max. 0.5) and Q and p are operator selected.

When F and Q are large the filter is 'sharp', improving spatial resolution at the expense of increased noise. If F and Q are low the filter is 'smooth', reducing noise and spatial resolution.

Figure 8 shows an example of the use of a Hanning Filter on a transaxial PET image.

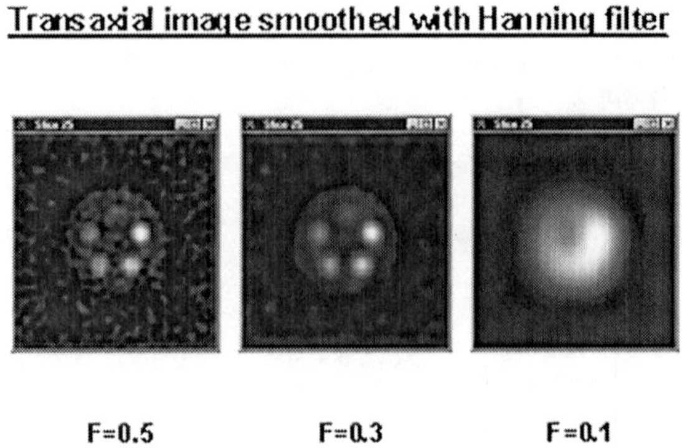

Figure 8. Effect of different values of cut off frequency values on an image.

ROI analysis & curve production is a very important application of computers in radiotracer imaging. In the acquisition process multiple frames of data are obtained and these can be analysed to produce time-activity curves for various regions of the images selected. In most cases several regions of interest (ROI) are planted on summed frames allowing the production of time-activity curves for each ROI. It is then possible to subtract background curves from specific organ data and then to fit the resulting curves with some form of physiological model. Finally it is possible to use the model to extract physiological parameters and produce parametric images.

An example of dynamic image analysis is shown in figure 9. Top left shows a summed image of the frames taken during a dynamic kidney function study (renogram). The image shows four ROI drawn around the kidneys, the heart and a representative background region. Bottom right shows the resultant time-activity curves for the kidneys after background subtraction. The curves clearly illustrate the difference in the function of the two kidneys and the data, top right, show that using a simple model, the time for the two kidneys to reach half- their activity level is a factor of three different. This type of data can be used to treat the malfunctioning organ as repeat studies can determine if therapy has produced any improvement in function.

Figure 10 shows similar data from a multi-gated dynamic cardiac study. Here the acquisition was triggered by the R-wave output from the patients' heart to allow the production of multiple frames each 28ms long, spanning the average heart beat. The graph plots the average activity in the heart

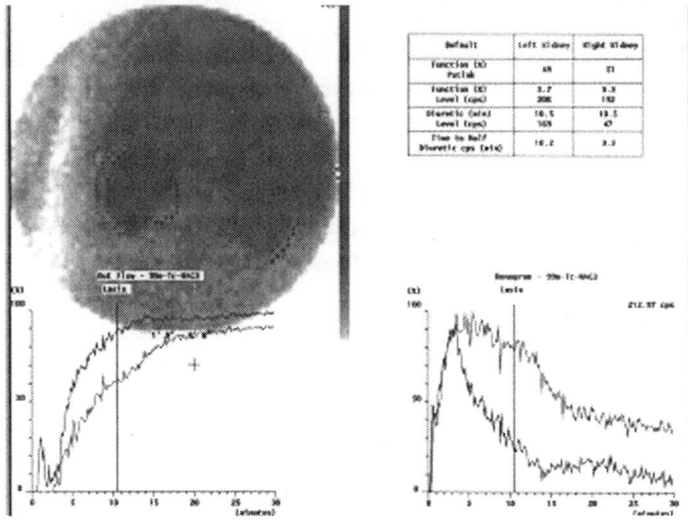

Figure 9. Summed image, time activity curves and extracted data from a renogram.

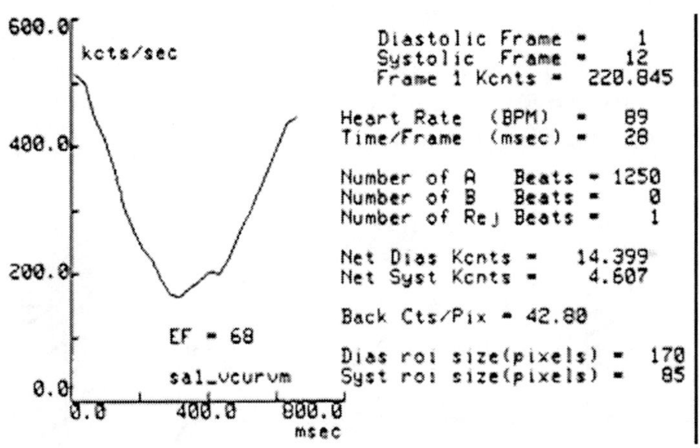

Figure 10. An example of the output from a multi-gated cardiac study.

summed over ~1250 heart beats A simple analysis measuring the minimum and maximum activities provides an assessment of the percentage cardiac output of the heart known as the ejection fraction (EF) which is measured as 0.68 in this case.

1.5. E) DATA STORAGE & ARCHIVE

This is a very important requirement as all data needs to be kept at least during the time a patient is undergoing treatment, for example. Modern computers have Gbyte hard discs for short term image storage and data can be backed-up or archived using a number of facilities including CD-ROM (>100 Mbytes), ZIP discs (>100 Mbytes), standard floppy disc (1.44 Mbyte) and Xabyte magnetic tapes (many Gbytes but slow). Some systems are now attached to a RAID architecture which allows many Gbytes of data to be stored locally to the computer. It is always essential to have some form of back-up or archive remote from the acquisition system in case of localised fire. Additionally it is common for most imaging systems to be attached to a local network or picture archiving system (PACS) allowing all types of images from different modalities to be stored under the same patient name.

1.6. F) SYSTEM CONTROL

All modern gamma camera gantries and heads are controlled by a computer to allow automatic head rotation and tilt, patient couch control for longitudinal scanning and lateral motion and a combination of movements for SPECT. For example by controlling the vertical movement of the patient couch and the rotation of the camera heads during a SPECT study it is possible to simulate an elliptical orbit as shown in figure 11.

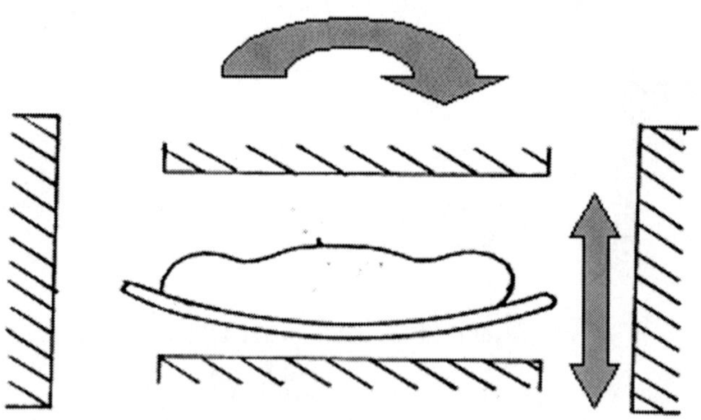

Figure 11. An illustration of how patient couch motion and camera rotation can minimise the distance between the camera heads and the patient at each angle. The vertical arrow shows the couch motion whilst the curved arrow shows the gantry rotation.

This provides the SPECT images with the best possible spatial resolution by keeping the patient as close as possible to the camera heads during rotation. In a similar manner the heads of a camera can be moved vertically during a whole body scan to ensure the minimum distance between the heads and the patient.

1.7. G) MULTI-MODAL IMAGE REGISTRATION

The use of multiple imaging modalities is becoming routine in patient diagnosis as each type of scan adds extra information for the clinician. If the scans are done at separate times there will be issues associated with registration of images from different modalities. Particular problems will be the use of different head/neck supports and bed shape for the different modalities. There are various methods for registering images from the different imaging techniques involving computer-based algorithms which might use information from external markers applied to the patient, mutual information within the studies or specific anatomical landmarks. The computer software and data formats are likely to differ between the modalities leading to difficult compatibility problems- some of these problems can be solved by use specific data transfer formats such as DICOM but there are still differences between these formats on different imaging machines. In any case it is now possible to register images from different modalities to ~ few mm in most cases although there are serious problems in cases where scans take different times and patient movement becomes important. A good example of this is the use of X-ray CT scans which are usually taken with breath holds and SPECT or PET scans which take 10s of minutes. Registration of thorax from these modalities can produces serious image artefacts due to patient breathing.

More recently the development of SPECT/CT and PET/CT system have improved the quality of multi-modal information but movement artefacts are still a potential problem.

2. Conclusions

Computer-based image acquisition and processing is a complex business requiring fast CPUs, large memories and hard disks plus connections to large-scale storage for archive and rapid access. In the future PACS systems, teleradiology and the internet will play a major role in multi-modal image utilisation.

References

1. The Physics of Medical Imaging. Ed. S Webb. IOP Publishing, ISBN 0-85274-349-1.
2. The Theory and Practice of 3D PET. Bendriem and Townsend, Kluwer Academic Publications,ISBN 0-7923-5108-8.
3. Basic Science of Nuclear Medicine, Parker, Smith & Taylor (Churchill/Livingstone).
4. Textbook of Nuclear Medicine. Vol. I - Basic Science. Vol II - Clinical Applications Harbert & Da Rocha, (Lea & Febiger).
5. Practical Nuclear Medicine. Sharp, Gemmell & Smith.(Oxford University Press).
6. Physics in Nuclear Medicine. Cherry, Sorenson and Phelps. (Saunders).
7. Positron Emission Tomography : Basic Science and Clinical Practice. Valk, Bailey, Townsend and Maisey. (Springer).
8. Positron Emission Tomography: Basic Sciences. Bailey, Townsend, Valk and Maisey. (Springer).
9. Computers in Nuclear Medicine: A Practical Approach. Kai H. Lee. (Society of Nuclear Medicine).

EVALUATION OF IMAGE QUALITY

PETER F. SHARP*
University Hospital NHS Trust, University of Aberdeen, Foresthill, Aberdeen, Scotland, UK

Abstract. In this paper we describe an overarching framework that allows us to interpret the information from an image. Starting with the imaging device, consideration will be given to the measurement of the quality of the raw data detected by the instrument. This uses Bayesian signal detection theory to combine the large area transfer characteristic, the modulation transfer function and the noise power spectrum in a single measure of quality.Then we will discuss how to assess the quality of the displayed image by measuring human performance directly. The most complete description of observer performance is provided by Receiver Operating Characteristic (ROC) analysis, which estimates all the combinations of sensitivity and specificity available from an imaging procedure. How subjective measures of image quality can be combined with the objective assessment of performance will be investigated. Finally, we will show how quality can be extended to the judgement of the influence of imaging technology on the clinical management of patients, including the quality of life of the patient.

Keywords: image quality; signal to noise ratio (SNR); modulation transfer function (MTF); receiver operating characteristic (ROC); noise equivalent quanta (NEQ); detective quantum efficiency (DQE); efficacy.

1. Introduction

Image quality is a measure of the effectiveness with which an image can be used for its intended task. In medical imaging this task has to be considered in terms of the influence that the information provided by the image has on the clinical management of the patient. Thus the challenge in assessing quality is to derive a process which leads from quantitative measures of imaging instrument performance, such as modulation transfer function, through an assessment of the ability of the clinician to interpret the information contained in the displayed image, to a judgement of the impact

* Peter F. Sharp, University Hospital NHS Trust, University of Aberdeen, Foresthill, Aberdeen, Scotland, UK; e-mail: p.sharp@biomed.abdn.ac.uk.

Y. Lemoigne et al. (eds.), Physics for Medical Imaging Applications, 311–319.

of the imaging modality on patient management as measured, for example, by the quality of life.

Since quality is task dependent, the assessment of image quality falls into two broad stages; defining the task and then measuring the ability of the person to carry out the task. The type of task can be considered as either classification or estimation. In the former the task is to decide to which one of a number of possible classes the image belongs. The most common classification task is that of detection where there are two classes; either the image is normal or abnormal. In contrast, estimation involves the measurement on a continuum of some parameter of the image such as the width of blood vessels, or the transit time of a radiopharmaceutical through the kidney. Most of the work in medical image quality has concentrated on the classification task and estimation will not be considered further.

The imaging process itself can be broken into two parts: data acquisition and image display. It is important to consider the two stages separately otherwise image quality may simply reflect the performance of the person viewing the image rather than the true potential of the imaging device.

2. Quality of the acquired data

The quality of the acquired data can be measured in terms of 3 parameters: large area transfer factor (K), modulation transfer function (MTF) and noise power spectrum (W_n).

The large area transfer factor describes the relationship between object and image contrast in situations where spatial information plays no part. An example would be the characteristic curve of X-ray film, linking X-ray exposure with the optical density of film. The modulation transfer function measures the reduction in an object's contrast when it is imaged, as a function the spatial detail in the image, as represented by the spatial frequency (Giger and Doi, 1984; Yip et al., 2001; Efstathopoulos et al., 2001).

Finally there is a requirement for a measurement of noise. While there are many examples in the literature of where noise is expressed in first order statistics, usually the standard deviation of intensity fluctuations, this is principally of value where the noise is white, i.e. has no spatial structure. To be of practical use, noise must be expressed in terms its spatial frequency content. The noise power spectrum or Weiner spectrum expresses noise by its autocorrelation function as a function of spatial frequency (Giger et al., 1984; Brettle et al., 1996).

While each of the above parameters is of value in exploring a specific aspect of device performance, the crucial question is how they combine to give an overall measure of quality. To answer that we need to look at how they contribute towards achieving the relevant clinical task

At this stage it is necessary to remember that we are only concerned with the acquired data, that is prior to it being formed into a displayed image that can be interpreted by a clinician. Thus we need to address the question of who is the decision maker, as clearly it is not a human.

The so-called ideal observer of Bayesian signal detection theory (Van Trees, 1968) provides a solution. This is a decision maker who, faced with a particular task, selects the cheapest or safest way of making a decision. The detection task can be presented as one in which the decision maker has an image, g, and must take one of two decisions. H_1 is the hypothesis that this image is normal and H_2 that it is abnormal. The decision is then based on the value of the odds ratio, or likelihood ratio (L), of the two probabilities

$$L = p(g \mid H_2)/p(g \mid H_1) \tag{1}$$

To make a decision the observer selects some threshold value, L_c, such that if $L<L_c$ then H_1 is chosen, if $L>L_c$ then H_2 is selected by the observer. Graphically this can be represented as in figure 1. The quality of the image can then be expressed in terms of the signal to noise ratio, SNR.

The likelihood ratios produced by normal and abnormal images are represented by probability density curves, figure 1. These simply reflect the inherent variability in the data from these two classes of images produced by such factors as photon noise, in the case of X-ray and nuclear medicine images, and biological variability.

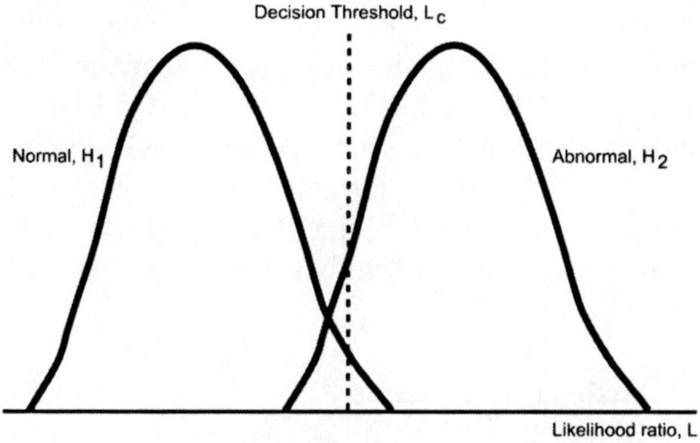

Figure 1. Decision theory paradigm. The ideal observer uses a threshold L_c such that if the value of the likelihood ratio for a particular image is greater than it, the observer classifies the image as abnormal.

In figure 1 the SNR is the distance between the means, $\langle L_I \rangle_1$ and $\langle L_I \rangle_2$, divided by the standard deviations (σ_1 and σ_2) of the distributions,

$$\mathrm{SNR}_I^2 = \frac{\left[\langle L_I \rangle_2 - \langle L_I \rangle_2\right]^2}{\frac{1}{2}\left[\sigma_1^2 + \sigma_2^2\right]} \tag{2}$$

This can be written (International Commission on Radiation Units and Measurements, 1996) in terms of the imaging parameters as

$$\mathrm{SNR}_I^2 = K^2 \int \frac{[\Delta S(\upsilon)]^2\, \mathrm{MTF}^2(\upsilon)}{W_n(\upsilon)}\, d\upsilon \tag{3}$$

where $\Delta S(\upsilon)$ is the difference in the spatial frequency spectra between the two alternative objects, i.e. under H_1 and H_2. For convenience the equations are written in terms of a single spatial frequency, υ.

It should be noted that equation (3) combines the three measurement parameters in a single equation reflecting quality and that this measure of quality is task dependent.

There are many circumstances in which it would be helpful to have a measure of the potential quality of the image without requiring the task to be specified. By removing the task dependent element from equation (3) we have a quality measure known as the Noise Equivalent Quanta (NEQ)

$$\mathrm{NEQ}(\upsilon) = K^2 \frac{\mathrm{MTF}(\upsilon)^2}{W_n(\upsilon)} \tag{4}$$

NEQ can be interpreted as the number of quanta at the input to a perfect imaging device that would give the same image noise as the real imaging device, or alternatively, the measured image noise transformed to input units, e.g. photons. The ratio of NEQ to the actual number of photons used to form the image is known as the Detective Quantum Efficiency, DQE. Examples of the use of NEQ and DQE can be found in Brettle et al., 1996.

The advantage of SNR as a measure of image quality is that it is independent of the imaging modality used and this will be discussed later. There is one important limitation; equation (3) is only valid for tasks in which the signal, that is the characteristics that make an image abnormal, are known exactly.

3. Quality of the displayed image

So far we have been interested in measuring the quality of the acquired data. It is now time to consider how that of the displayed image might be measured. In this case the observer is a human observer but this creates its own problems. Firstly it is necessary to specify the task that will be used to measure quality and, secondly, the criteria on which a decision is made need to be defined.

The link between quality and task has been emphasised a number of times. Clearly the choice of test pattern must reflect the type of task for which the imaging device is to be used. For example, one cannot expect quality based on the detection of "hot spots" in a test pattern to reflect the performance of the device in demonstrating fine spatial detail, such as vessel diameter. The choice must also depend on what it is practical to measure. The most realistic tasks would involve clinical images, but they may be difficult to obtain and it may not be a clear as to whether they are normal or abnormal. Hybrid test patterns involve inserting abnormal features into a clinically normal image. While avoiding the problem of knowing the answer, this approach may not be feasible for some pathologies. Other alternatives are computer generated patterns, but these depend on being able to model accurately the performance of the imaging device, or phantoms that are imaged with the device.

How one measures the effectiveness with which the observer interprets the image also needs to be addressed. The main issue is that the human observer is not consistent in the way he/she carries out a task. Variations in response can be misinterpreted as changes in imaging device performance. The approach that will be discussed here is the graded response, or Receiver Operating Characteristic (ROC) curve.

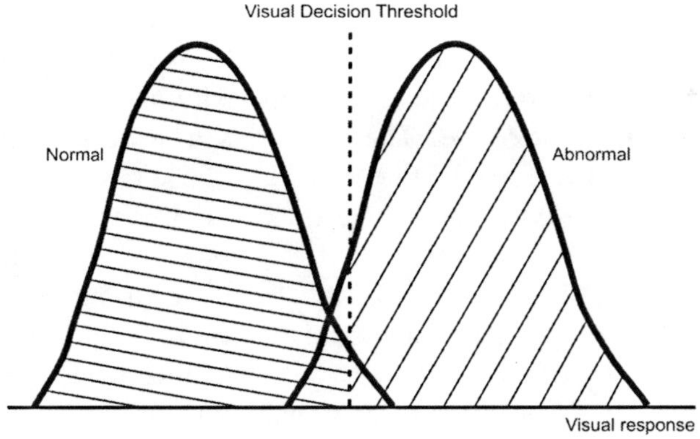

Figure 2. Representation of the visual detection task. If the appearance of the image produces an effect (visual response) on the human observer exceeding some threshold, then the observer report that the image is abnormal. The area under the abnormal curve, i.e. that representing the visual responses produced by abnormal images, to the right of the threshold (shaded as ///) equals the true positive response rate. The area under the curve for normal images to the left of the threshold (shaded as \\\) is the true negative response rate.

If we restrict ourselves to a simple two choice detection task, defined as identifying the image as being either normal or abnormal, then the task carried out by the human observer can be represented as in figure 2. This is similar to figure 1 but the x-axis is now visual response. This is a measure of the strength of the effect that the image has on the observer's ability to carry out the task. The images are again represented by two probability distributions one for normal images and the other abnormal.

The observer is assumed to adopt a visual decision threshold; a particular value of visual response which, if exceeded, leads to the observer responding that the image is abnormal. There are 4 possible outcomes as the image can either be normal or abnormal and the response can identify it as being either normal or abnormal. A response that correctly identifies the image as being abnormal is known as a True Positive Response. In figure 2 the True Positive Response rate is given by the area under the abnormal curve to the right of the visual threshold. Similarly the area under the normal curve to the left of the threshold represents the percentage of normal images correctly reported as being normal, the True Negative Response rate. However the observer can also give wrong responses either incorrectly identifying normal images as abnormal, the False Positive Response, or abnormal images as being normal, the False Negative response. These four types of response are not independent as:

True Positive Response Rate + False Negative Response Rate = 100%

True Negative Response Rate + False Positive Response Rate = 100%

Thus to define the observer's response to the two sets of images it is only necessary to measure one type of response from each pair, usually True Positive and False Positive. The terms sensitivity and specificity are often used in the medical literature. The former is true positive response rate while specificity is the true negative rate, i.e. 1-false positive rate.

As with the ideal observer, image quality is the measure of how well normal images can be distinguished from abnormal ones, the statistical difference between the two probability density distributions. The further apart they are, the easier it is to distinguish normals from abnormals, that is the better the quality of the image. As the visual threshold changes then the resulting true and false positive response rates will also vary. As the threshold moves from A, in the upper part of figure 3, to E, then the true and false positive rates decrease. The graph showing how true and false positive response rates vary, lower part of figure 3, is known as the Receiver Operating Characteristic (ROC) curve.

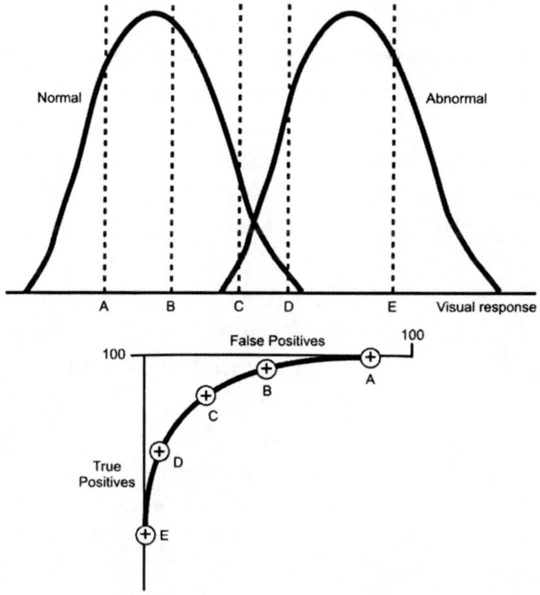

Figure 3. The Receiver Operating Characteristic (ROC) curve. The ROC curve, in the lower part of the figure, shows how True and False Positive responses vary as the visual threshold used by the human observer changes. The visual thresholds at positions A to E produce points A to E forming the ROC curve.

The position of the ROC curve reflects the overlap between the distributions and hence the quality of the image. If the variability of the two distributions are equal, then the signal to noise ratio, often referred to in the signal detection theory literature as d′, can be derived from the curves (Green and Swets, 1966). In a situation where they are not equal then the area under the ROC curve, Az, is often used to describe performance.

This approach is a very robust way of assessing displayed image quality and is described in detail in Metz, 1986. Examples of its application in medical imaging can be found in H MacMahon et al., 1988 and Yip et al., 2001.

It does have limitations and in the form described here is restricted to situations where the task is a binary one – the image can either be normal or abnormal- and where the observer is not required to report the location of the abnormality or abnormalities. The technique can be adapted for use in these other situations and further details can be found in Metz, 1986.

4. Quality and the patient

We have described an approach to measuring image quality which, in particular circumstances, allows us to measure both the quality of the raw data acquired by the imaging device and the quality of the same data displayed as an image, in terms of a dimensionless SNR. Thus we can, for example, measure any loss of quality in the display of the data by comparing the SNR of the acquired data with that for the displayed data. We can compare the performance of different imaging devices, even from different modalities, by comparing their respective SNRs.

Image quality, to be of real value in the medical setting, must reflect the effect that the imaging technology will have on the management of the patient. A new technique might give what appear to be wonderful images, but unless it provides either new clinical information that will influence the management of the patient, or produce information that is already available from existing techniques in a more effective, and perhaps cheaper way, then it has limited value. So how do our measures of quality fit into this broader picture?

TABLE 1. A model of efficacy (adapted from Fryback and Thornbury ,1991).

	Level	Output Measure
1	Technical	MTF, Noise
2	Diagnostic Accuracy	Sensitivity, Specificity
3	Diagnostic Thinking	Changes in clinician's diagnostic probability.
4	Therapeutic	Percentage of times therapy changes
5	Patient outcome	Changes in quality of life years.
6	Societal	Positive change in National Product

Fryback and Thornbury (1991) have proposed a six-level model of efficacy, table 1. In looking at the quality of the acquired data, in terms of measures such as MTF and Noise Power Spectrum, we have been addressing the technical efficacy of the instrumentation. Moving on to the evaluation of displayed data allows us to address the issue of diagnostic accuracy, in terms of sensitivity and specificity. Where we next need to engage in the process is with higher levels areas of diagnostic thinking, therapeutic and patient outcome. How does our imaging technique affect patient outcome – does it change the diagnosis, affect the treatment they receive and ultimately lead to them having a better quality of life? This takes the medical physicists into the area of health technology assessment, where the sensitivity and specificity associated with the technique are incorporated into the diagnostic pathway, including the cost of treatment and diagnosis, and finally an assessment of how this affects quality of life. An example of how such a process works can be found in Bradbury et al., 2002.

References

1. Bradbury, I., Bonell, E., Boynton, J., Cummins, E., Facey, K., Iqbal, K., Laking, G., McDonald, C., Parpia, T., Sharp, P., Single, A., and Walker, A., 2002, Positron emission tomography (PET) imaging in cancer management. Health Technology Assessment Report 2. Health Technology Board for Scotland, Glasgow.
2. Brettle, D.S., Workman, A. Ellwood, R.P., Launders, J.H., Horner, K., and Davies, R.M., 1996, The imaging performance of a storage phosphor system for dental radiography Br. J. Radiol. 69:256-261.
3. Efstathopoulos, E.P., Costaridou, L., Kocsis, O., and Panayiotakis, G., 2001, A protocol based evaluation of medical image digitizers, Br. J. Radiol, 74:841-846.
4. Fryback, D. G., and Thornbury, J.R., 1991, The efficacy of diagnostic imaging. Med. Decis. Making, 11:88-94.
5. Giger, M.L., and Doi, K., 1984, Investigation of basic imaging properties in digital radiography. 1. Modulation transfer function, Med.Phys.11:287-295.
6. Giger, M.L., Doi, K., and Metz C.E., 1984, Investigation of basic imaging properties in digital radiography. 2. Noise Wiener spectrum, Med. Phys. 11:287-295.
7. Green D.M., and Swets, J.A., 1966, Signal Detection Theory and Psychophysics, Wiley, New York, p.60.
8. International Commission on Radiation Units and Measurements, 1996, Report 54. Medical Imaging – The Assessment of Image Quality, ICRU, Washington, pp. 50-52.
9. MacMahon, H., Metz, C.E., Do,i K., Kim, T., Giger, M.L., and Chan, H.P., 1988, Digital chest radiography: effect on diagnostic accuracy of hard copy, conventional video, and reversed gray scale video display formats, Radiol. 168:669-73.
10. Metz, C.E., 1986, ROC methodology in radiologic imaging, Invest. Radiol. 21: 720-733.
11. Van Trees, H.L., 1968, Detection, Estimation, and Modulation Theory. Part 1: Detection, Estimation, and Linear Modulation Theory, Wiley, New York.
12. Yip, M., Pang, S.Y., Yim, W.S., and Kwok, C.S., 2001, ROC curve analysis of lesion detectability on phantoms: comparison of digital spot mammography with conventional spot mammography. Br. J. Radiol.74: 621-628.

ADVANCES IN PET AND SPECT

ANDREW TODD-POKROPEK*
University College London, Department of Medical Physics and Bioengineering, Gower Street, London WC1 6BT, UK

Abstract. There has been considerable progress in nuclear medicine instrumentation for both Positron Emission Tomography (PET) and Single Photon Emission Computerised Tomography (SPECT). The purpose of this article is to indicate areas of advance and progress and to make some predictions of what nuclear medicine acquisition and processing systems might look like in future. This subject includes new detectors and detectors systems, progress in tomographic reconstruction, in the correction of errors associated with the data such as uniformity, scatter, attenuation, motion and the partial volume effect.

Keywords: PET; SPECT; tomographic reconstruction; attenuation correction; scatter correction; partial volume effect; PET/CT; scintillators; solid state detectors.

1. Introduction

Historically the acquisition of images in nuclear medicine was performed with scanning devices (rectilinear scanners) and there afterwards with the so-called Anger gamma camera using NaI(Tl) scintillators[1], in the form of a large slab detector, an array of Photo-Multiplier tubes (PMTs) and electronics on one side, and a collimator on the imaging side. Positron images devices were developed shortly afterwards[2], also using conventional scintillators, and where typically arranges as a series of detectors arranged in a ring around the object to be imaged, PMTs etc on the outside, and very limited collimation (to limit the slice) on the inside.

Thus Single Photon Emission Computerised Tomography (SPECT or alternatively known as SPET) was initially developed in the 1960s[3], but was not widely used in clinical practice until the 1980s. It is now considered to be the standard method of acquisition of nuclear medicine image data, however increasingly in competition with Positron Emission Tomography (PET). Fig 1 illustrates a conventional SPECT system with a gamma camera rotating about an object (patient). The obvious advantage of

*Andrew Todd-Pokropek, University College London, Department of Medical Physics and Bioengineering, Gower Street, London WC1 6BT, UK; e-mail: A.todd@ucl.ac.uk.

Y. Lemoigne et al. (eds.), Physics for Medical Imaging Applications, 321–340.

SPET over conventional planar (or projection) images is the same as that of X-ray CT over conventional projection radiography, the important gain in contrast of small features. As an illustration, a lesion with an uptake to background ratio of 2:1 should show up on a SPECT images as twice a hot as the surrounding background, whereas in a planar images might show only a small (e.g. 2%) change in apparent uptake depending on the size of the lesion and its position in depth. Thus it is possible to gain an increase on contrast of about an order of magnitude. It is also hypothesised that in addition, the ability to quantitate the results should be considerably easier. The conventional SPET acquisition system is that of a rotating gamma camera where the detector performs a planar acquisition at a number of angles around the patient.

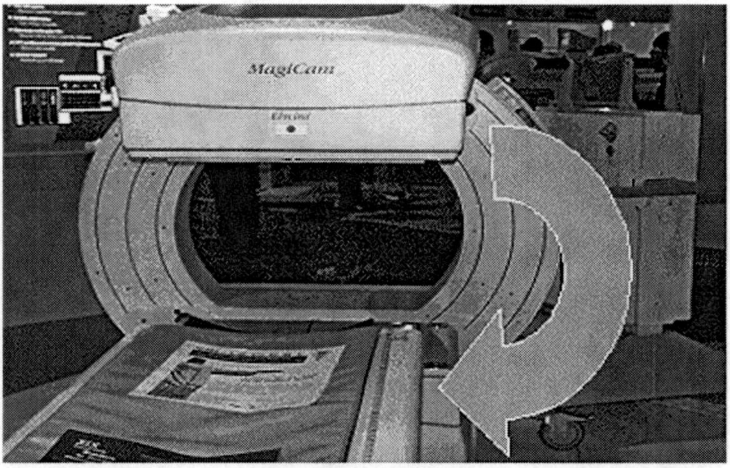

Figure 1. A standard rotating camera SPECT system.

Modern systems have evolved considerably, in the detector technology which is employed, in the electronics, and in the reconstruction algorithm and in the ensuring analysis and processing (computer) systems. One particular evolution is that of hybrid systems such as PET/CT4 and SPECT/CT5 where the X-ray Computerised Tomography (CT) component is used for both attenuation correction and additionally for image fusion to merge the anatomical (CT) data with the functional PET/SPECT images. Fig 2 shows an example of a current PET/CT system (basically a block with a cylindrical hole for the patient) and is very similar to comparable CT and MRI scanners). The key difference between SPECT and SPECT systems is that in single photon systems a collimator is required to eliminate photons not on a direct line between the emission source and the detector (normally perpendicular to the detector, whereas for a PET system coincident gammas are emitted along a so-called line of response (LOR) where as a result the collimation can be essentially omitted. Both types of system require the use of tomographic reconstruction algorithms.

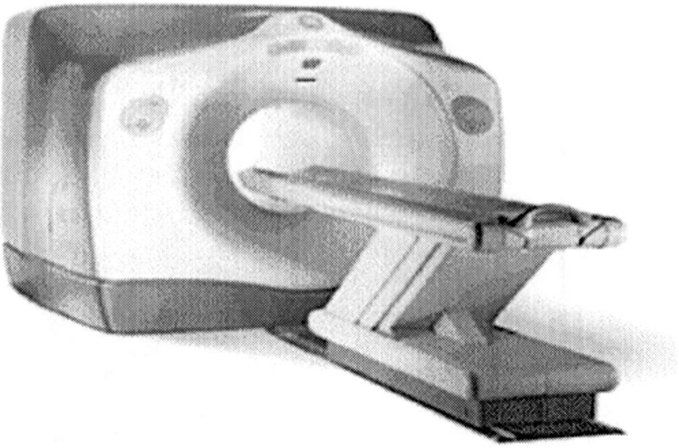

Figure 2. A standard PET/CT scanner.

2. Tomographic reconstruction

Tomographic reconstruction[6] is a classic inverse problem. We are trying to obtain values of the distribution of activity inside an object (the patient) which is an image **I,** from a set of observation of detected events in the form of a set of projections **P** from various positions and angles about the object. These can be expressed in the form of a matrix equation:

$$P = F.I$$

Where **F** is the so-called forward or system matrix indicating how each point in the image activity distribution can give rise to events detected on the recorded projection. Note that for convenience, the image is normally expressed as a vector as are the projections, such that the forward matrix **F** is two dimensional. The size of **F** can also be rather large. In an ideal situation however it is very sparse, that is, if we assume that the acquisition is such that each projection value is the line integral of a projection ray passing through the object, it will be both very sparse and only contain binary values, ones where the projection rays pass and zeros elsewhere. This is illustrated in Fig 3 which shows and ideal projection ray and an indication of a much larger system response function where data are acquired from a sensitive volume. For this reason a simple Fourier based decomposition may be performed, which results in the well known Filtered Backprojection (FBP) reconstruction technique.

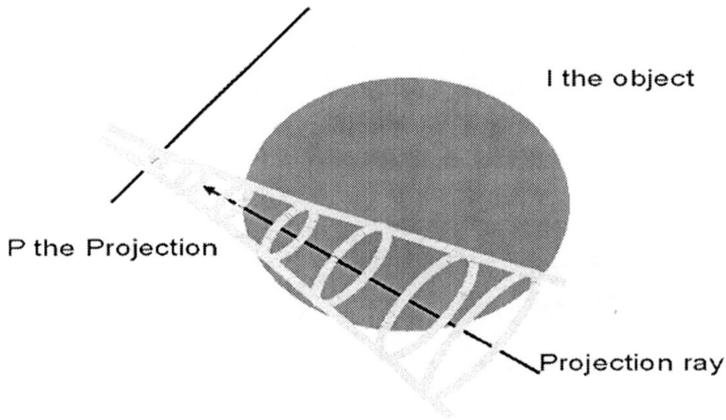

Figure 3. The relationship between P the projection and I the image to be reconstructed where the projection line indicates and ideal linear integral and the 3D shape indicated the more realistic region of response (the system response function).

In reality, in nuclear medicine, this simplification is not true[7]. Where scatter is present, and as a result of the finite system of the system response function, the forward matrix is neither sparse (nor binary). Thus alternative and improved reconstruction schemes are required, basically by the use of iterative algorithms[8], which can employ a more completely expressed form of the system response function and hence for the complete forward matrix. In addition, regularization to impose additional constraints to the solution can be very helpful. A typical form of regularization is to minimise the distance between observations **P** and the corresponding values generated from the suggested solution (**I**) by using the forward model (**F**) while at the same time minimising some value obtained by applying a regularising operator (**R**) to that solution, for example an estimate of smoothness.
Thus where

$$\mathbf{P} = \mathbf{F}\,\mathbf{I}', \text{ then we search for an } \mathbf{I}' \text{ to satisfy}$$

$$\min\{\ \|\mathbf{P} - \mathbf{F}.\mathbf{I}'\| + \mathbf{R}[\mathbf{I}']\ \}$$

Where a common form for the operator **R[ ]** is to look at a derivative, for example the first of second, and an L2 norm (least squares distance) is used. An additional development discussed below is in list mode reconstruction[9].

However the form of the forward matrix **F** needs to be well expressed and to be (reasonably) complete. This requires understanding the various corrections which need to be applied, for both SPECT and for PET.

3. Corrections required from a physical point of view

3.1. SPECT CORRECTIONS

In order to perform good SPECT acquisitions, a number of corrections must be performed, of which the first is that of uniformity. Any variation in the sensitivity of a detector system rotating about an object results in the presence of circular artefacts, as illustrated in Fig 4.

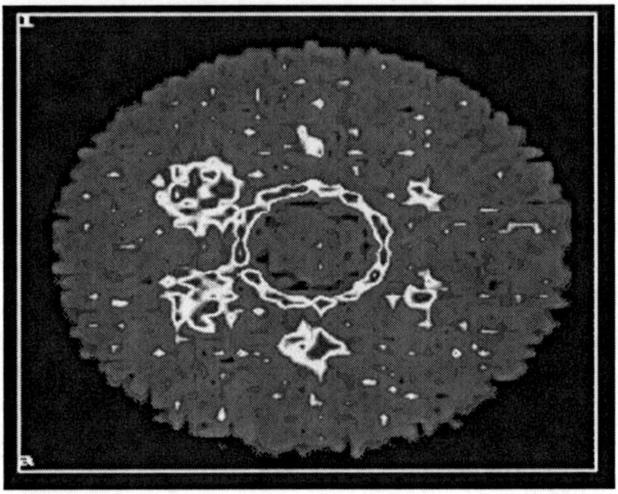

Figure 4. The effect of non-uniformity on a tomographic scan. The ring is an artefact caused by non-uniformity of the detector.

It is important to recall that variations in planar uniformity are amplified by the SPET process, such that the intensity of the circular artefact is dependent in particular on the distance of the area of planar non-uniformity from the centre of the field of view (which should correspond to the axis of rotation). This amplification can again be of about an order in magnitude that is 3% variations in planar uniformity can create circular artefacts of the order of 30% in amplitude[10]. It is for this reason that careful quality assurance of SPET detectors is required and it is not recommended to use detectors which have a planar non-uniformity of greater that 3%. To achieve such performance careful correction of spatial distortion, energy correction and residual planar uniformity need to be performed as described elsewhere. The improvement in gamma camera technology has enabled suitably uniform detectors for which the problems of tomographic non-uniformity (circular artefacts) has been reduced if not entirely eliminated. Such detectors have also become more stable requiring less frequent adjustment, although daily testing is still recommended.

A further obvious correction in the estimating of the centre of rotation and establishing correction factors for any error in motion of the detector(s) since any error in the perceived and real axis of rotation will lead rapidly to loss of resolution and eventually to the creation of artefacts. This subject is also well covered in the literature[11].

These difficulties having been resolved this leaves the following problems:

- Lack of sensitivity.
- Lack of resolution.
- Motion correction.
- Decay correction.
- The need for scatter correction and attenuation correction.
- The need for corrections of variations of spatial resolution.
- The need for partial volume correction.
- Model fitting and the extraction of physiological data.

3.2. PET CORRECTIONS

Corrections required in PET are similar with certain important differences and additions[12], which is also a good general reference). A uniformity of sensitivity correction is required, which is normally obtained by a so called blank acquisition (or a source without an object being imaged) where the relative sensitivities of each detector are measured, the results stored and then used as a straightforward multiplicative correction matrix. Scatter, Attenuation[13, 14] and Partial Volume correction[15, 16] are required as will be described. A key difference however between SPECT and PET is that the attenuation correction does not depend on the depth of the emission source and is independent of position along any line of response (LOR) or the PET system and so can be estimated by using an external emission source. However with the use of CT data to establish attenuation maps this is now long longer so important an advantage.

An additional important correction required in PET is that for random coincidences. Where a coincidence window with a certain time acceptance is used, random events can be detected as if they correspond to a true coincidence, a phenomenon which is linearly proportional to the width of the coincidence time window. A simple and reasonable solution is to use a second delayed coincidence window (where all observed coincidences must be randoms) and to use the count rate in the delayed window to correct to the coincident counts observed in the true coincidence window. Figure 5 indicates the different lines of acquisition for true, random and scattered events.

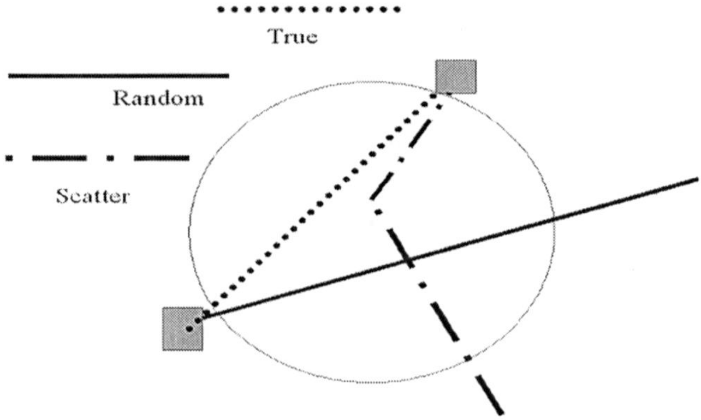

Figure 5. True, random and scattered events in a PET ring system.

In PET scatter is a serious problem. The simplest method to correct for scattered events is to use a second lower energy window, to use this to estimate the number of scattered events, and to correct for the coincidences observed in the upper window. This of course is only an approximation, in particular for the reason that the true coincident activity distribution and the scatter distribution are normally distinct and different.

3.3. ATTENUATION CORRECTING USING CT DATA

Whereas previously attenuation maps in both SPECT and PET where either estimated from the (boundaries) of the emission maps, or by using transmission maps obtained from external emission sources, the most popular current technique is to establish the transmission map for attenuation correction directly from CT data acquired on the hybrid PET and SPECT/CT scanner. A CT slab of data us acquired either before or after the emission scan without moving the patient. The time for acquisition is short (a few seconds) and with much better signal to noise ratios than for emission source transmission data. However the effective energy of the CT data is different being typically of the order of 70KeV average, and must be transformed in attenuation coefficients for the corresponding emission energy 140KeV for $^{99}Tc^{m}$ and 511Kev for PET[17]. The normal procedure is to establish a mapping function from Hounsfield Units (HUs) to values of the attenuation coefficient going from HU=-1000 up to about HU=1000. Normally a bilinear function is used for HU < or > zero. This of course is only an approximation. An alternative is to segment the CT image into tissue types (muscle fat air bone etc) and to use known values of the attenuation coefficient thereafter. An additional very significant problem is that the CT data does not correspond exactly to the emission data, in particular with respect to regions close to the diaphragm where breathing artefacts are common and can be significant.

An important caveat is that in emission scan attenuation correction, better results are normally obtained using the narrow beam attenuation coefficient (that values observed when both the source and detector are collimated and scatter is eliminated). This implies that the data are scatter corrected before attenuation correction is performed. Alternatively both attenuation and scatter information must be incorporated into the Forward matrix definition of the tomographic reconstruction method being employed. One very suitable method which can be used for this purpose is use to use Monte Carlo simulation method, a problem being that these are currently extremely slow.

4. New detectors

Some data about a selection of scintillators is included in the Table 1 below[18]:

TABLE 1. Selection of scintillators.

Material	Density	Light Output	Decay Time
NaI	3.67	40700	230
BaF	4.89	2000	0.7
BGO	7.1	5500	300
LSO	7.4	23600	40
GSO	6.71	~12000	60
YAP	5.58	14400	28
CsI	4.51	51000	1000
ZnSe	5.42	80000	3 - 70
Density is expressed as gm/cc, light output is given in photons/Mev, decay time in nanoseconds.			

Some data about solid state detectors (courtesy of H.Barrett[18, 19]) are included in the table 2 below:

TABLE 2. Some data about solid state detectors where *ehp* indicates number of electron hole pairs.

Material	*ehp/KeV*	Density (gm/cc)	$\mu\tau$ (cm^2/V)
Si	280	2.33	0.4 / 0.2
Ge	350	5.33	0.8 / 0.8
InP	240	4.79	10^{-5} / 10^{-5}
SI-GaAs	230	5.32	10^{-5} / 10^{-6}
CdTe	225	5.85	10^{-3} / 10^{-4}
CdZn0.2Te	200	6.0	10^{-3} / 10^{-4}
HgI2	240	6.4	10^{-4} / 10^{-6}
TlBr	170	7.5	10^{-4} / 10^{-5}
Diamond 80	3.51	11.85	10^{-6} / 10^{-6}

It is clear that many new detector materials are currently available. In general the best detectors have good stopping power (high density) good sensitivity (high light or electron hole pairs output) and are fast (in particular for PET applications. Thus NaI(Tl) for PET has largely been replaced firstly by BGO but more recently by LSO and GSO (and potentially YAP). SPECT is still largely dominated by the use of the conventional NaI(Tl) although some solid state pixilated detector arrays are commercially available. LSO is unfortunately unsuitable for SPECT since it is itself radioactive and generates an unsuitably large background count rate, but which a single photon events, is not a problem in PET.

The standard PET architecture involves rings of discrete detectors about the imaging volume. Resolution at the centre of the field of view is approximately equal to half the size of the detector cross-section, which is typically much less than the depth of the detector which should be long to improves sensitivity. Thus the ideal detectors (in addition to the above) should have a small cross section and pixelated arrays of detectors are the current solution. However, it is required to determine not only in which detector (which pixel of the pixelated detectors) the event was detected, but also the energy deposited and the time at which the event was detected. Traditionally this required a separate channel for each detector which for modern system with very large numbers of detectors (pixels) would be prohibitively expensive in terms of the electronics required. Thus hybrid systems have been developed using solid state light sensitive detectors for the scintillating crystals- Avalanche Photo Diodes (APDs) which are used to determine the pixel in which an event occurred, and more conventional PMT type systems for blocks of detectors to determine energy and timing for a group or block of individual detectors[11]. Figure 6 shows a diagram of a conventional block detector design as used in PET combining PMTs and APDs. Figure 7 shows as photo of a block APD,. Figure 8 illustrate how small current crystals and be while Fig. 9 shows how the electronic components (ASICs) are also being miniaturized.

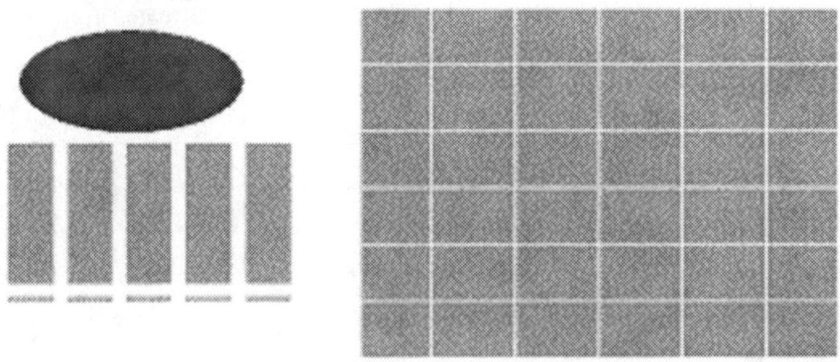

Figure 6. Diagram of a multi-channel PET block detector, side view left, top view right, where the circle (left) indicates a PMT and the small detectors at the bottom of the pixelated block detector shows the position of the APDs. Depth of Interaction Correction AND PHOSWich detectors.

Figure 7. Photo of a Hamamatsu Avalanche PhotoDiode (APD) detector.

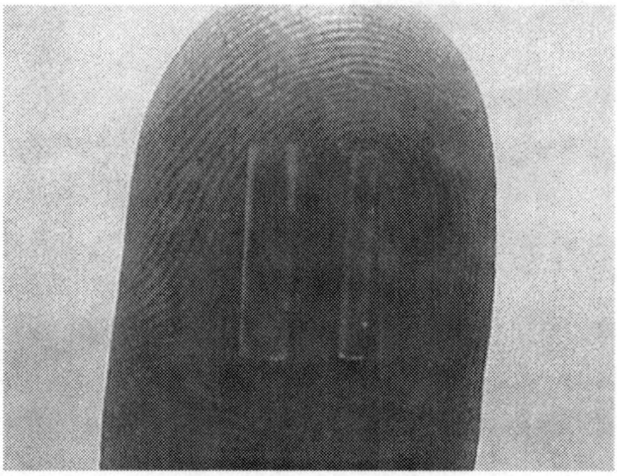

Figure 8. Example of some small crystals for use in new high performance PET systems.

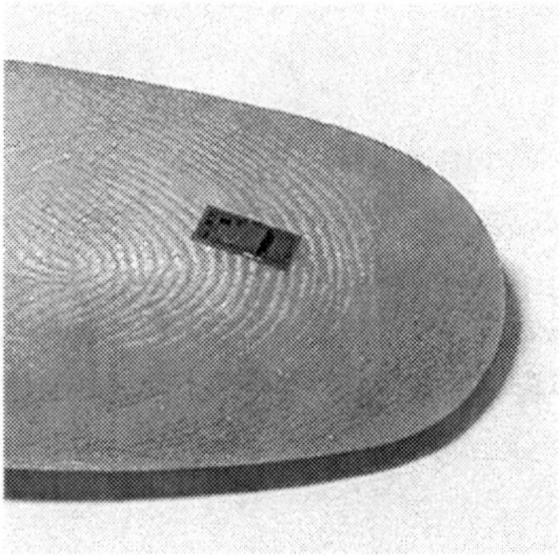

Figure 9. Example of an ASCI pre-multiplier to indicate the change of scale of Current micro-electrons.

In many cases it is of interest to determine the depth of interaction within a detector where an event was detected (or at least an estimate)[20]. Many schemes have been investigated. One such scheme, used with gamma camera (slab scintillating detectors) looks at the shape of the light output distribution which changes as a function of depth of interaction, but unfortunately is not very sensitive to this distribution. An alternative has been trials with dual thin crystals bonded together where an estimate of light output from each crystal is collected from the side of each crystal, the ratio of which is used to give the depth of interaction, whereas conventional Anger type electronics are used end-on, to give position and energy information. With discrete (pixelated) detectors, the light output from each end of the crystal can be collected (using suitable light guides) and the ratio can be used to estimate depth of interaction. The use of depth of interaction information in PET rings can be of particular value for events which are detected significantly off the axis of the crystal. Figure 10 indicated how the depth of interaction varies as a function of the position of the source within the detector ring.

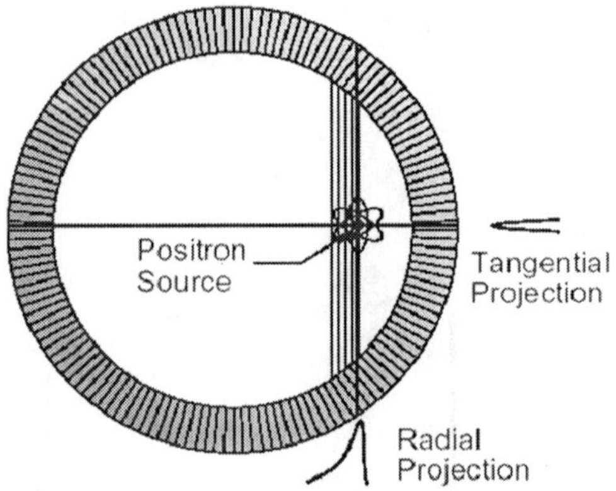

Figure 10. Illustration of the depth of interaction problem in a ring PET detector.

A related approach and more straightforward approach has been the use of crystals formed by bonding together multiple scintillators with different light emission decay times. Thus an analysis of the decay time can indicate in which crystal the event was detected. Typical combinations have been the use of LSO (40 nsec decay) with GSO (60 nsec decay). More complex multiple scintillators are being investigated for the purpose of developing hybrid SPECT/PET/CT detectors which rather than having separate sub-systems, the detector array is common for all modalities.

5. Small animal scanners

A sub-specialty in the development of nuclear medicine instrumentation has been that of systems for imaging small animals normally mice and rats[21]. Here there is a need for resolution of the order of a millimetre or better, not normally achieved with systems designed for adults. In PET there is an immediate benefit with respect to resolution when designing such systems in the effect of the non-colinearity of the pair of coincident gammas is reduced. The loss of resolution due to this effect is equal to (approx) 0.0022 D where D is the diameter of the detector ring. Thus where D is of the order of 10 cms this effect causes a blurring of only ~.2 mm. However in order to achieve the desired high resolution, very small crystal arrays are required using the approach previously described of blocks of pixelated detectors using APDs and at the other end PMT to establish overall position, energy and timing information. Many of the key development in this technology have been driven by the advances in small animal scanner instruments.

In SPECT however a collimator is still required, and this is still a considerable limitation with respect to resolution. One technique of interest has been the re-use of pinhole and multiple pinhole tomography, and other variations of coded aperture acquisitions[22]. A typical system acquires data with one or multiple such coded aperture detectors. Pinholes have the great advantage of geometrically magnifying the image of the activity distribution on the detector, and thus can obtain much higher resolution than conventional parallel hole, fan beam or cone beam collimators. The disadvantage is that they lack sensitivity as a result of poor solid angle. A further disadvantage of multiple pinhole and other coded aperture detectors is that the images need to be 'decoded' in order to be reconstructed (for example the images from individual pinholes will overlap) and therefore there is effectively a loss of sensitivity in terms of noise equivalent counts, when the data are reconstructed. However the amount of the injected dose to the animal and duration of acquisition time is less of a problem for a small animal than for an adults, and spectacular results have been obtained in a number of prototype systems.

6. Other advances

6.1. TIME OF FLIGHT PET

For the two annihilation gammas produced by a positron event, it is possible to estimate at what point on the line of response (LoR) the annihilation occurred by timing the difference in arrival times of the gamma to the two detectors[23]. I we had a temporal resolution of 0.1 nsec this would result in a spatial resolution along the LoR of 1.5 cm. Correspondingly a temporal resolution of 1nsec would give a spatial resolution of 15 cm. While neither of these would be enough to give adequate resolution for a PET system by itself, it can be used to improve the scatter rejection, and also as an additional component in the reconstruction algorithm where, instead of assuming uniform probability of the annihilation along the entire LoR, the known probability distribution from the time of flight information can be used to improve the signal to noise ratio (and perhaps the resolution itself). Clearly a sub-nano second temporal resolution would be required to be effective. This also requires a fast detector.

Time of flight systems have been tested for some time, and basically failed. While the required temporal resolution could be achieve, the electronics required were expensive and unreliable. However with current developments in electronics, this should now be achievable. Traditionally the detectors used for time of light systems was BaF (not a very convenient crystal mechanically), but recent developments using $LaBr_3$ crystals look very exciting.

6.2. COMPTON CAMERAS

Compton cameras have been under development for many years, but recent progress has been achieved. The basic idea for a Compton camera is to use two detectors, without collimation. In the first front detector, a Compton event occurs, while in the second rear detector the scattered gamma is stopped[24]. This is illustrated in Fig. 11. Thus we have the position of a scatter in the front detector, the position of the detection in the rear detector, and an estimate of energy loss between the two. This energy information can be used to define a cone the point of which is at the position of the scatter event, from which we know that the original event must have originated. This information can then used to reconstruct the original activity distribution. The sensitivity is potentially very high since no collimator is used. There is of course some loss of noise equivalent count rate due to the reconstruction process. Current developments have been based on the use of pixelated solid state detectors where the high energy resolution is an advantage in defining the shape (thickness) of the cone. It is also worth noting that performance improves as energy increases and while the system does not require coincident (positron) events in order to eliminate the collimator, is can be used at positron (511KeV) energies.

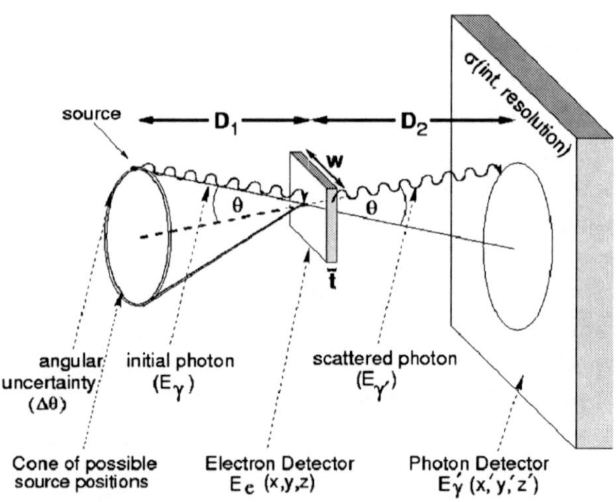

Figure 11. Diagram of how a Compton camera functions.

A theoretical development of the two layer Compton camera is the so-called imaging cube. For such a device, the time, position and energy of a series of events within a single detector would be measures, for example the first Compton scatter event would be measured, then any succeeding similar events, and the final photoelectric event. Thus in a similar manner the cone of initial event emission would be estimated and the activity distribution reconstructed in 3D. The field of view would be excellent and should be the sensitivity. Experiments have been performed with CsZnTe crystals. However the requirements for such a system to function are approximately a temporal resolution of say 1nsec, a spatial resolution of say 0.1mm of better in 3D, and an energy resolution of say 1KeV. Unfortunately, such a device cannot as yet be constructed. However with improvements in solid state detectors and electronics, who can say that such device in not feasible in the future?

6.3. 3D VOLUME, LIST MODE RECONTRUCTION AND OTHER IMAGE PROCESSING

Conventional tomographic reconstruction had been implemented by acquiring a series of projections about the object to be reconstructed. These are then, when required, sorted into a slice by slice data array (a sinogram) and then reconstructed, traditionally by Filtered BackProjection (FBP), and currently using (mostly) iterative algorithms such as Maximum Likelihood Expectation Maximisation (MLEM) and its faster variant, Ordered Subset Expectation Maximisation (OSEM)[8, 11]. These iterative algorithms have the advantage of being much less sensitive to the production of streak artefacts (always present in FBP) and normally produce better signal to noise ratios for a given reconstruction resolution. There are many other variants of such iterative algorithms which will not be discussed as such here.

A major change in PET resulted from acquisition and reconstruction of volumes rather than slices[25]. The original systems have lead slits between rings, to avoid by and large cross-talk between slices. For volume (3D) acquisition, these were removed. Coincidences between within a given slice and between slices were acquired. Thus data is fully sampled (in angle) perpendicular the central (longitudinal) axis, but limited angle sampling occurs in the other direction (along the central axis). The is also a considerable variation is sensitivity with respect to position, in particular between the central slice, and a slice at the edge of the field of view. If true 3D reconstruction is performed, this must be taken into account. Thus a series of fully 3D reconstruction algorithms have been developed, which achieve similar resolution to slice by slice reconstruction, but with improved sensitivity. Clinical evidence suggests that acquisition times can be reduced by a factor of two, achieving the same image quality. However, although volume acquisition achieves better sensitivity, it is also much more sensitive to activity outside the filed of view, and hence to scatter. Thus where substantial activity is present outside the field of view, reinsertion of the slice slits and slice by slice reconstruction may still be helpful. Figure 12 shows a diagram of the differences between a 2D and 3D PET system and indicated the additional sensitivity of the 3D system and its vulnerability to scatter.

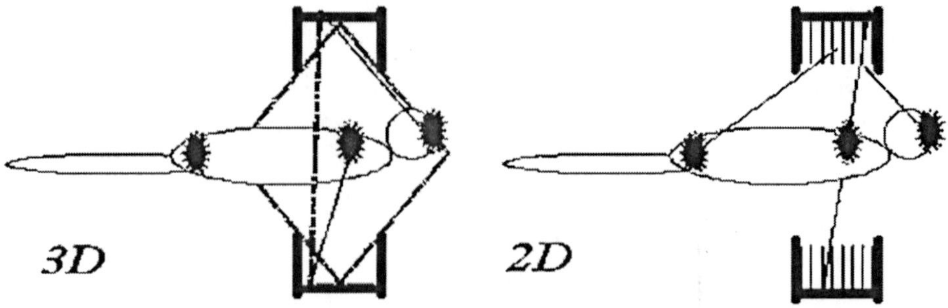

Figure 12. The difference between a 2D scanner system with slits (right) as contrasted with a 3D volume acquisition system (left).

When the total number of Lines of Response (LoRs) for a volume acquisition are estimated, it is observed that they may considerably exceed the number of events acquired. Since in principle the number of LoRs determines the input matrix for the reconstruction algorithm, event by event (list mode reconstruction) has been employed with considerable success. A number of corrections are easier to apply in list mode reconstruction, which has considerable additional flexibility[9].

In most cases, after reconstruction, regions of interest (ROIs) are defined, time activity curves (TACs) are generated, and after normalisation to produce activity (rather than count rate) estimates, a physiological model is fitted to determine physiological parameters such a blood flow, GFR, etc. An alternative approach being investigated is to avoid the generation of ROIs and TACs as such and to use the projection data (or the list mode data) to fit the models directly, without going through a reconstruction process as such. Temporal information for such an approach is better than for a conventional reconstruction where the time duration for the events included in the reconstruction must be sufficient for the resulting image to have an appropriate signal to noise ratio. A related topic is the of Dynamic SPECT reconstruction where instead of a single static image being reconstructed, a 4D (3D plus time) data set is generated on which similar physiological modelling processes can be performed. The data are acquired for example by a rotating gamma camera system, which can be performed at the rate of once per minute on many current systems.

Whole body acquisitions are also important in particular in PET. At present, in many cases these performed block by block, which are then fitted together. An alternative and almost certainly better technique (if the reconstruction can be performed) would be to acquire data as for a spiral CT system, translating the patient continuously though the ring.

Partial Volume (PV) correction is also an important additional correction of PET, SPECT and in reality all other similar modalities. It is observed that, as the size of the region where activity uptake has occurred becomes smaller with respect to the image resolution of the system, so the apparent uptake appears to be reduced[15, 16]. The from in which uptake is normally estimated in nuclear medicine images is expressed as the Specific Uptake Value (SUV), which is the ratio of uptake within an object, for example a lesion, and the surrounding so-called background activity. The SUV is used for example for tumours to indicate their metabolic activity and to assist in for example tumour staging. However at the size of the tumour decreases so does the apparent SUV decrease, staring from tumours or objects about twice the resolution of the system and becoming much worse when they are smaller than the resolution itself[26]. A common system to correct for this is to calculate a recovery coefficient which is just a value to multiple the SUV as a function of the size of the lesion. Lesions are therefore assumed to be spherical (or circular). A true PV correction would be some kind of deconvolution to re-sharpen the image to perform this correction for any shape or size of object. Two problems are encountered. Firstly the resolution and hence the correction required is not constant throughout the image[27]. Secondly such an enhancement process has bad noise amplification properties.

A second and related issue is when more than one tissue type is present in a pixel / voxel or region. This is more common on the edge of organs, tissues or objects. Fuzzy segmentation is one method used to address this where the voxel has a fuzzy membership value μ to each of several classes ($0<\mu<1$). An alternative is to segment the image into a number of regions **R**, from which we wish to estimate the true activities **T** for each region. We can then establish a straightforward matrix relationship **R=M.T** where **M** is the so-called mixture matrix.

Finally, at present by far the most common method of PV correction is to register the emission data with information from a higher resolution modality (for example CT or MRI) and to use the information from the high resolution segmentation to perform the PV correction or enhancement. There is a requirement for high accuracy of the registration in order to avoid the amplification and generation of errors using this method. Hybrid system can facilitate this process.

6.4. HYBRID SYSTEMS

The inclusion of a CT scanner together with the PET or SPECT system is now routine[5]. Very few PET systems are delivered currently which do not include a CT. While originally the CT component was low dose (obviously to reduce total dose to the patient) and single slice, many new systems are fully functional multi-slice CT systems (64 slices are available) and CT image quality is as for stand alone CT scanner. SPECT/CT systems are not as common at present probably for cost reasons. Such multi-slice PET/CT scanners have an obvious application in cardiac studies. However there are still significant problems associate with motion (respiratory, cardiac etc) and with handling dynamic information such as contrast injection for the CT data.

Other hybrid devices are being explored of which possibly the most interesting is PET/MRI[28]. Here the PET detectors are inserted within a (conventional) MRI system, and several prototypes of small animal scanners are being tested. The obvious and major problem is that components such as PMTs do not work in a high magnetic field. The solution to this problem has been the use of optical fibres to extract the light from scintillators away from the MRI system and to have the electronics outside and shielded from the magnetic field. Another interesting development (at present small animal scanners) is the use of hybrid optical (fluorescent). Software fusion of data from separate MR, CT and PET scanners is a very interesting alternative[29]. Figure 13 shows the (so-called) image of the year (SNM 2006) of a SPECT/CT image fusion, while Fig. 14 illustrates and example of image fusion between CT and PET in the liver, a striking illustration of the additional information conveyed by a functional imaging modality to that from anatomical images as obtained by CT.

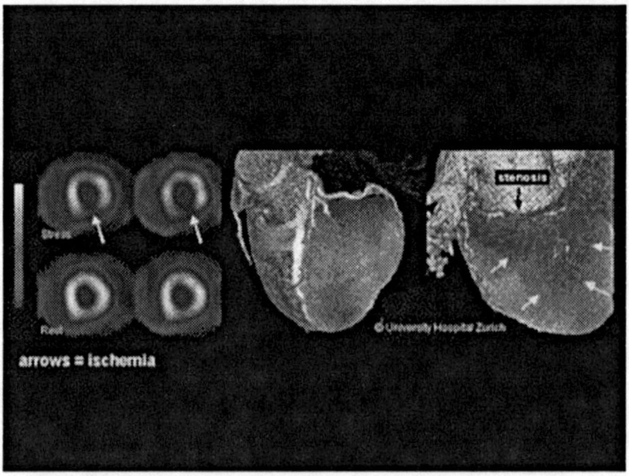

Figure 13. SPECT/CT Society of Nuclear Medicine image of the year (Courtesy JNM/, Von Schulthess).

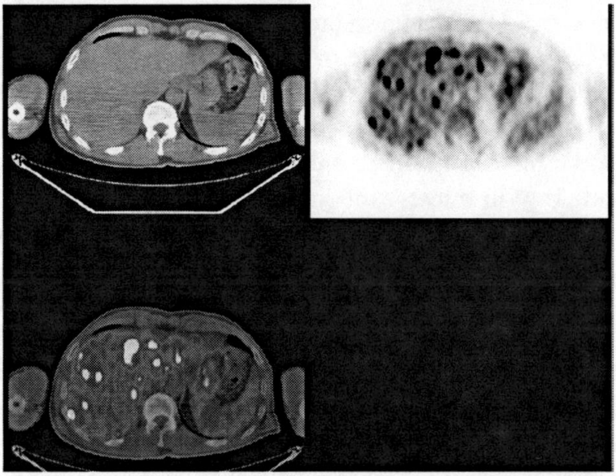

Figure 14. An example of a fusion of PET/CT (Courtesy of: Liselotte Højgaard, MD DMSc, Annika Eigtved, MD Ph.d., Anne Kiil Berthelsen, MD. PET & Cyclotron Unit, Dept. Nuclear Medicine, Rigshospitalet, University of Copenhagen).

7. Summary and Conclusions

A brief overview of some of the developments of nuclear medicine instrumentation, including processing has been given. While PET has been evolving rather rapidly, it would appear the conventional Anger type detectors still provide the work horse for much nuclear medicine imaging. However some advances have been made, and energy resolution in particular is considerably better is current detectors than in the past. Systems are still limited by collimators.

To quote Harry Barrett[19], in three years time we will have:

- Defect-free, high-resistivity CZT crystals of large area (25-30 cm^2)
- Shaper-amp ASICs with on-chip clock and bias
- Reliable bonding methods
- Affordable 128×128 or 256×256 arrays with good photopeak efficiency
- Routine use of semiconductor arrays in small-animal imaging and specialized clinical applications
- Demonstrated NanoSPECT (100 μm linear resolution or 1 nanolitre volume resolution) for mice and rats
- Methodology for systematic task-based optimization of image quality in semiconductor systems

The long term aim is for detectors arrays with 10 μm linear resolution and picolitre volume resolution. Some of the facilitators are already available: better solid state detectors, ASICs and micro-electronics components, high performance computing. Image registration and fusion with other modalities should also continue to enhance the value of the images obtained by tracer technology in nuclear medicine. However underlying the value of and the advances achievable in nuclear medicine is the continuing interest in the use of tracers, where the sensitivity of nuclear medicine imaging systems in terms of the molecular concentration of such tracers is orders higher than most competing modalities, and in the development of new tracers (radiopharmaceuticals) and in the inclusion of such methods within the expanding knowledge of genomics, proteomics and in general in molecular imaging.

References

1. H.O. Anger, Radioisotope camera, In G.J. Hine (ed), Instrumentation in nuclear medicine, Vol 1, 485-552 (Academic Press, New York, 1967).
2. M.M. Ter-Pogossian, M.E. Phellps, E.J. Hoffmann, and N.A. Mullani, A positron-emission transaxial tomography for nuclear imaging (PETT), *Radiology*, 114, 89-98 (1975).
3. D.E. Kuhl, R.Q. Edwards, Cylindrical and section radioisotope scanning of the liver and brain. *Radiology* 83, 926-935, (1964).
4. T. Bayer, D.W. Townsend, T. Brun, et al, A combined PET/CT scanner for clinical oncology, *J. Nucl. Med.* 41, 1369-1379 (2000).
5. G.K. Von Schulthess, Clinical Molecular Anatomic Imaging, PET, PET/CT and SPECT/CT, (Lippincott, Philadelphia, 2003).
6. H.H. Barrett, and W. Swindell, Radiological imaging, the theory of image formation, detection and processing, (Academic Press, New York, 1981).
7. A.Todd-Pokropek, Advances in computers and image processing with applications in nuclear medicine, *Quarterly J. of Nucl. Med.*, 46, 62-69, (2002).
8. L.A. Shepp, and Y. Vardi, Maximum likelihood estimation for emission tomography, *IEEE Trans. Med. Imaging*, 1, 113-121, (1982).
9. C. Byrne, Likelihood maximization for list-mode emission tomographic image reconstruction, *IEEE Trans. Med. Imaging*, 20, 1084-1092, (2001).
10. NEMA NU1-2001, NU2-2001 www.nema.org.
11. A. Todd-Pokropek, Non-circular orbits for the reduction of uniformity artefacts in SPECT. *Phys. Med. Biol.*, 28, 309-313, (1983).
12. S. R. Cherry, J.A. Sorenson, and M.E. Phelps, Physics in Nuclear Medicine, 3rd Edition, (Saunders, Philadelphia, 2003).
13. G. Glatting, M. Wuchenaur, and S.N. Reske, Iterative reconstruction for attenuation correction in positron emission tomography: Maximum likelihood for transmission and blank scan, *Medical Physics* 26(9), 1838-1842, (1999).

14. C.M. Tonge, M. Manoharan, R.S. Lawson, R.A. Shield, and M.C. Prescot, Attenuation correction of myocardial SPECT studies using low resolution computed tomography images, *Nucl. Med. Commun.* 26(3): 231-237, (2005).
15. C.C. Meltzer, J.K. Zubieta, J.M. Links, P. Brakeman, M.J. Stumpf, and J.J. Frost, MR-based correction of brain PET measurements for heterogeneous gray matter radioactivity distribution. *J. Cereb. Blood Flow Metab.* 16:650-658, (1996).
16. O.G. Rousset, Y. Ma, and A.C. Evans. Correction for partial volume effects in PET: principle and validation, *J Nucl. Med.* 39(5), 904-11, (1998).
17. C. Burger, G. Goerres, S. Schoenes, A. Buck, A. H. R. Lonn, and G. K. von Schulthess, PET attenuation coefficients from CT images: experimental evaluation of the transformation of CT into PET 511-keV attenuation coefficients, *Eur. J. Nucl. Med.*, 29, 922-927, (2002).
18. H.H. Barrett, and L. Furenlid, State of the art in gamma-ray detectors, Short Course, IEEE Nuclear Science Symposium and Medical Imaging Conference, Rome (2004).
19. H.H. Barrett, and H. Bradford Barber, Semiconductor arrays for nuclear medicine: From materials to devices to systems. IEEE Medical Imaging Conference, Rome, Italy, on CD, (2004).
20. Y. Yang, P.A. Dokhale, R.W. Silverman, K.S. Shah, M.A. McClish, R. Farrell, G. Entine, and S.R. Cherry, Depth of interaction resolution measurements for a high resolution PET detector using position sensitive avalanche photodiodes, *Phys. Med. Biol.* 51(9):2131-42, (2006).
21. S.R. Cherry, Y. Shao, R.W. Silverman, A. Chatziioannou, K. Meadors , S. Siegel , A. Boutefnouchet, T. Farquhar, J. Young, W.F. Jones, D. Newport, C. Moyers, M. Andreaco, M. Paulus, D. Binkley, R. Nutt , M.E. Phelps ME., MicroPET: a high resolution PET scanner for imaging small animals, *IEEE Trans. on Nucl. Sci.* 1997; 44: 1161-1166, (1997).
22. N. Schramm, G. Ebel, U. Engeland, M. Behe, T. Schurrat, and T.M. Behr, Multi-pinhole SPECT for small animal research, *J. Nucl. Med.*, 43, 5, Suppl., 913, (2002).
23. T.K. Lewellen, Time of flight PET, *Semin. Nucl. Med.*, 28(3), 268-75, (1998).
24. M-G. Scannavini, R.D. Speller, G.J. Royle, I. Cullum, M. Raymond, G. Hall and G. Iles, Design of a small laboratory Compton camera for the imaging of Positron Emitters, *IEEE Trans. Nuclear Science*, 47, 1155-1162, (2000).
25. B. Bendriem, and D.W. Townsend, The theory and practice of 3D PET, (Kluwer, Dordrecht, 1998).
26. J. Qi, R.M. Leahy, A theoretical study of the contrast recovery and variance of MAP reconstructions from PET data, *IEEE Trans. Med.. Imaging*, 18, 293--305, (1999).
27. J.A. Fessler, and W.L. Rogers,Spatial resolution properties of the penalized-likelihood image reconstruction methods: Space invariant tomography, *IEEE Trans. Image Proc.* 5(9) 1346-1358, (1996).
28. P.K. Marsden, D. Strul, S.F. Keevil, S.C. Williams, and D. Cash. Simultaneous PET and NMR. *Br J Radiol.*, 75 Spec No:S35–S39, (2002).
29. J.V. Hajnal, D.G.L. Hill, and D.J. Hawkes, Medical Image Registration (CRC Press Boca Raton, 2001).

POSITRON EMISSION TOMOGRAPHY APPLICATION TO DRUG DEVELOPMENT AND RESEARCH

PIERO A. SALVADORI*
PET, Cyclotron and Radiopharmaceutical Chemistry Department, CNR Institute of Clinical Physiology, Via Moruzzi 1, 56124 Pisa, Italy

Abstract. The research for the identification and development of new drugs represents a very complex process implying long times and massive investments. This process was not able to parallel the rate of discoveries made in the field of genomic and molecular biology and a gap created between demand of new drugs and the ability of pharmaceutical companies to select good candidates. Positron Emission Tomography, among the different Molecular Imaging modalities, could represent a new tool for the early assessment and screening of new drug candidates and, due to its physical performances and the characteristics of positron-labeled tracers, gain the role of "Biomarker" accepted by the Companies and the Regulatory Bodies of Drug Agencies. To fulfil this task PET has to exploit all of its special features such as data absolute quantification and modelling, high spatial resolution and dynamic imaging. Relevant efforts need to be directed to the careful design and validation of experimental protocols with the main goal of achieving consistency in multi- centric trials.

Key words: positron emission tomography; drug development; molecular imaging; *in vivo* pharmacology.

1. Introduction

The concept of drug development has deeply changed over the time, clearly being a function of actual scientific knowledge and technological discoveries. The available know-how has impacted on this field, changing the tools and the research strategies, modifying therapeutic targets and healing concept as well. Although a long way has been made from serendipity through classical pharmacology to computational approaches and "*omics*", still the complexity of the multivariable process of pathogenesis and disease progression remains a challenge. The conflicting

* Piero A. Salvadori, PET, Cyclotron and Radiopharmaceutical Chemistry Department, CNR Institute of Clinical Physiology, via Moruzzi 1, 56124 Pisa, Italy; e-mail: Salvador@ifc.cnr.it.

Y. Lemoigne et al. (eds.), Physics for Medical Imaging Applications, 341–351.

relationship existing between the task of demonstrating drug safety and efficacy and Company revenues added further complexity to the problem.

The strong expectations from the use of high-throughput chemistry and screening coupled to computational approaches determined a quick alignment of drug research strategy in all major Pharmaceutical Companies, mostly based on the generation of a wide number of tentative active compounds and the *in silico* prediction of their biological activity [1]. The bottleneck of the process has emerged from the poor performances or behaviour of the tested compounds *in vivo*. Somewhere in the process the link was missed between the nature of the compound and the expected action in vivo. Interspecies differences and poorly characterised animal models strongly contribute to such a problem.

In other words, the need to revert attention to physiology was perceived, and the limits of a clinical trial system remained unchanged for decades, Figure 1.

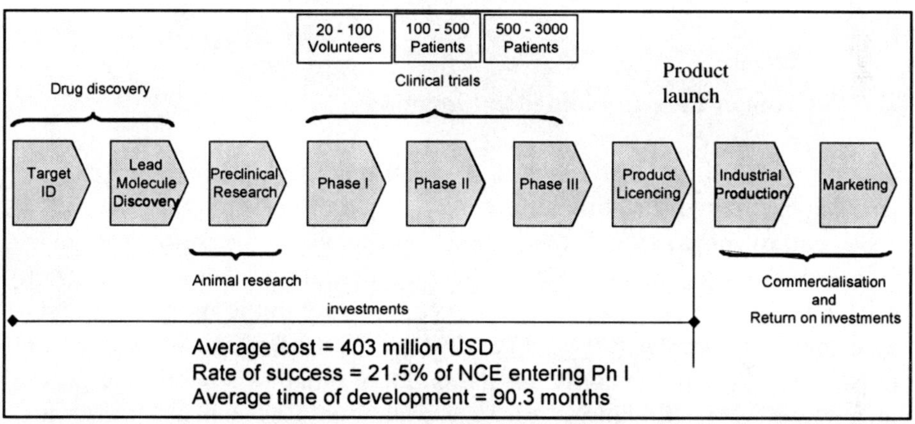

Figure 1. Drug development Pathway (economics and performance data from JA Di Masi et al., J. Health Economics, 2003).

Biomedical imaging represents the latest promising tool in DDR. This is strictly related to the ability demonstrated by imaging modalities of providing measurements of a wide range of *in vivo* parameters such as anatomical variations, changes in physiology and, particularly with those modalities belonging to Molecular Imaging (MI), biochemical processes.

Positron Emission Tomography (PET) may be considered at the leading edge of MI. In fact, the sound link between PET and *in vivo* biochemistry has emerged from over than thirty years of scientific papers, which have come to the general attention during the last decade as a consequence of the massive number of publications on the clinical use of PET as a diagnostic and therapy control tool in oncology.

The urgent need to improve the capacity of Companies and Health Authorities to screen New Chemical Entities (NCE) and the consequent search for new strategies triggered the discussion on the use on biomedical imaging and MI on DDR.

The hinging point of this new approach is on translational research: this delicate passage lying between preclinical phase of drug development, performed in experimental animals, and the "First time in man" (phase I). MI is expected to play a major and specific role on this issue which is starting to be referred to as "*phase 0*" or "*exploratory study*"[2]. The feeling of the possible impact of PET on DDR became stronger as the understanding grew among drug developers that therapeutic intervention had to address the earliest phases of the disease, and subjective (phenotypic) response to drugs should not be necessarily regarded as a failure but could be further explored under the perspective of "*personalised medicine*".

2. The role of PET in biomedical imaging

Without pretending to give a unique definition of Molecular Imaging, it is generally assumed that this term can be applied to a modality that relates molecular events to signals that can be represented in the form of an image. However, very often biomedical imaging modalities that are not strictly related to phenomena occurring at molecular level but to morphology, such as Computed Tomography (CT) and Magnetic Resonance Imaging (MRI) mostly reporting on anatomy and function, are included within the general meaning of MI. Modalities such as Magnetic Resonance Spectroscopy (MRS), Single Photon Emission Tomography (SPET) and PET have the prominent role in MI even though other imaging techniques are emerging (*e.g.* Optical Coherence Tomography, Fluorescence Imaging). The different performances and output data of imaging modalities, Table 1, have suggested their integrated use (multimodality imaging).

TABLE 1. Core information from different imaging modalities.

	US	CT	MRI	MRS	SPET	PET	PET/CT
Anatomy	+/-	++	+	-	-	-	++
Function	+	+/-	+	-	+/-	+/-	+
Biochemistry	-	-	+/-	+	+	++	++

For quite a long time all these modalities have been used, or their efficacy has been claimed, in the field of clinical therapy monitoring. Well known examples refer to the reduction of tumour mass during treatment ascertained by CT or imaging of cardiac function recovery by Echocardiography or cardiac MRI.

It is obvious to infer that these changes, which are usually related to the clinical evaluation of the patient, under a different approach could give partial answer or evidence to the question: does the drug have an effect? Useful results and data can be produced following such an approach. However, some further considerations should be made: it is well known that anatomical modifications occur as the final effect of a chain of events starting from biochemical aberrations: i.e. the elapsed time from disease (or therapy) onset to evident effect may be quite long.

Similar conclusions, even though with better temporal sensitivity, can also be reached evaluating functional parameters. Furthermore, only pharmacodynamics (drug-induced effect) is addressed and poor or no information can be obtained on pharmacological parameters (Absorption, Distribution, Metabolism, Excretion, Toxicology = ADMET) which may each influence drug performance.

Under these assumptions nuclear medicine modalities can express an added value that is strictly related to their ability to track the fate of labelled compounds. However, this would be of limited importance unless the radiotracers are of high specific activity, i.e. the amount of radioactivity needed to the acquisition of statistically significant information is carried out by a very small (typically lower than the μg scale) amount of substance. Such a characteristic implies that interaction between the radiotracer and the pertinent biochemical targets can be "observed" with low interference from mass effect, quite often even at physiological concentration, and without the induction of massive, if any, biological response.

PET shows some relevant advantages over other nuclear medicine modalities: beyond higher spatial resolution and sensitivity than SPET, the possibility of attenuation correction plays a fundamental role. PET makes it possible to obtain quantitative data or, in other words, to convert the counts/pixel, at the basis of the image, into Bq/pixel. Each pixel of the PET image represents a unit volume of the scanned object, indeed we should refer to them as "*voxel*", and this means that the image itself can be looked at as representing a regional distribution of radioactive concentrations.

The most common utilisation of this information is the comparison of averaged values of radioactivity between Regions Of Interest (ROIs), which can be defined on the image dataset according to anatomical references. Indeed, the high resolution of PET should be regarded to as a relative concept in nuclear medicine, since it is not sufficient to fully define organ substructures (e.g. some cerebral structures or trans-mural differences of cardiac walls). In addition, it should not be forgotten that PET can image any structure having the tracer inside, but tracer deposition or presence may not be ubiquitous. To overcome this limitation PET has evolved into PET/CT. Such a "hybrid" machine represents the result of the coupling of a PET and a CT scanner. The CT image is acquired just before the PET emission image but during the same imaging session and without moving the subject from the imaging bed. It is used for both displaying the anatomy of the observed region and, as a density map, correcting the PET emission scans for regional densities [3]. The single acquisition geometry of PET and CT images yields major advantages on image post processing such as image superimposing and fusion. Obviously, the ability to relate very precisely radiotracers information to organ anatomy greatly improves its clinical added value [4].

A further consideration on PET fortune concerns the nature of the mostly used PET nuclides used to prepare radiotracers [5]. First of all they are the radioisotopes of the elements making up the living matter: the substitution of a stable atom (e.g. ^{12}C) by its positron-emitting isotope (^{11}C) will not produce any change into the chemistry of a biologically active molecule, neither into its biological acceptance (negligible isotopic effect).

TABLE 2. Physical characteristics of the most common PET radionuclides.

Nuclide	Half-life (min)	Decay mode	E_{max} (β^+) (MeV)	Specific activity (Ci/mol)
^{15}O	2	100% β^+	1.72	$9 \cdot 10^{10}$
^{13}N	10	100% β^+	1.19	$2 \cdot 10^{10}$
^{11}C	20	99% β^+	0.96	$9 \cdot 10^{9}$
^{18}F	110	97% β^+	0.64	$2 \cdot 10^{9}$

Due to this reason, they are often named "physiologic tracers"; furthermore, they can be prepared at high specific radioactivity, Table 2, and their positron ranges compatible with good imaging performances. Thus, these radioisotopes, with the limitation represented by their half-lives, can be used to synthesize radiolabelled biological molecules that in turn can be used as tracers. The short half-life may be at the same time a disadvantage or not. Certainly, long-lasting processes, like chemical syntheses during preparation or slow biochemistry upon their utilisation in vivo, may be prevented by half-life; however, this feature may turn out to play a positive role on radiation protection (lower environmental impact, lower radiation dose to subjects) and on research protocols requiring serial administrations, such as test/re-test study during physiological or pharmacological stimuli.

3. PET pharmacological application

The opportunities arising from the use of PET radiotracers in DDR are twofold:

- To use PET radiotracers (often called "radiopharmaceuticals") as reporter probes to account for the interaction of the "*druggable*" molecule with the expected biological and/or biochemical process.

- To synthesize a labelled form of the "*druggable*" molecule and make use of such a probe to assess its behaviour *in vivo*;

DDR might define these two approaches, respectively, as detection of a *surrogate end-point* and a *direct end-point*.

More than 500 different compounds have been labelled with positron-emitting nuclides and used for pharmacological and metabolic studies [6]. Some of them have proved or are about to prove their value in diagnostic and clinical application, Table 3.

TABLE 3. Radiopharmaceuticals having applications as reporter probes.

Radiopharmaceutical	Targeted biochemical process
^{18}F-2-fluoro-2-deoxyglucose	glucose consumption and metabolism
^{11}C-palmitate	fatty acid metabolism
^{18}F-fluorothyrosine	amino acids transport
^{11}C-methylmethyonine	protein synthesis
^{18}F-fluorodopamine	dopamine pool
^{11}C-raclopride	receptor ligand
^{13}N-ammonia	regional blood flow

The use of radiotracers to label active molecules is not new: there is a long history of pharmaceutical research making use of ^{14}C and ^{3}H to address the questions "*how much*" and "*where*" in the body. Unfortunately these radionuclides are beta emitters and they are not suitable for direct *in vivo* imaging but only using *ex vivo* specimens or samples. Major drawbacks with the use of beta emitters for imaging were the cumbersome manipulation of the specimen (e.g. autoradiography) and the loss of dynamic information. This latter point was approached using point-by-point measurements (each point representing the average value of repeated experiments, e.g. at a given time from tracer injection) and hence required the use of a relevant number of experimental animals. Additional problems such as costs, workload and possibility of error introduction during manipulation should also be considered.

The advantage of getting all this information *in vivo* into intact subjects (non invasively) is well evident but the possibility of exporting these experiments directly into humans is indeed the real breakthrough.

This is possible, due to the non-invasivity of the modality, the low radiation dose and the pharmacological safety of the radiotracers. However, this application requires the use of dynamic PET studies; the addition of the time scale to imaging (4D images) moves up the complexity of the study, in particular with moving organs [7], and the amount and size of data sets, but may be considered as the enabling factor towards in vivo pharmacokinetics.

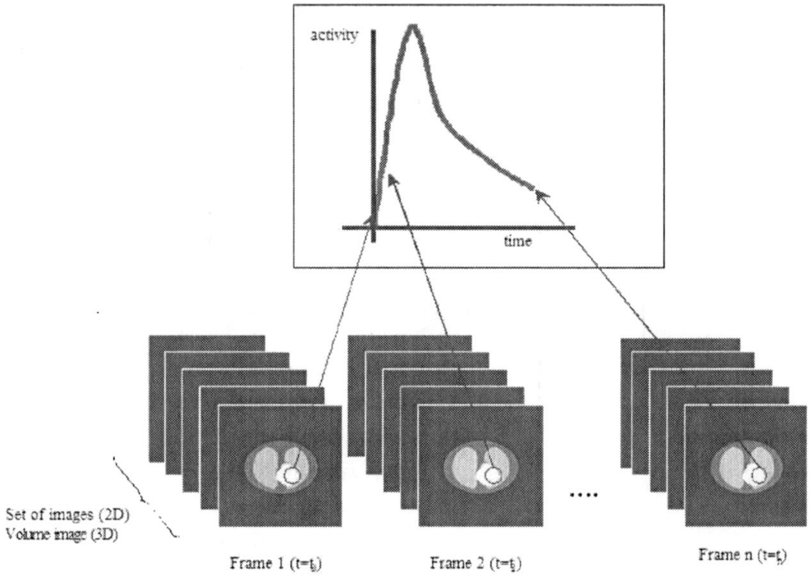

Figure 2. 4D imaging set the bases for dynamic evaluation of regional concentrations.

Whole-body scanners can be used to explore any relevant anatomical district, Figure 3. The pattern of distribution and the kinetic behaviour of the tracer will then depend upon the regional interaction with the tissue, i.e. existence and intensity/specificity of interacting mechanisms, but also upon a number of additional parameters, first of all those influencing the transport of the tracer to the target site such as blood-flow, blood volume, distribution volume, carriers. This implies a careful design of the study protocol that may require multi-tracer imaging and additional measurements (e.g. blood sampling and analysis) to address all variables in the system [8].

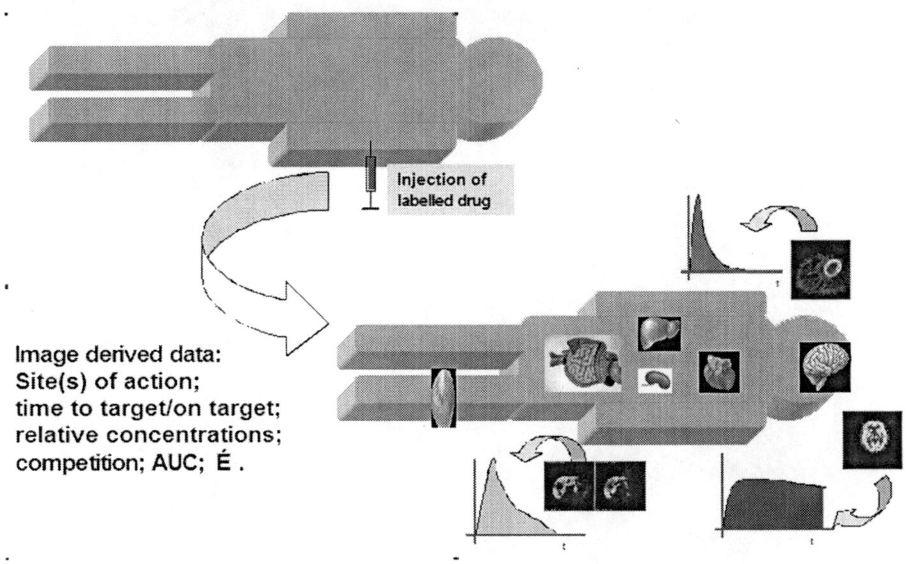

Figure 3. Regional pharmacology using PET Molecular imaging.

Data obtained from PET images can then be used to perform pharmacological calculations, ranging from simpler graphical analysis to full complex compartmental analysis. Validation of image-derived data (protocol hypothesis) versus gold standard is strongly recommended for any application oriented to demonstrate absolute performance of a study protocol and its ability to produce results that in turn may be used to compare different cohorts. Indeed this is a very delicate issue and robustness of protocol should be assessed with specific regards to the targeted goal and the experimental conditions.

Hundreds of thousands patients are scanned worldwide per year, mostly using a "radiopharmaceutical", [^{18}F]-2-fluoro-2-deoxy-D-glucose (FDG), able to measure the regional rate of glucose consumption [9]. About 95% of PET scans belong to oncology medical queries: and response is given on the basis of the different rate of glucose consumption between normal and proliferating tissue, demonstrated by the accumulation of FDG.

Thus FDG is itself a tracer having a perspective place in DDR, as a *surrogate end-point*; its differential accumulation over repeated tests can be used to monitor therapeutic treatment and therapeutic agent efficacy as well [10]. This has proven to be a very important clinical utilisation of FDG because the usual patient follow-up has based on tumour mass shrinking (usually by CT) that requires weeks/months to be detectable, while metabolism changes can be assessed much earlier, Figure 4.

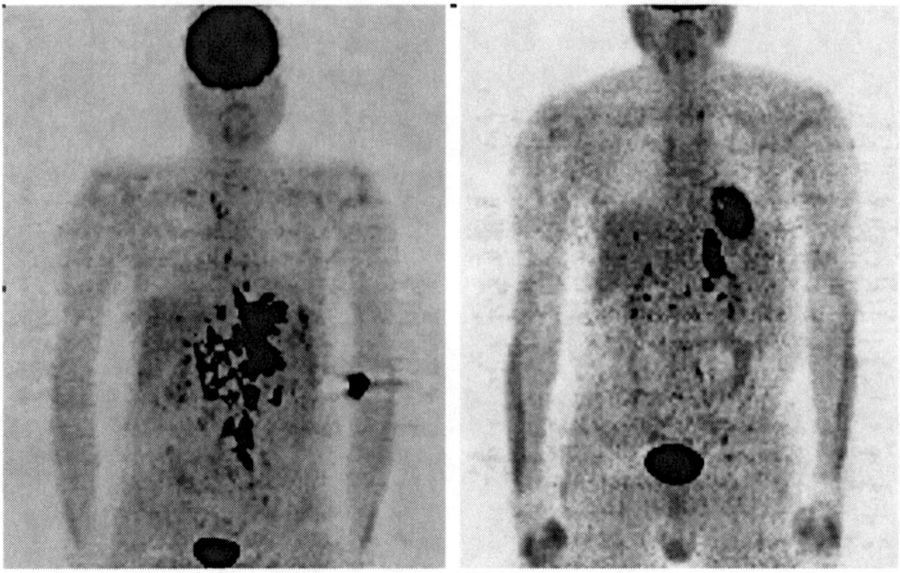

Figure 4. Reduction of metabolic activity in tumour after first cycle of chemotherapy (2 weeks delay).

Regional rate of glucose consumption has been also used to detect response to therapeutic treatment in different pathologies such as inflammatory diseases, brain functional diseases.

The success of FDG has triggered a very important change into the PET community: first, it drew the attention of biomedical companies into this field; second, it reinforced the belief that PET had to deal with numbers more than pictures. It should be regarded as a major progress in the use of PET/FDG imaging as "*biomarker*" in DDR (oncology) the recent publication in the literature of consensus and position papers dealing with the relevant rules to be applied [11]. Although the use of full quantitative modelling of FDG metabolism may not be applicable on large clinical trials the use of semi-quantitative approaches, based on Standard Uptake Value (SUV) validated on a single pathology, is strongly recommended. PET scanner manufacturers made a strong effort in developing algorithms directly on board of scanners for image processing for specific tasks, often incorporating correction and modelling tools or available as upgrading options. In terms of becoming a perspective pharmaceutical tool, PET demonstrated not only to have good physical sensitivity (radioactivity detection) but also high biological sensitivity: this is exactly what DDR was looking for [12].

4. Micro-imaging

Although shortening the time to first-time-in-man stands as the ultimate goal, animal studies maintain their value and represent a first step that cannot be avoided prior to human experiments. Therefore, the great attention on PET studies in humans in not limiting the attention to imaging applications in preclinical phase[13]. On the contrary, small animal scanners (μPET, μMRI, μCT) are becoming more and more common in preclinical pharmaceutical research, both in Academia and in pharmaceutical industries.

The smaller dimensions have a positive impact on spatial resolution (1.8-2 mm vs. 4-6 mm) and result into lower costs than human scanners (about 6/7-fold reduction). At the same time the reduced dimension of the "subject" poses additional methodological challenges, such as input function measurement and quantification.

It is interesting to note that the perspective use of molecular imaging into this field might also produce additional benefits such as changing the concept of "control group" during a preclinical study. A control group might be determined not only on the base of conventional approaches, such as breeding or run-in based on nutrition or environmental conditions, but defined via metabolic and/or physiological studies. This perspective could strongly reduce the number of animals needed to have statistical consistency data and would allow more precise data to be gathered. This new field of "micro-imaging" is thus triggering the development of new technologies, such as hybrid small scanners (μSPET/CT), and opening up new and challenging research issues[14].

References

1. JM Collins, Imaging and other Biomarkers in early clinical studies: One step at a time or re-engineering drug development, J. Clin. Oncol., 2005; 23(24): 5417-5419.
2. S Ekins, Y Nikolosky, T Nikolskaya, Application of Systems biology to absorption, distribution, metabolism, excretion and toxicity, Trends Pharmacol. Sci., 2005; 26(4):202-209.
3. DH Townsend, T Bayer, TM Blodgett, PET/CT scanners: a hardware approach to image fusion, Seminars Nucl. Med. 2003: 193-204. PE Kinahan, BH Hasegawa, and T Beyer, X-ray Based Attenuation Correction for PET/CT, ibid. 166-179.
4. I Sachelarie, K Kerr et al., Integrated PET-CT: evidence-based review of oncology indications, Oncology (Williston Park). 2005;19(4):481-90; discussion 490-2, 495-6.
5. GB Saha, Basics of PET Imaging: Physics, Chemistry, and Regulations; Springer Verlag, New York, 2005.

6. ME Phelps, PET Molecular imaging and its biological applications, Springer Verlag, New York, 2004.
7. SA Nehmeh, YE Erdi, et al., Four-dimensional (4D) PET/CT imaging of the thorax Med Phys. 2004; 31(12):3179-86.
8. P Iozzo, A Gastaldelli, et al., 18F-FDG assessment of glucose disposal and production rates during fasting and insulin stimulation: a validation study, J Nucl Med. 2006; 47(6):1016-22.
9. SS Gambhir, J Czernin, et al. A tabulated summary of FDG PET literature, J.Nucl.Med. 2001; 42(suppl.5):1S-93S.
10. SM Larson, LH Swartz, 18F-FDG PET as a candidate for "qualified biomarker": Functional assessment of treatment response in oncology, J.Nucl.Med., 2006; 47:901-903.
11. LK Shankar, JM Hoffman et al., Guidelines for the use of 18F-FDG PET as an indicator of therapeutic response in patients in national Cancer Institute trials, J.Nucl.Med., 2006; 47: 1059-1066.
12. M Rudin, R Weissleder, Molecular imaging in drug discovery and development. Nature Rev Drug Disc. 2003;2:123-131.
13. SR Cherry. Use of positron emission tomography in animal research. ILAR J. 2001;42:219-232.
14. SR Cherry, In vivo molecular and genomic imaging: new challenges for imaging physics, Phys. Med. Biol., 2004; 49:R13-R48.

Part V

Digital Imaging

PHYSICAL PRINCIPLES OF MAMMOGRAPHY

DAVID R. DANCE*
Joint Department of Physics, Institute of Cancer Research and The Royal Marsden NHS Foundation Trust, London SW4 6JJ, UK

Abstract. An outline is given of the underlying physical principles that govern the selection and use of systems for X-ray mammography. Particular attention is paid to screen-film mammography as some aspects of digital mammography are considered in another lecture. The size and composition of the compressed female breast and of calcifications are described and the magnitude of photon interaction processes in breast tissues discussed. The physical performance measures contrast, unsharpness, dose, noise and dynamic range are outlined and used in a treatment of the various components of the mammographic system. The selection of photon energy is a compromise between contrast and/or signal-to-noise ratio on the one hand, and breast dose on the other. For screen-film imaging the contrast achieved is considered to be the most important image measure and the performances of different mammographic target/filter combinations (including Mo/Mo, Mo/Rh, Rh/Rh and W/Rh) are compared on this basis. For digital imaging, the signal-to-noise ratio is the most important image measure, and the optimal X-ray spectra are then different to those for screen-film mammography. The relationship between image unsharpness and focal spot size and image magnification is explored. The importance of breast compression is stressed and the advantages of compression listed. The contrast in the image is degraded by scattered photons recorded by the image receptor and the magnitude of this effect and the reduction achievable using mammographic anti-scatter grids considered. The performance of mammographic screen-film receptors is described and analyzed, paying attention to unsharpness, noise and receptor DQE.

Key words: mammography; screen-film systems; spectra, mean glandular dose.

1. Introduction

X-ray mammography is an important examination widely used for breast screening and for the examination of symptomatic women. This lecture provides an introduction to the physical factors that govern the design and

*David R. Dance, Joint Department of Physics, Institute of Cancer Research and The Royal Marsden NHS Foundation Trust, London SW4 6JJ, United Kingdom; e-mail: drdance@f2s.com.

Y. Lemoigne et al. (eds.), Physics for Medical Imaging Applications, 355–365.

use of mammographic imaging systems. More attention is paid to screen-film systems than digital systems as the latter systems are covered in a companion lecture. A more extended treatment of the physical principles of mammography can be found for example in IPEM (2005) and Säbel and Aichinger (1996).

2. Properties of the female breast

There is a large variation in the size and composition of the female breast. In infancy, it is largely composed of adipose tissues but at puberty fibroglandular tissue starts to develop and this continues until maturity. Thereafter the fibroglandular tissue is gradually replaced by adipose tissue. However, for women of a given age, there is wide variation in the glandularity of the breast and hence of the radiographic density and appearance of the mammogram. In general breast glandularity decreases with increasing breast size, but again there can be large variations.

The breast is compressed when imaged and the compressed thickness may vary from less than 20 mm to more than 110 mm. This variation in both breast composition and thickness prevents a significant challenge to the X-ray imaging system, which must achieve adequate quality at a low dose for a wide range of conditions. Breast abnormalities may appear on the mammogram as a soft tissue lesion which may be rounded or spiculated, but sometimes the only sign of abnormality is one or more calcifications or a distortion in the breast architecture. Calcifications are specks of calcium hydroxyapatite or phosphate which range in size from extremely small to several mm. It is considered desirable to be able to detect calcifications as small as 100 μm, though this again presents a significant challenge to the imaging system.

3. Photon interactions

In the mammographic energy range, the important photon interactions are the photoelectric effect and scattering processes. The photoelectric effect is the dominant interaction below about 22 keV, and gives rise to most of the energy absorption from the incident X-rays and hence to the breast dose. There are two scatter processes which must be considered: coherent (or Rayleigh) scattering and incoherent scattering. The former process is most significant in the forward hemisphere, giving rise to a change in photon direction but no transfer of energy. Incoherent scattering gives rise to a small transfer of energy, which increases with scattering angle.

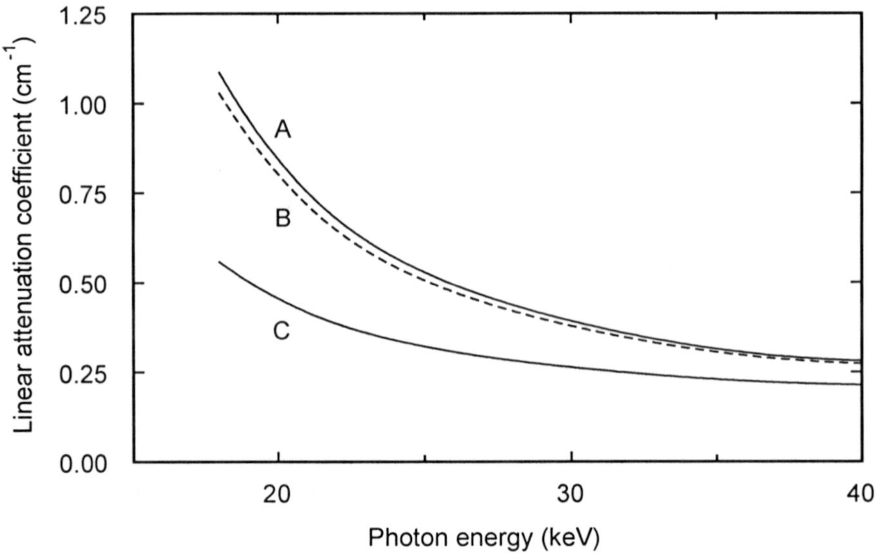

Figure 1. Variation of the linear attenuation coefficient for breast tissues with photon energy and breast tissue type. Curve A infiltrating duct carcinoma; curve B fibrous tissue; curve C fatty tissue. Data taken from Johns and Yaffe (1987). Figure taken from IPEM (2005). Copyright Institute of Physics and Engineering in Medicine 2005. Reproduced with permission.

Figure 1 shows the dependence of the linear attenuation coefficient for photon interactions on breast tissue type and photon energy. It follows that the highest contrast will be achieved at the lowest photon energy. There is an important difference in the attenuation offered by fibrous tissue and that for fatty tissue, but only a small difference between infiltrating duct carcinoma and fibrous tissue. The lines plotted are averages from a series of tissue samples, and in fact the data for carcinoma and fibrous tissue overlap. It follows that some cancers may be difficult to distinguish from fibrous tissue in terms of their attenuation properties; they are identified because of differences in breast architecture.

4. Important physical parameters

In mammography it is important to important to visualize at a low dose both subtle differences in soft tissue density and small calcifications. Figure 2A shows that the highest contrast is achieved at the lowest photon energy, and that if the energy is not low enough, subtle differences in breast density

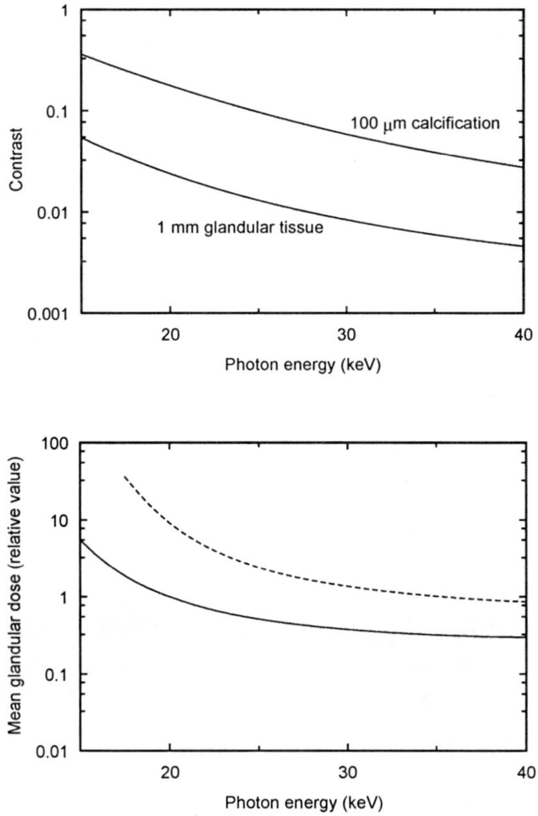

Figure 2. A (Top) Variation of contrast with photon energy. The contrast has been calculated for a simplified model that neglects image unsharpness. Contrast is shown for a 100 μm calcification and I mm of glandular tissue. B (Bottom) Variation of mean glandular breast dose with photon energy. Upper curve 80 mm thick compressed breast, lower curve 20 mm thick compressed breast. Figure taken from IPEM (2005). Copyright Institute of Physics and Engineering in Medicine 2005. Reproduced with permission.

and small calcifications will not be visualized. Figure 2B shows how the breast dose increases as the photon energy decreases and hence that a compromise must be reached between these two performance requirements. Dose is also strongly dependent on breast thickness.

Various dose specifications have been suggested for mammography (Dance et al., 1999): incident or surface dose, mid-breast dose, mean breast dose and the mean dose to the glandular tissues within the breast, which are the tissues at highest risk of radiation-induced carcinogenesis. The latter quantity is the quantity of choice and has been recommended by the ICRP and generally adopted in national dosimetry protocols.

In order to visualize small calcifications, it is important to have good resolution and low noise. Resolution is specified here in terms of *unsharpness*. Its magnitude depends upon the size of the focal spot of the X-ray tube and the imaging geometry used (penumbra or 'geometrical' unsharpness), the properties of the image receptor (receptor unsharpness) and patient movement during the exposure. Generally, patient movement will be very small because the breast is compressed during the exposure. Image noise receives contributions from fluctuations in the number of photons detected per unit area (quantum mottle), and from variations in the response of the receptor itself. The related performance of the image receptor is expressed in terms of the detective quantum efficiency (DQE). For a given imaging task, the signal-to-noise ratio (SNR) may be employed. As shown below, the relationship between dose, contrast or SNR may be used as the basis for the selection of the X-ray spectrum to be used in mammography.

The amount of tissue traversed by X-ray photons during mammography ranges from zero at the edge of the breast to the full thickness of the breast. As a consequence the image receptor must be able to respond adequately to signals varying significantly in magnitude, and the dynamic range of the image receptor can be an important limitation of performance.

5. The mammographic X-ray imaging system

The mammographic unit is designed to allow easy examination of the breast from a range of directions. The patient may be examined standing or sitting, with her breast resting on a support plate. The X-ray tube and breast support plate are mounted on an assembly that may be rotated to achieve the desired radiographic projection. An anti-scatter grid is placed between the breast support and image receptor. The receptor itself may be a screen-film cassette or a digital device. Firm compression is applied to the breast using a plastic compression plate. An automatic exposure device terminates the exposure so that the required signal level is achieved in the image.

5.1. THE X-RAY TUBE

5.1.1. *X-ray spectrum*

In mammography a low energy X-ray spectrum is used which is often achieved by using a molybdenum or rhodium target and a K-edge filter of molybdenum (with a Mo target) or rhodium (with a Mo or Rh) target, although a tungsten target is also used in some situations. Because Rh has a higher atomic number than Mo, spectra using an Rh target are the harder than those for a Mo target.

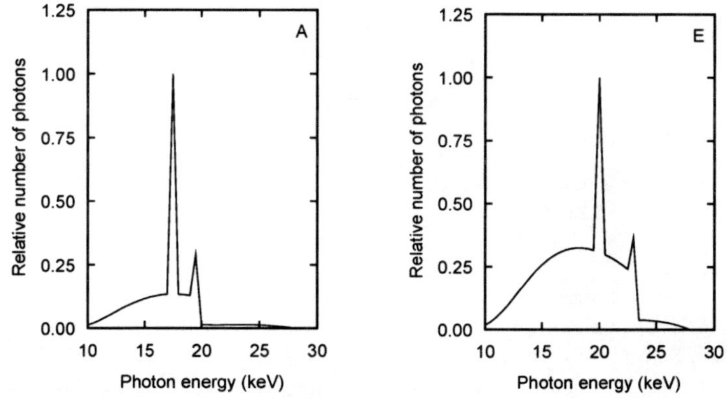

Figure 3. (A) X-ray spectrum from a Mo target at 28 kV filtered by a 30 μm Mo filter. (E) X-ray spectrum from a Rh target at 28 kV filtered by a 25 μm Rh filter Figure taken from IPEM (2005). Copyright Institute of Physics and Engineering in Medicine 2005. Reproduced with permission.

It is common for X-ray tubes to have more than one target and/or a filter wheel so that the X-ray spectrum (target, filter and tube voltage) can be varied automatically according to the size and density of the breast to be examined. Typical spectra are shown in figure 3.

Figure 4 shows the relationship between contrast (of a calcification) and mean glandular dose for different target/filter combinations and an 80 mm thick compressed breast of 10% glandularity. Each curve shows the variation as the tube voltage is changed from 25-32 kV (Dance et al., 2000). No choice of spectrum can match the contrast achieved using the Mo/Mo combination at 25 kV. Only one spectrum can match the contrast for Mo/Mo at 28 kV (Mo/Rh) and some dose saving is then possible. If the contrast requirement is relaxed (say to match Mo/Mo at 30 kV), a larger dose saving is possible and other target/filter combinations can then be used. In general, for larger breasts the contrast requirement is relaxed so that harder spectra are used.

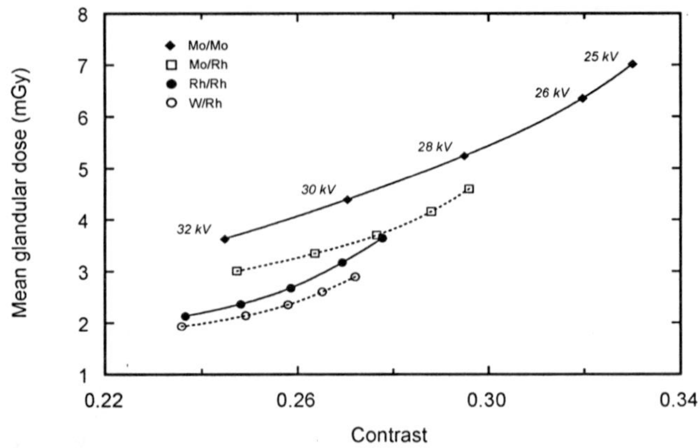

Figure 4. Relationship between contrast and mean glandular dose for imaging a particular image detail; 80 mm 10% glandularity breast and a typical screen-film combination. Curves for four choices of target/filter are shown. Figure taken from IPEM (2005). Copyright Institute of Physics and Engineering in Medicine 2005. Reproduced with permission.

Figure 5 shows the relationship between the mean glandular dose and the choice of X-ray spectrum in **digital mammography** for the task of imaging a particular detail at a signal-to-noise ratio of 5 (Dance et al., 2000).

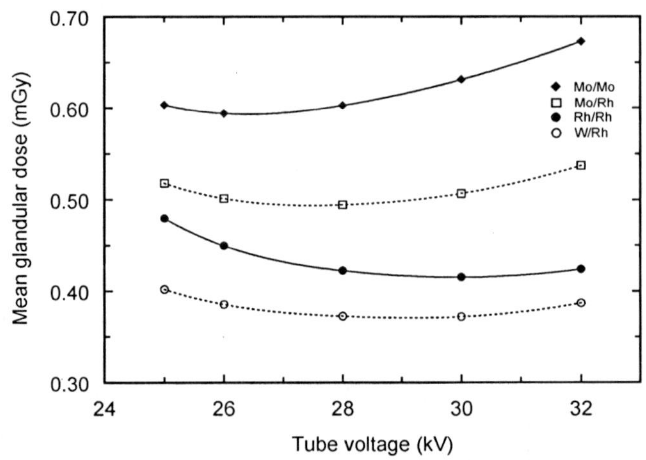

Figure 5. Relationship between tube voltage and mean glandular dose for imaging a particular detail at an SNR of 5; 80 mm breast 10% glandularity. Curves for four different target/filter combinations are shown. Figure taken from IPEM (2005). Copyright Institute of Physics and Engineering in Medicine 2005. Reproduced with permission.

Results are again for an 80 mm thick breast of 10% glandularity. The W/Rh spectrum is the best of those illustrated. It is important to note that the choice of optimum spectrum is dependent on breast thickness.

5.1.2 Focal spot-size

For conventional (contact) mammography, unsharpness increases with image magnification and a measured focal spot of 0.6mm or less and a focus-film distance (FFD) of at least 60cm are recommended. Increasing the FFD or decreasing the object-film distance decreases magnification and improves unsharpness.

In magnification mammography, an air gap is introduced between breast and image receptor. If a sufficiently small focal spot size is used, magnification will decrease unsharpness but will increase dose. Because of the air gap, there will be a decrease in scatter (no grid is normally used for magnification), and a larger image on the same receptor means the effect of noise is reduced.

5.2 BREAST COMPRESSION

Breast compression is essential in mammography because:

- It reduces dose
- It reduces scatter and hence improves contrast
- It enables a softer X-ray spectrum to be used, and hence contrast to be improved
- It reduces geometric unsharpness
- It reduces movement unsharpness
- It reduces the dynamic range requirement
- It reduces the amount of tissue overlap and hence allows better visualization of the breast tissues.

5.3 ANTI-SCATTER GRIDS

The contrast in the image is degraded by the scattered radiation recorded by the image receptor. The amount of degradation varies with photon energy, breast size and image receptor. It can be quantified using the *contrast degradation factor* CDF which is the ratio of the image contrast with and without the effects of scatter. The CDF decreases with increasing breast thickness, and for a 50 mm thick breast imaged without a grid is 0.65, indicating the desirability of using a grid to improve contrast. When a grid designed for mammography is used for imaging a large breast, a significant improvement in contrast is achieved but at the cost of increased dose (necessary for screen-film mammography to maintain the density on the film). For example for an 80 mm compressed breast, the contrast is improved by around 54% but the dose is increased by a factor of 2.1 For a 20 mm thick breast the corresponding figures are 17% and 1.7 (Dance et al., 1992).

5.4 SCREEN-FILM RECEPTORS

Mammographic screen-film receptors make use of a rare earth phosphor, usually gadolinium oxysulphide, in a single layer which is placed in contact with, but behind the photographic emulsion. This configuration brings the point of production of the light photons emitted by the phosphor as close as possible to the emulsion, thereby reducing the lateral spread of these photons as they pass from phosphor to emulsion, and improving resolution.

The design of the screen is a compromise between resolution, dose and noise. As the phosphor thickness increases, it absorbs more X-ray photons, but the unsharpness becomes worse because of the greater lateral spread of the light fluorescent photons. Lateral spread can be reduced by the addition of dyes to the phosphor, but at the cost of increased dose because the number of light photons detected per incident X-ray then decreases.

The performance of the screen-film receptor is strongly limited by the performance of the film emulsion. This has a non-linear response to light photons, showing a very flat variation with exposure at both high and low exposures. Figure 6A shows the film characteristic curve for two mammographic screen-films.

The contrast achievable is proportional to the slope of this curve (film gamma), which is shown in figure 6B. The region of high gamma or good contrast is very limited and correct exposure is very important.

In addition, the noise arising from variations in the number of developed silver grains per unit area of the emulsion is particularly important at low and high doses, and at high spatial frequencies, leading to an important reduction in the DQE of the screen-film receptor. This is illustrated in figure 7, which shows the DQE for the MinR2000 screen-film combination. For this combination, the highest DQE value is about 0.4. The highest values of the DQE and noise equivalent quanta occur at optical densities higher than 1.0. In addition to the loss of performance at high and low optical densities there is a significant loss at high spatial frequencies.

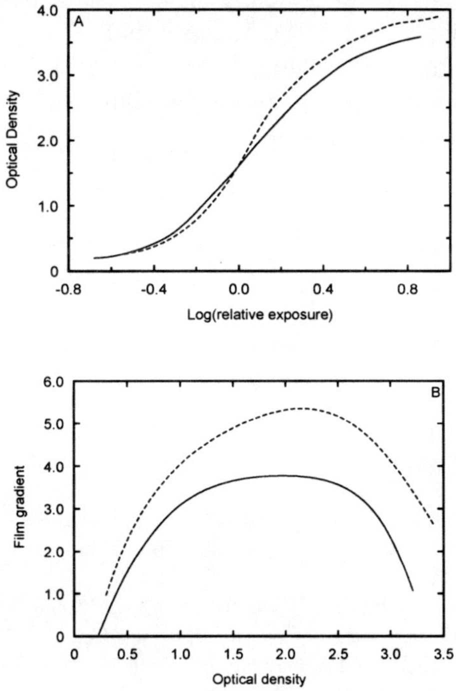

Figure 6. Film characteristic curve (A, top).; and film gamma (B, bottom) for two typical mammographic screen-film combinations Data based on Meeson et al. (2001). Figure taken from IPEM (2005). Copyright Institute of Physics and Engineering in Medicine 2005. Reproduced with permission.

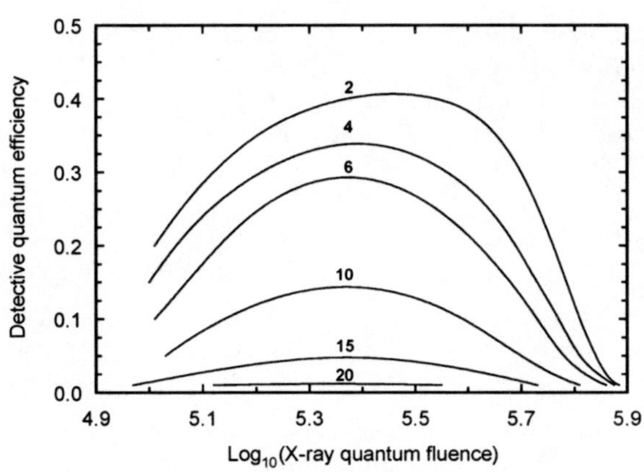

Figure 7. DQE for Kodak MinR screen-film combination. Data based on Bunch (1999) . Figure taken from IPEM (2005). Copyright Institute of Physics and Engineering in Medicine 2005. Reproduced with permission.

The desirability of using an optical density greater than 1.0 is consistent with work by Young et al. (1994). These authors found that that the cancer detection rate in the UK National Breast Screening Programme was dependent upon the optical density on the film, with the higher detection rates found at higher optical densities.

References

Bunch, P.C., 1999, The effects of reduced film granularity on mammographic image quality, *SPIE* **3032** 302-314.

Dance, D.R., Skinner, C.L., and Alm Carlsson G., 1999. Breast dosimetry, in: *Subject dose in radiological imaging*, K-H Ng, D A Bradley and H M Warren-Forward. Eds, Elsevier, Amsterdam, pp. 185-203.

Dance, D.R., Thilander Klang, A., Sandborg, M., Skinner, C.L., Castellano Smith, I.A., and Alm Carlsson G., 2000, Influence of anode/filter material and tube potential on contrast, signal-to-noise ratio and average absorbed dose in mammography: a Monte Carlo study, *Brit. J. Radiol.* **73** 1056-1067.

Dance, D.R., Persliden, J., and Alm Carlsson, G. 1992, Monte Carlo calculation of the properties of mammographic anti-scatter grids, *Phys. Med. Biol.* **37** 235-248.

IPEM (Institute of Physics and Engineering in Medicine), 2005, Physical principles of mammography, in: *The commissioning and routine testing of mammographic X-ray systems*. IPEM report 89, 3rd Edition, IPEM, York pp. 10-30.

Johns P.C. and Yaffe, M.J., 1987, X-ray characterisation of normal and neoplastic breast tissues, *Phys. Med. Biol.* **32** 675-695.

Meeson, S., Young, K.C., Rust, A., Wallis, M.G., Cooke, J., and Ramsdale, M.L., 2001, Implications of using high contrast X-ray film screen combinations, *Brit. J. Radiol.* **74** 825-835.

Säbel, M., and Aichinger, H., 1996, Recent developments in breast imaging, *Phys. Med. Biol.* **41** 315-368.

Young, K.C., Wallis, M.G., and Ramsdale, M.L., 1994, Mammographic film density and the detection of small breast cancers, *Clin. Radiol.* **49** 461-465.

COMPUTED TOMOGRAPHY

ISABEL CASTELLANO*
Royal Marsden Hospital, Fulham Road, London SW4 [1]6JJ, UK

JACOB GELEIJNS
Leiden university centre, PO box 9600, NL-2300 RC, Leiden, The Netherlands

Abstract. After its clinical introduction in 1973, computed tomography developed from an x-ray modality for axial imaging in neuroradiology into a versatile three dimensional imaging modality for a wide range of applications in for example oncology, vascular radiology, cardiology, traumatology and even in interventional radiology. Computed tomography is applied for diagnosis, follow-up studies and screening of healthy subpopulations with specific risk factors. This chapter provides a general introduction in computed tomography, covering a short history of computed tomography, technology, image quality, dosimetry, room shielding, quality control and quality criteria.

Keywords: computed tomography; image quality; patient dose; quality control.

1. Introduction

"Computerized transverse axial tomography is a new and fundamentally different diagnostic x-ray method. Modern electronic and computer technology have been allied to x-ray detection and measurement to analyse content of a beam of x-ray photons passing through tissue, in such a way that remarkable tissue differentiation becomes possible" (J. Ambrose[1] in 1973).

"As the exposure of the patient to x-rays must be restricted, there is an upper limit to the number of photons that may be passed through the body during the examination, and so to the amount of information that can be

---

* Isabel Castellano, Royal Marsden Hospital, Fulham Road, London SW4 16JJ, UK; e-mail: *Elly.Castellano@rmh.nhs.uk*

Y. Lemoigne et al. (eds.), Physics for Medical Imaging Applications, 367–397.

obtained. It is, therefore, of great importance that the method of examination ensures that all the information obtained is fully utilized and interpreted with maximum efficiency" (G.N. Hounsfield[2] in 1973).

Two citations from pioneers in the development of computed tomography, one radiologist (J. Ambrose) and one physicist (G.N. Hounsfield), illustrate the expectation in 1973 for unique opportunities in x-ray imaging that would become available by computed tomography but at the same time Hounsfield already stressed the need for optimising the use of ionizing radiation in clinical applications. More than 30 years after its introduction, CT plays a key role as a versatile technique in diagnostic radiology and numerous technical innovations contributed to improvements of image quality and to the optimisation of the delicate balance between image quality and patient dose.

2. Technology

Conventional radiography has provided images of the human body since 1895, however a major limitation of this technique is the superimposition of tissues and organs in the resulting two-dimensional radiographs. This limitation was overcome by the technique of tomography (also called planigraphy) that was developed by Bernard Ziedses des Plantes in 1931. This technique allows for imaging a two dimensional plane by mechanical movement of the x-ray tube and detector (film-screen combination). The volume outside the plane of interest is blurred by this movement whereas the plane of interest is represented more or less accurately [REF]. Tomography generally yields coronal images, but special techniques allow for imaging curved planes (e.g. orthopantomography) and axial planes.

The theoretical work by J.H. Radon on the reconstruction of cross sectional images from transmission measurements in 1917 and by A.M. Cormack on the mathematical analysis of the conditions for acquiring a correct radiographic cross-section in a biological system, together with the introduction of computers, provided the preconditions for the development of computer assisted tomography (from the Greek tomos, a cut, and graph, written). Computed tomography provides two dimensional cross sectional images of the body. G.N. Hounsfield realized the clinical introduction of computed tomography in 1972. A.M. Cormack and G.N. Hounsfield shared the Nobel Prize for medicine and physiology in 1979 for the "development of computer assisted tomography". Computer assisted tomography is nowadays referred to as computed tomography (CT). The first CT scanners only allowed scanning of the head but they provided unique opportunities for brain imaging. New applications became available with the clinical introduction of whole body CT scanners in 1975.

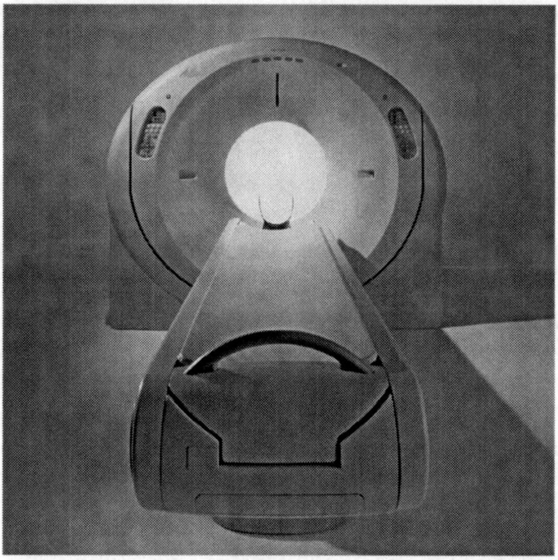

Figure 1. A multi slice CT scanner, the figure shows the patient table and the donut shaped housing containing the gantry. The x-ray tube and detector are mounted on a rotating part of the gantry. A spiral CT scan is achieved by linear transport of the p patient table through the gantry during the acquisition (Courtesy Toshiba Medical Systems Europe).

Computed tomography requires measurement of x-ray transmission profiles through cross sections of the body under a large number of angles. To accomplish this, an x-ray tube and a detector rotate around the patient. Modern CT scanners make use of a rotating fan shaped x-ray beam and arrays of hundreds of detectors. Figure 1 shows a state-of-the-art multislice CT scanner. The reconstruction of cross sectional images is achieved by the technique of filtered backprojection of the measured transmission profiles[3]. The pixels in the reconstructed images are correlated to the linear attenuation coefficient of the tissues (μ_{tissue}) in the corresponding cross section of the body. The pixel values in the reconstructed images are referred to as hounsfield units (HU). The hounsfield unit is normalized to the linear attenuation coefficient of water (μ_{water}):

$$HU = \frac{\mu_{tissue} - \mu_{water}}{\mu_{water}} \times 1000$$

The hounsfield unit representing air is by definition -1000 HU and water is by definition associated with 0 HU. Fat tissue has a hounsfield unit slightly below zero (-80 HU to -100 HU), lung has a hounsfield unit in the range -600 HU to -950 HU, soft tissues are represented by hounsfield units in the range 20 HU to 70 HU and hounsfield units for compact bone may exceed 1000 HU. CT images are generally visualized in grey values that correspond to a certain window width (WW) and window level (WL). hounsfield units are usually visualized in an eight bit grey scale; this is achieved by linear mapping of the hounsfield units in a 'window'. The window width (WW) defines the range of hounsfield units that are represented by the grey values (ranging from white to black) and the window level (WL) defines the central hounsfield unit within the selected window width. Optimal representation of the tissues of interest in the CT image is achieved by selecting the most appropriate window width and window level. Different settings of the WW/WL are used to visualize for example soft tissue, lung tissue or bone.

Computed tomography allows in principle for a quantitative analysis of tissues. This has led to applications such as bone densitometry (osteoporosis), lung densitometry (lung emphysema), calcium scoring (calcium scoring is used to quantify the degree of calcification of coronary arteries) and perfusion studies (to measure blood flow in regions of interest such are brain lesions or tumours).

The main acquisition parameters in computed tomography are tube voltage, tube current and rotation time. A relatively high tube voltage (120 kV - 140 kV) is used in computed tomography to achieve good x-ray transmission and sufficient detector signal. This is necessary since the effective use of the generated x-ray beam is limited in computed tomography due to the tight collimation of the x-ray beam. The tube current used in computed tomography is limited by the required long scan time and the heat capacity of the x-ray tube. The rotation time is the time required for a 360° rotation of the x-ray tube. The first CT scanners were operated in a sequential mode, meaning that each cross sectional image was reconstructed from the transmission profiles measured during a 360° rotation of the x-ray tube. After the acquisition the patient table could be translated over a short distance, next the acquisition of another axial image was performed.

The introduction of spiral CT in 1989 was a major step in the development of CT technology. During the acquisition in spiral CT a continuous rotation of the x-ray tube is combined with linear transport of the patient table through the gantry. This allows for the acquisition of a large volume interest within one breathhold and for example the development of CT angiography. The table translation is generally expressed relative to the (nominal) beam width (in single slice CT this equals the slice width): the ratio of table translation per 360° tube rotation relative to the nominal beam width is in spiral CT referred to as the pitch factor. The rotation time of single slice CT scanners was 1-2 s, the slice thickness (and nominal beam width) in most clinical applications was 5-10 mm.

Ten years after the introduction of spiral CT a next step in CT technology, that provided the preconditions for new clinical applications, was made: the introduction of multislice fast rotating CT scanners. In single slice CT scanners only one linear array of detectors was used, in multislice scanners this was extended to respectively 4, 16 and 64 adjacent active arrays of detectors, enabling the simultaneous measurement of a corresponding large number of transmission profiles. At the same time the rotation time dropped well below 1 s, to 0.3-0.4 s. Consequently, with the latest multislice CT scanners it is possible to scan almost the entire body of an adult within one breathhold at a slice thickness well below 1 mm. Figure 2 shows the geometry of a spiral CT acquisition, note the circular trajectory of the x-ray tube, which from the perspective of a patient transforms into a spiral course.

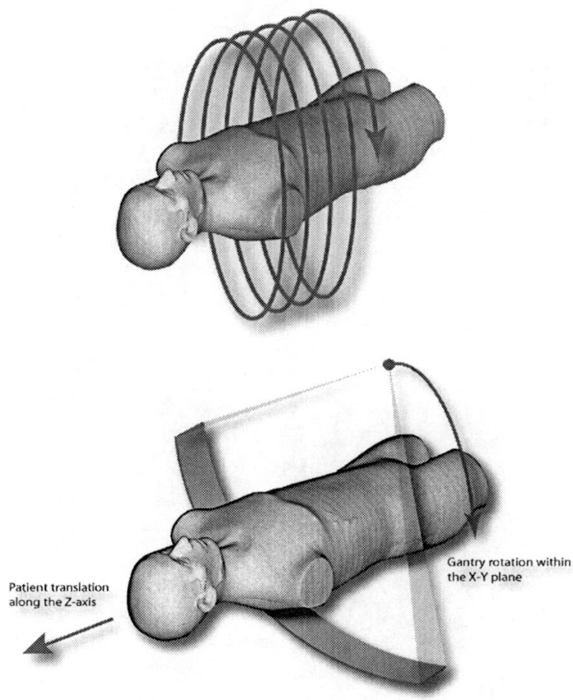

Figure 2. The combination of a rotating x-ray tube and linear transport of the patient table along the Z-axis during the acquisition results, from the perspective of the patient, in a spiral path of the x-ray tube around the patient. The axis of rotation is referred to as the z-axis; the patient table moves along this Z-axis; the focal spot rotates within a plane perpendicular to the Z-axis (the X-Y plane).

The latest developments in computed tomography technology are the introduction of dual source CT and cone beam CT. The dual source CT scanner mainly aims at improved temporal resolution which is particularly beneficial in applications of cardiac CT[4]. The dual source CT might also be used for dual energy imaging. The cone beam CT features 256 detector arrays that allow scanning a volume containing 256 slices, each only 0.5 mm thick, within one tube rotation[5]. The cone beam CT offers new opportunities for CT angiography in general and particularly for cardiac CT since it allows 'isophasic' imaging, i.e. imaging of a volume of interest at one and the same phase of contrast enhancement. Cone beam CT might also be beneficial for dynamic studies, such as evaluation of the movement of joints. Hybrid systems have also become available; they integrate a PET

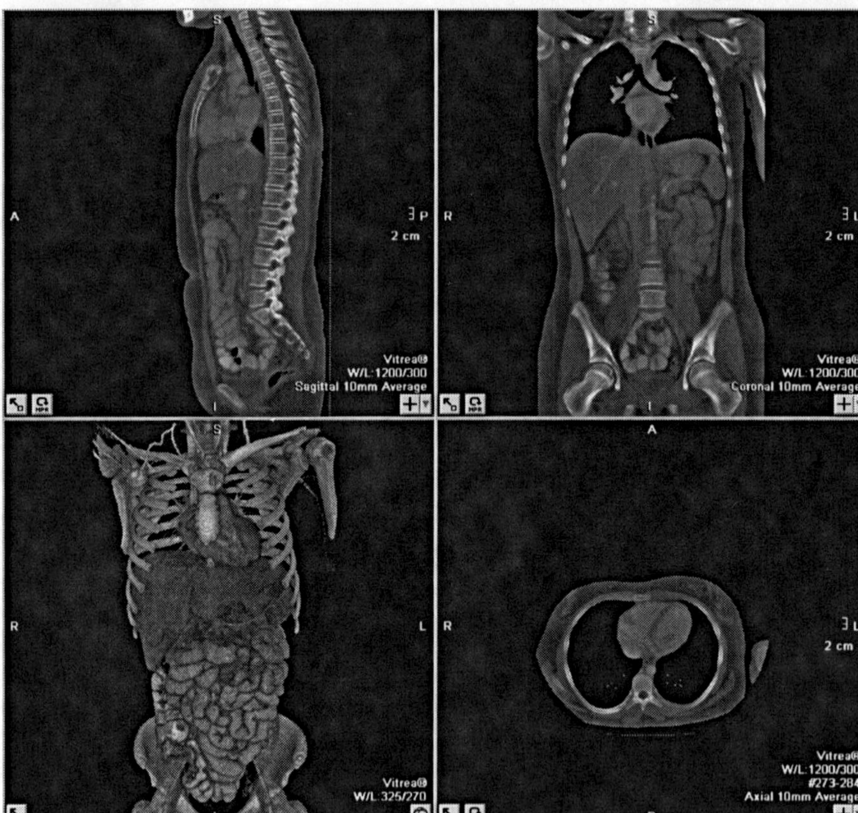

Figure 3. The CT scan is usually evaluated on a dedicated workstation that allows for assessment of the scanned volume in any plane and in threedimensional display; under any desired viewing angle (Vital Images, Inc., Vitrea). The figure shows a sagittal view (above, left), a coronal view (above right), a three-dimensional display (below, left) and an axial view (below, right).

scanner or a SPECT scanner with a CT scanner (PET-CT, SPECT-CT). Besides advantages for image interpretation (image fusion) these hybrid systems also offer improved quality of PET or SPECT images since more accurate attenuation corrections can be made during the reconstruction of PET or SPECT images.

Multislice CT scanners yield hundreds or thousands of cross sectional images that have to be evaluated on a dedicated workstation. The workstation allows for visualization of the scanned volume in any plane or in three-dimensional display (Figure 3). Advanced options of the workstations include maximum intensity projection (MIP), curved multiplanar reformat (CMR), volume rendering, and shaded surface display.

3. Image quality

At its introduction excellent low-contrast resolution of computed tomography was the most prominent feature that distinguished computed tomography from radiography and conventional (film-screen) tomography. With computed tomography it became possible for the first time to distinguish between grey and white matter in the brain. Low-contrast resolution is the ability to detect structures that demonstrate only a small difference in signal (expressed in hounsfield units) compared to their direct environment. Image noise is the main limitation for low-contrast resolution. Image noise may be decreased either by raising tube current (mA) at the cost of patient exposure, or by increasing the reconstructed slice thickness, at the cost of spatial resolution. In addition, low-contrast resolution depends on tube voltage, beam filtration and the reconstruction algorithm. Physicists usually assess low-contrast resolution with phantoms that contain different sized inserts. With a phantom low-contrast resolution can be assessed either subjectively by an observer that has to decide whether an insert is visible or not, or objectively by calculating the contrast to noise ratio. The noise power spectrum would provide an objective measure of scanner performance but is not yet applied on a large scale. Sometimes native tissue contrasts are not sufficient to differentiate between structures, such as the vessel wall and its lumen, or the colon and its filling. In these cases the image contrast can be enhanced by injection of iodine based contrast in a small peripheral vein (CT angiography) of by colonic air insufflation after full bowel cleansing (CT colonography).

Spatial resolution, or high-contrast resolution, is the ability to observe contours of small objects within the scanned volume. Small objects can only be resolved well when they exhibit a rather large difference in signal (hounsfield units) compared to their direct environment. Voxel size is often used as an indicator of spatial resolution, however, it should be noted that voxel size should be interpreted with care since smaller voxel size does not necessarily imply better spatial resolution. Spatial resolution is preferably expressed as a response of a delta-function, in computed tomography this response is either called a point-spread-function (spatial resolution in the axial plane) or a slice sensitivity profile (spatial resolution along the z-axis), the response is often quantified as the full-width at half-maximum. Spatial resolution is limited by the acquisition geometry of the CT scanner, the reconstruction algorithm and the reconstructed slice thickness. The performance of current 64-slice scanners with regard to spatial resolution, expressed as the full-width half-maximum of the response of a point spread function, is within the range 0.6-1.0 mm in all 3 dimensions. Temporal resolution is the ability to resolve fast moving objects in the displayed CT image. Good temporal resolution avoids motion artefacts and blurring of the image. A good temporal resolution in CT is realized by fast data acquisition (fast rotation of the X-ray tube). Reconstruction algorithms that are used for general CT applications provide in principle a temporal

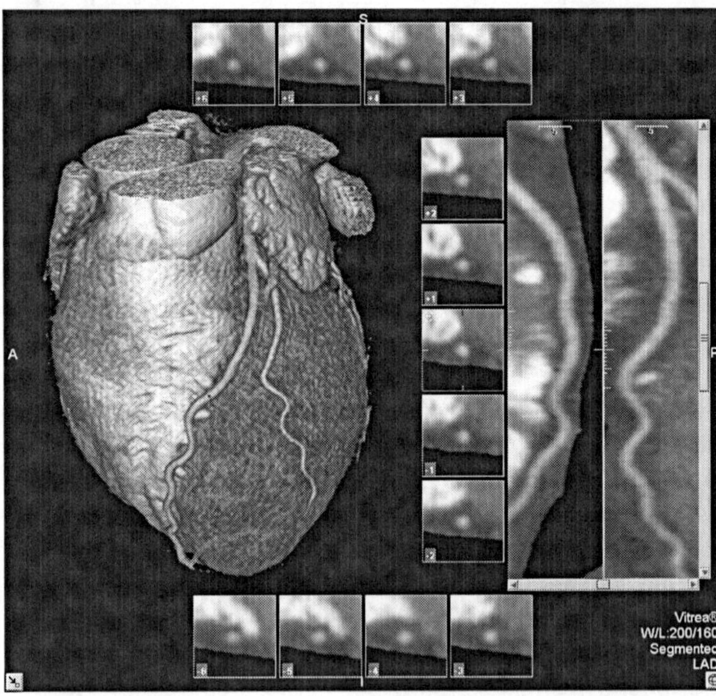

Figure 4. Multislice CT angiography provides sufficient temporal and spatial resolution to freeze the motion of the heart and to visualize the small coronary arteries. Administration of intravenous contrast is required to enhance the coronary arteries. The figure shows a three-dimensional image of the heart, small reformats of planes perpendicular to the coronary artery and two curved multiplanar reformats of the coronary artery[6].

resolution equal to the rotation time (360° rotation, full reconstruction), the best routinely achievable temporal resolution is slightly longer than 50% of the rotation time (180° rotation, half reconstruction). Current 64-slice scanners provide a rotation time of 330-400 ms. These typical rotation times are not short enough for achieving a 100 ms or shorter snap-shot of the beating heart that is required in cardiac CT.

Therefore, dedicated reconstruction algorithms have been developed for cardiac CT (segmented reconstruction). In combination with ECG synchronization, these algorithms provide a motion free three-dimensional image of the beating heart and the coronary arteries (Figure 4).

4. Dosimetry

Concepts have been developed to define the dosimetric characteristics of CT scanners. They are used for quality control and evaluation of acquisition protocols; to assess patient dose; and used for risk assessment. The computed tomography dose index is measured during one 360° rotation of the x-ray tube and is defined as the integral of the dose profile along the z-axis (D(z)) divided by the nominal beam width (T). The integral length is usually defined by the length of the pencil ionization chamber that has been developed for CT dosimetry (in general this is 100 mm).

$$CTDI = \frac{1}{T} \int_{-50mm}^{+50mm} D(z).dz$$

The unit of CTDI is absorbed dose in air (mGy). The CTDI can be measured in standardized cylindrical PMMA (Perspex) phantoms of 16 and 32 cm diameter representing respectively the attenuation and scatter of the head and body. A derivate of the CTDI has been defined that provides for weighting of CTDI's measured at central and peripheral positions of the PMMA phantoms yielding the weighted CTDI ($CTDI_{weighted}$). For spiral (volume) acquisitions it is common practice to correct the weighted CTDI by dividing this value by the pitch factor yielding the volume CTDI ($CTDI_{volume}$). CTDI can also be measured free in air at the centre of rotation of the CT scanner ($CTDI_{air}$). A clinical CT acquisition always involves more that one 360° rotation, so the dose length product (DLP) is a derivate of the CTDI that takes into account the total number of rotations and it is calculated by multiplying the volume CTDI by the actual used scan range. The unit of DLP is mGy.cm. A total DLP for more than one spiral acquisition may be calculated by adding the respective DLP's; however the DLP is not an additive quantity for body and head scans due to the use of different sized phantoms. The DLP for a scan of the head and the body can thus not be added to yield a total DLP. Modern CT scanners provide an indication of patient dose by display of the actual used volume CTDI and DLP on the scanner console.

The effective dose[7] for the patient can be derived from acquisition parameters, such as scan range, tube voltage, tube current, rotation time, pitch factor and beam collimation. This is usually done with software tools such as the ImPACT CT Dosimetry Calculator[8]. A rough estimation of effective dose can also be derived from the dose length product, which is generally displayed on the scanner console after each clinical acquisition. Table 1 shows conversion factors for assessment of effective dose (mSv) from dose length product (mGy.cm) for different parts of the body.

TABLE 1. Conversion factors for assessment of effective dose (mSv) from dose length product (mGy.cm) for different parts of the body[9].

Region of the body	Normalizes effective dose E/DPL (mSv mGy^{-1} cm^{-1})
Head	0.0023
Neck	0.0054
Chest	0.019
Abdomen	0.017
Pelvis	0.017
Legs	0.0008

Table 2 shows the median volume computed tomography dose index, dose length product and effective dose for different clinical indications for computed tomography. These values were derived in a large European field study[9].

When repeated CT scans are being made at one level the peak entrance skin dose becomes of interest. In this case it has to be verified that cumulative peak entrance skin dose remains well below the threshold dose for inducing deterministic skin effects such as erythema and epilation. Peak entrance skin dose should be evaluated for some special applications of computed tomography such as CT fluoroscopy and CT perfusion studies. Deterministic skin effects have been observed incidentally[10].

5. Shielding and quality control

Prior to the installation it must be verified that appropriate shielding of the CT room is available. With the increasing tube capacity and increasing beam width, together with the decreasing rotation time it has become possible to scan larger numbers of patients and to acquire long scan ranges within one breathhold.

This resulted in relative high demands on the room shielding design for modern multislice CT scanners compared to the previous single slice CT scanners. The required shielding can be easily assessed by information about the distribution of scattered radiation around the scanner (isodose curves of scattered radiation are made available for this purpose by the manufacturers) and the expected workload. Practical guides are available for establishing the required shielding[11].

An acceptance test has to be performed after installation of a CT scanner. Requirements for the acceptance test have been established by the International Electrotechnical Commission[12] and practical guidance is provided in numerous documents including the ImPACT Information Leaflet on CT scanner acceptance testing[13]. Once put into clinical use, the scanner needs periodic maintenance and periodic quality control by means of status and constancy testing. Quality control of CT scanners addresses technical aspects such as the accuracy of the hounsfield numbers, the uniformity of the image of a homogeneous object, image noise, spatial resolution, low-contrast resolution, slice thickness, computed tomography dose index, geometric accuracy, table position and movement, and accuracy of the laser light localizers.

TABLE 2. Effective dose (mSv) for different clinical indications for computed tomography. These values were derived in a large European field study[9].

Clinical indication	$CTDI_{vol}$ mGy	DLP mGy cm	Effective dose mSv
Cranium			
Acute stroke	53	746	1.7
Sinusitis	20	151	0.3
Hearing loss	48	233	0.5
Chest			
Interstitial lung disease	3.0	117	2.5
Pulmonary embolism	11.4	302	5.9
Pulmonary metastases	9.4	309	5.6
Abdomen			
Abcess	11.3	551	9.3
Follow up lymphadenopathy	11.6	518	9.0
Urolithiasis	6.8	428	10.6
Livermetastases	12.5	643	9.5
Cranium, paediatric			
Trauma (1-12 months old)	25	236	1.8
Hydrocephalus (4-6 year old)	27	262	1.3
Chest, paediatric			
Congenital abnormality (1-12 months old)	3.1	62	4.7
Bronchiectasis (4-6 year old)	2.1	31	1.4

6. Quality criteria for multislice computed tomography

The clinical application of CT has to be optimised. Adequate protocols for the acquisition of CT scans have to be implemented and the resulting patient dose has to be evaluated against diagnostic reference levels. Practical guidance for proper use of CT is provided in the European Guidelines on Quality Criteria for Multislice Computed Tomography (www.msct.info). European guidelines for multislice CT have been established for 26 clinical indications. These guidelines address the preparatory steps, including advisable preliminary or alternative investigations and patient preparation. Next, the acquisition objectives are described, including the target volume, and the image weighting (contrast resolution, exposure, beam collimation, pitch, CTDI volume and tube voltage). Guidelines for image reconstruction involve the primary reconstruction section thickness, overlap of primary reconstructions, the reconstruction algorithm, field of view, viewing slice thickness and the overlap of viewing slices. Image quality criteria are provided separately for both, visualization and critical reproduction. Finally guidance on the use of contrast media and advice on modification to technique are provided. The guidelines are preceded by a general introduction to Multi Slice Computed Tomography (MSCT), including technical and clinical principles of MSCT, and including paediatric CT, and an extensive discussion of the clinical and associated performance parameters for MSCT. The guidelines are followed by a glossary and list of abbreviations. Three appendices deal with respectively a detailed description of CT dosimetry, a report on a European field survey of multislice CT and a report on original developments in CT dosimetry, including paediatric CT dosimetry.

7. Further reading

Comprehensive textbooks offer excellent information about the general principles of medical imaging[14] computed tomography technology[15,16] and the clinical application of computed tomography[17].

References

J. Geleijns, E. Castellano // Computed Tomography // European School on Medical Physics, ESI, Archamps (France), *Proceedings Week 2: Medical Imaging,* http://www.cur-archamps.fr/esi .

1. Ambrose J. Computerized transverse axial scanning (tomography).. Clinical application. *Br J Radiol.* 1973 Dec;46(552):1023-1047.

2. Hounsfield GN. Computerized transverse axial scanning (tomography). 1. Description of system. *Br J Radiol.* 1973 Dec;46(552):1016-1022.
3. A.C. Kak and M. Slaney, *Principles of Computerized Tomographic Imaging, IEEE Press*, 1988 (http://www.slaney.org/pct/).
4. Flohr TG, McCollough CH, Bruder H, Petersilka M, Gruber K, Suss C, Grasruck M, Stierstorfer K, Krauss B, Raupach R, Primak AN, Kuttner A, Achenbach S, Becker C, Kopp A, Ohnesorge BM. First performance evaluation of a dual-source CT (DSCT) system. *Eur Radiol.* 2006 Feb;16(2):256-68.
5. Mori S, Endo M, Obata T, Tsunoo T, Susumu K, Tanada S. Properties of the prototype 256-row (cone beam) CT scanner. *Eur Radiol.* 2006 Mar 28.
6. C.B. Higgins, A. de Roos; *Cardiovascular MRI and MRA*, ISBN: 0781734827.
7. ICRP publication 60, 1990 recommendations of the International Commision on Radiological Protection, *Annals of the ICRP*, Vol. no. 1-3, Pergamon Press, Oxford (UK), 1991.
8. ImPACT CT Patient Dosimetry Calculator Version 0.99 (2006) (www.impactscan.org).
9. Bongartz G., Golding S.J., Jurik A.G., Leonardi M., van Persijn van Meerten E., Rodríguez R., Schneider K., Calzado A., Geleijns J., Jessen K.A., Panzer W., Shrimpton P.C., Tosi G., 2004 *CT Quality Criteria*, www.msct.info.
10. Imanishi Y, Fukui A, Niimi H, Itoh D, Nozaki K, Nakaji S, Ishizuka K, Tabata H, Furuya Y, Uzura M, Takahama H, Hashizume S, Arima S, Nakajima Y. Radiation-induced temporary hair loss as a radiation damage only occurring in patients who had the combination of MDCT and DSA. *Eur Radiol.* 2005 Jan;15(1):41-6.
11. D.G. Sutton and J.R. Williams Radiation Shielding for Diagnostic X-rays: *Report of a Joint BIR/IPEM Working Party London, England: British Institute of Radiology,* 2000. ISBN 0-905749-44-8.
12. *IEC 61223-3-5*, International Electrotechnical Commission, Evaluation and routine testing in medical imaging departments, Part 3-5: Acceptance tests – Imaging performance of computed tomography X-ray equipment (2004).
13. *ImPACT Information Leaflet* No. 1: CT Scanner Acceptance Testing Version 1.02 (2001) (www.impactscan.org).
14. H.H. Barrett, K.J. Myers, *Foundations of Image Science*, ISBN: 0-471-15300-1, Wiley, 2003.
15. J. Hsieh, Computed Tomography: Principles, Design, Artifacts, and Recent Advances, *SPIE PRESS Monograph* Vol. PM114, 2003.
16. Willi A. Kalender, *Computed Tomography: Fundamentals, System Technology, Image Quality, Applications,* ISBN: 3-89578-216-5, 2005.
17. Mathias Prokop, M. Galanski, A.J. van der Molen, C. Schaefer-Prokop, *Spiral and Multislice Computed Tomography of the Body,* ISBN 0-86577-870-1, 2003.

IMAGE RECONSTRUCTION ALGORITHMS IN POSITRON EMISSION TOMOGRAPHY

CLAUDE COMTAT*
Service Hospitalier Frédéric Joliot, CEA, 4 Place Marechal Leclerc, 91406 Orsay CEDEX, France

Abstract. The principles of image reconstruction in positron emission tomography will be presented in the lecture. The filtered backprojection algorithm will be explained in detail for 2D reconstruction. The generalization of the algorithm in 3D will be described. A brief introduction to iterative reconstruction methods will be given, and their advantages and disadvantages against the filtered backprojection algorithm will be discussed.

Keywords: PET; tomographic reconstruction; filtered back-projection; EM-ML.

1. Data acquisition

Data acquisition in positron emission tomography (PET) consists in the recording of the number of coincident events for each pair of detectors in coincidence. The virtual line going through the center of two detectors in coincidence is called a *line of response* (LOR). Its location is defined by the radial position x_r, the azimuthal angle ϕ, and the indices (r_1, r_2) of the pair of detector rings in coincidence (see Figure 1).

The number of events per line of response is stored in histograms called sinograms. A sinogram is a 2D matrix containing all (x_r, ϕ) pairs. There is one sinogram $s(x_r, \phi)$ per pair of rings (r_1, r_2) in coincidence. For a system with N_R rings, up to $N_R \times N_R$ sinograms can be acquired. In practice, sinograms are indexed by their axial position $z = (r_1 + r_2)/2$ and their ring difference $\Delta z = r_1 - r_2$. Sinograms are classified as *direct* ($|\Delta z| \leq 1$) and *oblique* ($|\Delta z| > 1$) sinograms. During a 2D acquisition, only direct sinograms are recorded.

* Claude Comtat, CEA-SHFJ, 4 Place Marechal Leclerc, 91406 Orsay CEDEX, France; e-mail: comtat@ieee.org.

Y. Lemoigne et al. (eds.), Physics for Medical Imaging Applications, 381–387.

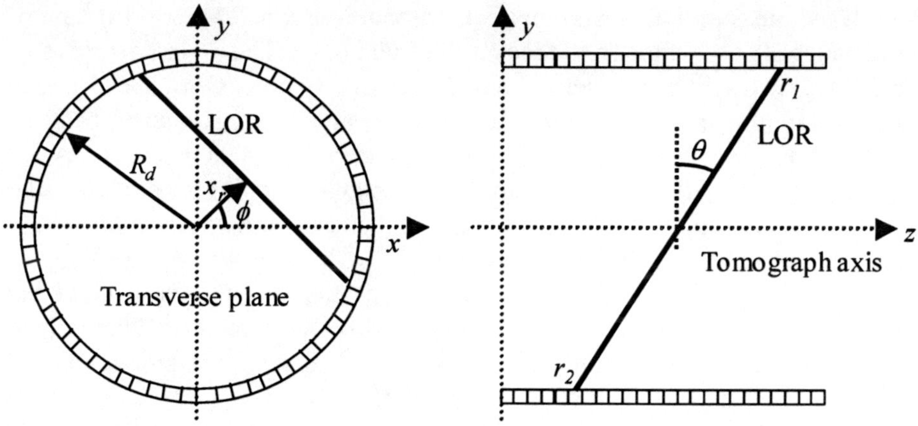

Figure 1. Acquisition geometry.

2. Principles

Tomographic reconstructions allow to produce a three-dimensional image of the radiotracer distribution $f(x, y, z)$ inside the body. In practice, $f(x, y, z)$ is described as a discrete 3D matrix, where each element is called a voxel. For a system with N_R rings with a thickness of Δr, $2 \times N_R - 1$ image slices with a thickness of $\Delta r/2$ are reconstructed.

All reconstruction techniques are based on a mathematical model (**Model**) that links the measured data (*sinogram*) to the image (*image*)

$$Sinogram = \textbf{Model}\{\ image\ \}$$

and a mathematical algorithm to invert the model and reconstruct an image. We distinguish between two classes of reconstruction techniques:

1. analytical reconstructions, when the model can be analytically inverted;
2. iterative reconstructions, when the model cannot be easily analytically inverted and the solution needs to be iteratively found.

3. 2D and 3D reconstruction

In a 2D reconstruction, only direct sinograms are processed. Each direct sinogram allows reconstructing a single image slice, and each image slice is reconstructed independently from the others. The three-dimensional image is obtained by stacking the 2D image slices. In a 3D reconstruction, all image slices are reconstructed simultaneously, using all direct and oblique sinograms.

If the measured data were perfect, without noise, a 2D reconstruction of the image slices is enough to recover the true image. The main advantage of a 2D reconstruction of a set of image slices is its processing speed, much faster than a 3D reconstruction of the same image slices. However, because of the Poisson nature of radioactive decays, measured data are noisy. A 3D reconstruction allows increasing the signal-to-noise ratio in the reconstructed image by accounting for the oblique sinograms. An alternative that preserves the advantages of both reconstruction techniques (fast processing in 2D and increased sensibility in 3D) consists in the acquisition of the data in 3D mode, followed by a rebinning of the oblique sinograms into a stack of direct sinograms. The rebinned sinograms are reconstructed with a 2D method. The two main rebinning techniques are the Single Slice ReBinning (SSRB) (Daube-Witherspoon and Muehllehner, 87) and the FOurier REbinning (FORE) (Defrise et al, 97) algorithms. The first algorithm is very fast, but it degrades the spatial resolution, whereas the second algorithm preserves the spatial resolution, but its processing speed is somehow slower. FORE is frequently used in clinical routine.

4. Analytical Reconstructions

Analytical reconstructions are based on the line-integral model. This model states that the measured data $s(x_r,\phi,z,\Delta z)$ describe the 2D analytical projection $p(x_r,y_r,\phi,\theta)$ of the image $f(x,y,z)$

$$s(x_r,\phi,z,\Delta z) = p(x_r,y_r,\phi,\theta) = \int_{\mathrm{LOR}(x_r,y_r,\phi,\theta)} f(x,y,z)\,\mathrm{d}z \qquad (1)$$

The direction of a 2D projection is given by the azimuthal angle ϕ and the co-polar angle θ (see Figure 2). One element of a 2D projection is parameterized by the radial coordinate x_r and the axial coordinate y_r. The link between the sinogram indices $(z, \Delta z)$ and the projection coordinates (y_r, θ) is given by

$$\begin{aligned} y_r &= z \cdot \cos\theta \cdot \Delta r \\ \tan\theta &= \frac{\Delta z \cdot \Delta r}{2 \cdot \sqrt{R_d^2 - x_r^2}} \end{aligned} \qquad (2)$$

where R_d is the detector ring radius. When the size of the field-of-view along the scanner axis is small compared to the radius R_d, which is true for most whole-body scanner, the co-polar angle value is approximated by

$$\tan\theta \approx \frac{\Delta z \cdot \Delta r}{2 \cdot R_d}. \quad (3)$$

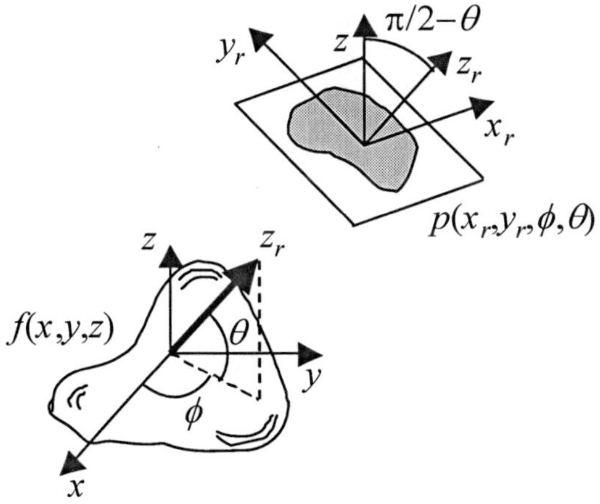

Figure 2. The 2D projection $p(x_r, y_r, \phi, \theta)$ of the object $f(x, y, z)$.

Several analytical algorithms allow to invert (1) and reconstruct $f(x, y, z)$ from $s(x_r, \phi, z, \Delta z)$. The most commonly used is the Filtered BackProjection algorithm (FBP) (Natterer, 86). This algorithms requires a complete sampling of $p(x_r, y_r)$ for all measurement directions (ϕ, θ). However, because of the limited axial size of the tomograph, this requirement is not satisfied for oblique sinograms (see Figure 3). To overcome this limitation, the 3D reprojection algorithm (3DRP) (Kinahan and Rogers, 90) has been developed. This algorithm is divided into four steps: 2D reconstruction with FBP of a first image from the direct sinograms only, estimation of the oblique sinograms that are not measured by numerical projection of the first image, merging of the estimated projections with the measured projections, 3D reconstruction with FBP of the merged sinograms. 3DRP is considered as the *gold standard* of 3D reconstructions. For many PET systems, however, this algorithm is not implemented. To speed-up the reconstruction process, oblique sinograms are rebinned into direct sinograms with FORE, and the rebinned sinograms are reconstructed in 2D with FBP. This FORE+FBP-2D technique introduces slight spatial resolution degradation when compared to 3DRP.

Analytical algorithms assume that the measured data are not noisy. In practice, because of the count-limited statistics of PET acquisitions, measured data are characterized by a high level of statistical noise. Analytical algorithms perform poorly when the data are noisy. To limit the effect of the noise, an apodization window is often applied during the filtering step of FBP. Its effect is to smooth the projections, hence degrading the spatial resolution in the reconstructed image. From a practical point of view, the user of a FBP technique shall specify the type of apodization window and the level of filtering.

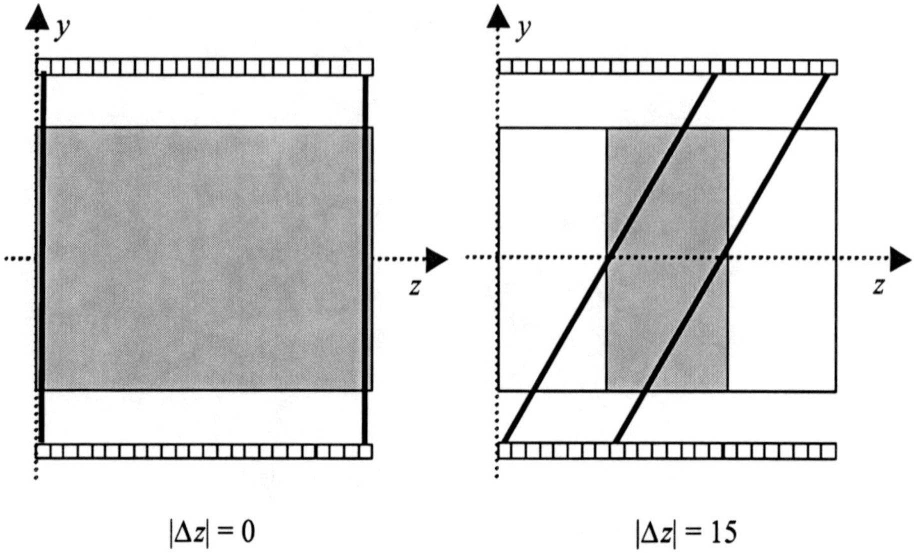

Figure 3. Segment of the field-of-view that is measured (in gray) as a function of the ring different Δz.

5. Iterative reconstructions

Iterative reconstruction techniques allow to use a more complex, hence more realist model than the line integral model. Usually, a statistical noise model is added to the line integral model. Iterative algorithms intrinsically incorporate the discrete nature of the sinograms ($\mathbf{s} = \{s_i \mid i = 1,\ldots,N\}$) and the image ($\mathbf{f} = \{f_j \mid j = 1,\ldots,M\}$), written here as a vector with a length equal to the total number of sinogram elements (N) or image voxels (M).

Given a matrix model $A_{i,j}$ that defines the probability to detect in the LOR i a decay occurring in voxel j, the expectation of the number of detected events s_i is equal to

$$\mathrm{E}\{s_i\} = \langle \mathbf{A}_i, \mathbf{f} \rangle = \sum_{j=1}^{m} A_{ij} \cdot f_j \,, \tag{4}$$

where f_j is the intensity of voxel j. The measurement s_i is a random variable that follows a Poisson distribution

$$p\left(s_i \middle| \langle \mathbf{A}_i, \mathbf{f} \rangle\right) = \frac{e^{-\langle \mathbf{A}_i, \mathbf{f} \rangle} \cdot \langle \mathbf{A}_i, \mathbf{f} \rangle^{s_i}}{s_i!}. \tag{5}$$

Most iterative algorithms used in PET seek at maximizing the log-likelihood $L(\mathbf{f})$ (ML, for *maximum likelihood*)

$$L(\mathbf{f}) = \log\left(\prod_{i=1}^{N} p\left(s_i \middle\| \langle \mathbf{A}_i, \mathbf{f} \rangle\right)\right) = \sum_{i=1}^{N} \left(s_i \log\left|\langle \mathbf{A}_i, \mathbf{f} \rangle - \right|\langle \mathbf{A}_i, \mathbf{f} \rangle - \log(s_i!)\right) \tag{6}$$

EM-ML (Shepp and Vardi, 82) (*Expectation Maximization – Maximum Likelihood*) is the likelihood maximization algorithm most commonly used because of its implementation simplicity. The image at iteration $(p + 1)$ is given by

$$f_j^{(p+1)} = \frac{f_j^{(p)}}{\sum_{i=1}^{N} A_{ij}} \sum_{i=1}^{N} A_{ij} \frac{s_i}{\left\langle \mathbf{A}_i, \mathbf{f}^{(p)} \right\rangle}. \tag{7}$$

However, the convergence rate of the EM-ML algorithm is very slow. An accelerated version has been developed (OSEM, *Ordered Subset EM-ML*) (Hudson and Larkin, 94), where the measured data **s** are distributed among K subsets. Rather than updating the image once per iteration, after the back-projection of all N elements of **s**, the image is updated K times during one iteration, after the back-projection of all elements of one subset of **s**. The convergence rate of OSEM is approximately K times faster than for EM-ML. The RAMLA algorithm (Browne and de Pierro, 96) (*Row Action Maximum Likelihood Algorithm*) converges also more rapidly to the ML solution than EM-ML.

From a practical point of view, the user of an iterative reconstruction technique shall specify the number of iterations. The statistical noise in the reconstructed image increases with the number of iterations, but the spatial resolution gets better. Usually, the algorithm is halted before convergence. To further decrease the statistical noise, the reconstructed image is sometime smoothed.

In many implementations, the ML iterative algorithm is applied to data pre-corrected for the random and scattered coincidences, attenuation and normalization. These pre-corrected data do not follow a Poisson distribution, in contradiction with the Poisson model of EM-ML. In order to partially restore the Poisson properties of the data, one solution is to weight the data according to the attenuation (AW, *Attenuation Weighted*) (Comtat et al., 98).

In general, the computing time for one iteration is of the same order than for an analytical reconstruction. To keep a total computing time compatible with a clinical use when the data are acquired in 3D mode, the sinograms are sometime rebinned with FORE prior to the use of a 2D iterative algorithm.

References

Browne J, De Pierro AR. A row action alternative to the EM algorithm for maximizing likelihood in emission tomography. IEEE Trans Med Imag 1996;15:687-99.

Comtat C, Kinahan PE, Defrise M, Michel C, Townsend DW. Fast reconstruction of 3D PET with accurate statistical modeling. IEEE Trans Nucl Sc 1998;45:1083-9.

Daube-Witherspoon ME, Muehllehner G. Treatment of axial data in three-dimensional PET. J Nucl Med 1987; 28 :1717-24.

Defrise M, Kinahan PE, Townsend DW, Michel C, Sibomana M, Newport DF. Exact and approximate rebinning algorithms for 3D PET data. IEEE Trans Med Imag 1997;16: 145-58.

Hudson H, Larkin. Accelerated image reconstruction using ordered subsets of projection data. IEEE Trans Med Imag 1994; 13: 601-9.

Kinahan PE, Rogers JG. Analytical three-dimensional image reconstruction using all detected events. IEEE Trans Nucl Sci 1990; 36: 964-8.

Natterer F. The mathematics of computerized tomography. New York, NY: Wiley; 1986.

Shepp LA, Vardi Y. Maximum likelihood reconstruction for emission tomography. IEEE Trans Med Imag 1982; 1:113-22.

QUALITY CONTROL IN MAMMOGRAPHY

ALAIN NOEL*
Centre Alexis Vautrin, Unité de Radiophysique Médicale, Avenue de Bourgogne – 54511 Vandoeuvre lès Nancy, France
Centre de Recherche en Automatique de Nancy (UMR 7039 CRAN CNRS-UHP-INPL), 2 Avenue de la Forêt de Haye - 54516 Vandoeuvre lès Nancy, France

Abstract. In mammography to obtain the best image quality at the lower possible dose to minimize the risk, implementation of an effective quality control protocol is of the utmost importance. This paper will mainly review quality control (QC) procedures for screen-film mammography and will briefly present how screen-film and digital differ. Finally, the possibility to automate some tests in digital mammography will be introduced.

Key words: quality control; screen-film mammography; digital mammography.

1. Introduction

Nowadays, breast cancer is very frequent and will affect about 1 out of 8 women during her lifetime. Mammography screening programs have been set up to examine asymptomatic women, in order to detect cancers in a very early stage when treatment is most effective and the survival rate is still high. Mammography which is the x-ray examination of the breast is a very demanding radiological task. So, image quality is of paramount importance in mammography as we need at the same time, high spatial resolution to detect micro-calcifications, low contrast detestability to visualize masses and low dose to minimize risk due to the use of ionizing radiation.

Quality Control (QC) is an important part of a quality assurance program in mammography. The quality control protocols designed primarily for film/screen units have defined the critical image quality criteria that reflect the overall performance of a given system. Quality Control is even more important when dealing with digital techniques that

* Alain Noel, Centre Alexis Vautrin, Unité de Radiophysique Médicale, Avenue de Bourgogne – 54511 Vandoeuvre lès Nancy ; e-mail : a.noel@nancy.fnclcc.fr.

Y. Lemoigne et al. (eds.), Physics for Medical Imaging Applications, 389–395.

are more complex and more exposed to failures due to electronic components drifts. In addition, viewing conditions and printing are also highly demanding.

This paper will focus mainly on the quality control of screen-film mammography before addressing the quality control of digital mammography stressing on how screen-film and digital differ. Finally, we will address the possibility with digital systems, to implement data processing tools to simplify and automate quality control tests. The main advantage of such an approach is that it reduces the time spent in tasks required when dealing with quality controls.

2. QC in screen-film mammography

2.1 QUALITY CONTROL GUIDELINES

Mammography quality goal is to obtain the best possible image at an acceptable dose for each patient, make sure that the equipment performs in a consistent manner and with an acceptable level of artefacts in the image. For doing so, many countries have implemented mammography quality control programs[1,2] mainly based on the protocols developed and published by The European Commission[3] and the American College of Radiology[4, 5]. For example, quality control guidelines adopted in Belgium, France, Germany, Greece, Ireland, Israel, Italy, Luxembourg, Norway, Spain and Sweden are based on European guidelines, while in Uruguay and USA they are based on ACR guidelines. On the other hand, Australia, Canada, Denmark, Finland, Hungary, Iceland, Japan, The Netherlands, Portugal and UK have developed country-specific QC guidelines. Finally, in Finland, France, Hungary, The Netherlands, Spain, Sweden and USA QC is required by law.

2.2 QC PROTOCOL OF THE MAMMOGRAPHY EQUIPMENT

QC of mammography equipment allows evaluating and tracking performances, to identify problems and to verify the results of a corrective action. A quality control protocol should consider the automatic processor, the Screen-Film system, the mammography X-ray unit, the view boxes and the viewing conditions. We can distinguish different mammography surveys. The first one is the initial equipment evaluation which is done on new equipment before it is used on the patient to ensure that the equipment meets all the performance specifications. The second one which could

constitute an annual survey includes the equipment evaluation plus the QC review.
A standard annual survey should include these tests:

- Darkroom: to assure that darkroom will not be responsible for dust artefacts and that darkroom safelights do not fog mammography films.

- Processor: to assure that the film processor-chemical system is working in a consistent manner. This is done by checking the operating levels and the control limits and by performing an artefacts evaluation.

- Mechanical safety and function of the mammography unit : this is just a physical inspection to make sure that :

    - The unit is mechanical rigid and stable
    - All moving parts are free-running
    - All locks and detent work properly
    - All indicators lights and displays function
    - The receptor assembly is free from vibration
    - Image receptor is hold securely in the receptor assembly
    - The compression plate functions correctly, with a uniform compression and a force not exceeding the maximum force allowed
    - The breast thickness indicator is accurate

- Collimation assembly :

    - Check the radiation to light-field alignment
    - Check that the radiation beam is confined to the image receptor except at chest wall edge
    - Look at the alignment of the compression paddle to the chest wall edge of the film

- Image receptors :

    - Verify the light tightness of cassette
    - Screen-film contact: to assure that optimal contact is maintained between screen and film in each cassette

- Relative uniformity of screen speed : to assure that uniformity of relative sensitivity of image receptors routinely used is adequate

- kVp performance : to assure that the actual kVp is accurate and reproducible

- Focal spot size : to determine if the focal spot size is within acceptable limits

- Beam quality : to assure that the HVL of the X-ray beam is adequate

- Automatic Exposure Control : to assess the AEC system with regard to performance capability (kVp and thickness compensation) and density control selector function

- Image Quality: to assess that under clinical exposure settings mammography image quality of the phantom[7] is adequate and to evaluate artefacts.

- Entrance Surface Dose: to calculate the average glandular dose (AGD) based on the half value layer by measuring the standard entrance surface dose on the mammography phantom under clinical conditions.

- Specific Radiation Output Rate: to assure that under routine exposure conditions the radiographic exposure time will not be too long.

- Viewing boxes : to assure that viewing boxes provide sufficient luminance to allow visualization of all clinical information

- Viewing Conditions : to assure that viewing conditions are optimized and maintained at an optimal level

Thanks to the implementation of such comprehensive quality control protocol screen-film mammography quality has dramatically increased over the years. Today, we can state that screen-film is really quite optimized. What about digital mammography?

3. QC in digital mammography

QC protocol designed for screen-film is now well established. The demand is not different for digital mammography and QC protocols[5, 6] need to address this issue. For example, due to the introduction of digital mammography, the fourth version of the European Protocol for the Quality Control of the Physical and Technical Aspects of Mammography Screening incorporates both screen-film and digital mammography. It is going to be released soon. Let's see how QC for screen-film and Digital differ.

3.1 HOW SCREEN-FILM AND DIGITAL DIFFER?

First, is that image acquisition and display are decoupled in digital. So, both must be evaluated. The second very important thing is that due to the large linear dynamic range of digital detectors, it is easier for problems to be masked. For example, the relation between optical density and dose with screen-film no longer applies in digital with a risk of overexposure. Thirdly, image in digital form allow quantitative analysis. On the other hand, many components between screen-film and digital mammography units are still the same. So when regarding digital mammography QC protocols, we can distinguish between tests that are the same, similar or different to screen-film and those which are specific to digital.
Here are some examples:

1. Same QC test:

- Unit assembly evaluation
- Viewing boxes and Viewing conditions

2. Similar QC tests:

- Image quality: can be assessed with a phantom as for screen-film mammography or from contrast-detail threshold measurements completed with Signal to Noise Ratio (SNR) and Contrast to Noise Ratio (CNR) measurements.

3. Different QC tests:

- Automatic Exposure Control system performance. AEC algorithms are more sophisticated for digital and need to be tested under varying conditions. The AEC system should perform adequately such that image quality is sufficient and dose is within an acceptable range regardless the clinical settings.
- Geometric distortion and Artefacts evaluation.
- System resolution: Modulation Transfer Function (MTF) of the detector can be determined by using appropriate test tool and software.

- Noise properties: Noise Power Spectrum (NPS) can be determined by imaging a standard test block and by using appropriate software.

4. Specific QC tests:

- Image presentation on Monitors and Printers. These tests make an extensive use of the so-called TG18 test pattern designed by the American Association of Physicists in Medicine, Task Group 188.

- Monitor display: To assure a visual consistency in how a mammogram will appears whether viewed on different viewing stations or as a film on a light-box. To that purpose, DICOM 3, part 14 specifies a Grayscale Standard Display Function9-11 that relates pixel values to displayed Luminance levels.

3.2 POSSIBILITY TO AUTOMATE SOME TESTS

Image in digital form offers some possibilities to automate some tests. Using dedicated Phantom and appropriate software tools some quality criteria such as Modulation Transfer Function (MTF), Noise Power Spectrum (NPS), Detective Quantum Efficiency (DQE) and Uniformity measurements can be easily performed. These criteria characterise the intrinsic imaging system performances but is not linked in a straightforward manner to the Radiologist detection task. Contrast-Detail phantom (CDMAM) scoring[12-14] is widely used to assess the detection performance of human readers from images acquired with a given imaging system and review on a given image display equipment. Replacing human observers with mathematical observers would therefore make the automated scoring of phantom a potential alternative to the assessment and the comparison of system imaging performance, while getting rid of observer variability and fastidious review process.

4. Conclusion

Quality control protocols have been established to monitor the quality of each step in the image chain generation and display for screen-film and digital mammography. For digital many features are similar to screen-film and some are different such as Automatic Exposure Control, Artefacts and display of the image.

Finally, it is worth mentioning with digital the potential to automate some tests which could help in the regularly image quality control in mammography.

References

1. Shapiro S, Coleman EA, Broeders M et al. Breast cancer screening programs in 22 countries: current policies, administration and guidelines. Int J Epidemiol 1998; **27**: 735-742.
2. Hendricks RE, Klabunde C, Grivegnee A, Pou G, Ballard-Barbash R. technical quality control practices in mammography screening programs in 22 countries. International Journal for Quality in Health Care 2002; **volume 14, Number 3**: pp. 219-226.
3. The European Protocol of the Quality Control of the Technical Aspects of Mammography Screening. CEC Report EUR 14821, 3rd edn 2001.
4. Hendrick RE, Bassett LW, Dodd GD et al. Mammography Quality Control: Radiologist's Manual, Radiologic Technologist's Manual, Medical Physicist's Manual. Reston, VA: American College of Radiology, 1st edn, 1990; 2nd edn, 1992; 3rd edn 1994, 4th edn, 1999.
5. Hendricks RE, Berns EA. Full-Field Digital Mammography Quality Control Manual. Phantom Image Publishing. Chicago, Illinois. Version 1.1 - May 2001-10-01.
6. Kimme-Smith C, Lewis C, Beifuss M, Bassett LW. Establishing minimum performance standards, calibration intervals and optimal exposure values for a whole breast digital mammography unit. Med. Phys. 25 (12), December 1998.
7. Noel A, Stines J, Zerhouni M. *Développement d'un fantôme en matériau équivalent-sein pour la mammographie,* RBM, Vol. 15-7 (1993) 412-423.
8. Samei E, Badano A, Chakraborty D, Compton K, Cornelius C, Corrigan K, Flynn MJ, Hemminger B, Hangiandreou N, Johnson J, Moxley M, Pavlicek W, Roehrig H, Rutz L, Shepard J, Uzenoff R, Wang J, Willis C. Assessment of Display Performance for Medical Imaging Systems, Report of the American Association of Physicists in Medicine (AAPM) Task Group 18, *Medical Physics Publishing*, Madison, WI, AAPM On-Line Report No. 03, April 2005.
9. Digital Imaging and Communications In Medicine (DICOM), *Grayscale Standard Display Function*, PS 3.14, January 1998, National Electric Manufacturers Association.
10. Barten PG. *Physical model for the Contrast Sensitivity of the human eye*, proc. SPIE 1666, 57-75, 1992.
11. Barten PG. Spatio-temporal model for the Contrast Sensitivity of the human eye and its temporal aspects, proc. SPIE 1913, 01, 1993.
12. Thijssen MAO, Veldkamp W, Van Engen R, Swinkels M, Karssemeijer N, Hendricks J Comparison of the detectability of small details in a film-screen and digital mammography system by the imaging of a new CDMAM-phantom. Proceedings of IWDM 2000, pp. 666-672, Yaffe M ed, Medical Physics Publishing, Madison, WI, Toronto, 2000.
13. Barret HH, Yao J, Rolland JP, Myers KJ. *Model observers for assessment of image quality*, Proc. of the Nat. Acad. of Sc. (USA), 1993.
14. Myers KJ. *Ideal Observer Models of Visual Signal Detection*, Handbook of Medical Imaging, Vol. 1, pp 559-592, J. Beutel, H.L. Kundel, R.L. Van Metter Editors, SPIE Press, 2000.

PATIENT DOSE IN DIAGNOSTIC RADIOLOGY

ALAIN NOEL*
Centre Alexis Vautrin, Unité de Radiophysique Médicale, Avenue de Bourgogne – 54511 Vandoeuvre lès Nancy, France
Centre de Recherche en Automatique de Nancy (UMR 7039 CRAN CNRS-UHP-INPL), 2 Avenue de la Forêt de Haye - 54516 Vandoeuvre lès Nancy, France

Abstract. One of the basic principles, stated explicitly in Article 4 of the EC Council Directive 97/43 Euratom, is optimization. This means that all radiological examinations should be performed with a dose that is As Low As Reasonably Achievable (ALARA principle applied to the protection of the patient) in order to obtain the required diagnostic information. Therefore, dose needs to be determined with the relationship between image quality and dose always kept in mind. In this paper, radiation quantities and units to report patient doses in diagnostic radiology will be identified.

Keywords: patient dose; diagnostic radiology.

1. Introduction

It is very important to assess medical exposure as ionizing radiation is potentially harmful, capable of causing serious and lasting biological damage even at low doses.

The collective population exposure is the sum of the exposure from natural and artificial or man-made radiation sources. From UNSCEAR[1], the annual average dose to the French is about 2.4 mSv from natural radiation exposure and 1.1 mSv from artificial exposure. Then, artificial exposure represents a 50% increase of the annual contribution of the natural sources of radiation. Not only for France but all over Europe, man-made radiation is dominated by medical exposure and the majority of this can be attributed to diagnostic radiology[2].

Dose evaluation is also an important quality indicator for the optimisation of the medical examinations.

* Alain Noel, Centre Alexis Vautrin, Unité de Radiophysique Médicale, Avenue de Bourgogne – 54511 Vandoeuvre lès Nancy, France; e-mail : a.noel@nancy.fnclcc.fr.

Y. Lemoigne et al. (eds.), Physics for Medical Imaging Applications, 397–404.

2. Medical exposure

We should not overlook the fact that for the radiological examination of symptomatic patients, unlike natural exposure most of the benefit and possible detriment accrue to the exposed individual. On the other hand, when assessing the risk, it is usual to compare the effective dose delivered by radiological examinations with the dose due to natural radiation exposure. But an important point to consider is the huge difference between dose rate of the natural radiation and dose rate of the diagnostic radiology exposure. The dose rate of medical exposure is about 1 million times higher and we cannot conclude that there is no dose rate effect

2.1 WHY MEASURE PATIENT DOSE

The first reason could be due to legislation. Indeed, one of the basic principles, stated explicitly in article 4 of the EC Council Directive 97/43/EURATOM, is optimisation. This means that all examinations should be performed with a dose that is As Low As Reasonably Achievable (ALARA principle applied to medical exposure) in order to obtain the required diagnostic information.

But, it is assumed that the goal of achieving acceptable image quality to get adequate diagnostic information, consistent with the medical imaging task, is the overriding clinical objective. Therefore, dose needs to be determined with the relationship between image quality and dose always kept in mind.

Another important point to consider is that in our modern society there an increasing request to information on all aspects of health problems.
The main objectives for dosimetry in diagnostic radiology are :

- To establish the approximate risk from a particular examination
- To assess the risk to an individual patient
- For comparison with diagnostic reference level agreed nationally and internationally
- To monitor the collective dose to the population
- To assess equipment performance as part of a quality assurance programme for diagnostic radiology

2.2 PRIORITIES

Patient doses have to be measured for a wide range of different types of X-ray examination. The situations in which dose control is most important are :

- High risk examinations e.g. paediatric radiology, mammography
- High dose examinations e.g. interventional procedures, CT, barium enema
- High frequency/low dose examinations e.g. dental radiology

3. Patient dosimetry

3.1 BASIC DOSIMETRY QUANTITIES[3]

Absorbed dose relates to where secondary electrons deposit their energy in the medium.

$$D = dE/dm$$

dE is the mean energy imparted to mass dm of material. The special unit of absorbed dose is the gray.

$$1 \text{ Gy} = 10^{-2} \text{ J kg}^{-1}$$

KERMA is increasingly used in radiation dosimetry. This stands for Kinetic Energy Released per unit Mass and must specify the material concerned. Kerma places the emphasis on removal energy from the beam of indirectly particles in order to create secondary electrons.

$$K = dE_{tr}/dm$$

dE_{tr} is the sum of the initial kinetic energy liberated by uncharged particles in mass dm of material. The special unit of kerma is the Gray.

$$1 \text{ Gy} = 10^{-2} \text{ J kg}^{-1}$$

Kerma in air may differ from dose in air for 2 reasons :

1. First, secondary electron may be radiated as bremsstrahlung
2. Second, the point of energy deposition in the medium is not the same as the point of removal energy from the beam because of the range of secondary electrons.

In diagnostic radiology bremsstrahlung is negligible and the ranges of secondary electrons are so short that:

$$K_{air} = D_{air}$$

3.2 MEASUREMENT METHODOLOGY[4-6]

The entrance surface dose is defined as the absorbed dose in air at the point of intersection of the X-ray beam axis with the entrance surface of the patient including back-scattered radiation. Note that in diagnostic radiology the back-scattered factor (BSF) can be quite high and will typically be in the range 1.2-1.4. It is recommended that direct measurements of entrance surface dose should be made using small, independently calibrated, dosimeters (TLDs).

However, when a direct measurement has not been made indirect measurement should not be overlooked. Knowing tube output measured under reference conditions, the entrance surface dose can then be calculated from the knowledge of the exposure factors applying any necessary corrections.

Applications specific quantities to be used :

- Air kerma free in air (mGy) measured at reference distance :

$$K_{air,dref} = {}_{n}K_{air,\,dref} \,.\, mAs$$

where ${}_{n}K_{air}$, d_{ref} is the normalized kerma (mGy/mAs) in air at the reference distance d_{ref}.

- Entrance Surface Air Kerma (mGy) which is the kerma in air on the X-ray beam axis at the patient's skin.

$$ESAK = K_{air,\,dref}.(d_{ref}/FSD)^2$$

where FSD is the focus to skin distance.

- Entrance Surface Dose (mGy)

$$ESD = ESAK.BSF.(\mu a/\rho)_{tissu,\,air}^{depth=0}$$

To convert Entrance Surface Air Kerma (ESAK) to Entrance Surface Dose (ESD) of the medium you have to take into account the back scatter factor (BSF) and to recall that the energy absorbed per unit mass depends only on the mass absorption coefficient of the medium relative to air at the surface $[(\mu a/\rho)_{tissu,\,air}^{depth=0}]$.

- Depth dose

$$D_d = ESAK.(\mu en/r)_{tissue,\ air}^{\ depth=0}\ .\ (FSD/FSD+d)^2.TAR(d,S)$$

where $(\mu en/r)_{tissue,\ air}^{\ depth=0}$ is the mass energy absorption coefficient of the medium relative to air at the surface, and TAR is the tissue air ratio at depth d for a beam area S.

Finally, it is also possible to make a crude estimation using published data on similar types of examination.

4. Practical techniques of measurements

4.1 RADIOLOGY

4.1.1 *Measurement techniques*

Recommended dosimetric quantities to measure patient dose have to:

- Be defined unambiguously.
- Be simple and precisely measured with dosimeters easily available.
- Give a typical value in relation with the dose delivered.

Two practical dose quantities have been recommended:

- entrance surface dose (ESD) for individual radiographs in other word for simple examination with a limited number of projections such as chest, pelvis, abdomen, skull, spine dose area product (DAP) meter for complete or more complex examinations involving radiography with many projections and fluoroscopy such as IVU, Barium enema, Barium meal, Barium swallow, interventional procedure. These quantities are essential to derive organ doses and effective dose to estimate the risk[7, 8].
- For interventional radiology, there have been several reports of entrance skin doses exceeding 2 Gy which can cause deterministic effect. So, it could be of interest to determine, along with the DAP the maximum Entrance Skin Dose (ESDmax) to be aware of the possibility of deterministic effects for those types of procedures.

4.1.2 *Choice of dosimeters*

Two types of dosimeter are commonly used for estimating entrance surface dose to patients during X-ray examination, namely ionisation chambers (IC) and thermoluminescent dosimeters (TLDs). It is worth mentioning that scintillation detectors can also be used to perform ESD.

Ionisation chambers are mainly used to perform measurements of air kerma on the axis of the X-ray beam (beam calibration) and/or absorbed dose on phantom. They are not well suited for *in vivo* measurements because they are bulky and require connecting cable, they are difficult to attach in close contact to the patient's skin to measure the BSF and moreover you get cast interfering shadow on radiographs.

Properly calibrated, TLDs are recommended[9] for direct measurement of ESD. These dosimeters have some advantages. They are physically small and available in a variety of physical forms and different materials. Moreover, TLDs can be easily stuck directly on patient's skin to ensure full measure of BSF from the patient and they are not likely to obscure useful diagnostic information.

The dose-area-product (DAP) is most conveniently measured with specially designed dose-area product meters. They consist of flat, large area parallel plate ionisation chambers connected to suitable electrometers.

DAP meters are mounted on the diaphragm housing to intercept the entire cross-section of the X-ray beam. The response of DAP meter in terms of charge collected is proportional to both the area of the beam and the averaged dose over the area of the X-ray beam. DAP is usually measured in $Gy.cm^2$ or $cGy.cm^2$ and radiation back-scattered from the patient is excluded. If reset to zero at the beginning of each examination they can provide a single measurement of the total amount of radiation used even in the most complex examination involving radiography and fluoroscopy.

4.2 CT SCANNERS

CT and Mammography are two examinations where strict dose control is essential. CT has been considered in the Council Directive 97/43 as involving high doses to the patient. It has been indicated in some papers[10] that in our modern society dose delivered from CT examination account for about 50% of the collective dose delivered by medical exposure.

EU[11] has recommended two reference dose quantities to report doses in CT examinations. First, by determining the CT Dose Index (CTDI). In practice, a convenient assessment of CTDI can be made by using the standard 16- and 32-cm-diameter perpex dosimetry phantoms and a pencil ionisation chamber with an active length of 10 cm to provide a measure of $CTDI_{10}$ expressed in term of absorbed dose in air given in mGy. For a complete examination, the Dose-Length-Product (DLP expressed in mGy.cm) is used. The total DLP for the entire procedure is calculated from the CTDI value as a function of the scanning protocol used and the sum of the contribution from each sequence performed.

4.3 MAMMOGRAPHY

Because mammography is used for screening programme and because the high radio sensitivity of the tissues involved attention should be paid to the dose delivered to the breast.

In mammography, the attenuation as the X-ray beam passes through the breast is very important with no irradiation to other organs. Thus, EU[12, 13] recommends that the relevant dose estimation for mammography should be the dose to the breast gland which is called the average glandular dose (AGD) rather than ESD. A complete procedure for the determination of average glandular dose to typical breasts simulated with PMMA is presented in the European guidelines[13] for quality assurance in mammography screening.

5. Conclusion

The determination of patient dose in diagnostic radiology is very important but is not a simple matter. Patient doses have to be measured for a wide range of different types of X-ray examinations. Nowadays, routine measurements are increasing, mandatory by law in some countries. There is clear evidence that under careful attention patient doses are coming down. EC requires optimization for all radiological procedures to get high quality image and low dose keeping in mind that too low dose image of non-diagnostic quality is useless.

References

1. UNSCEAR 2000-Sources and Effects of Ionizing Radiation. United Nations Scientific Committee on the Effects of Atomic Radiation.
2. NRPB-R289-Hart D, Hillier MC, Wall BF, Shrimpton PC, Bungay D. Doses to Patients from Medical Examinations in the UK – 1995 Review.
3. International Commission on Radiation Units and Measurements. Radiation Quantities and Units. ICRU Report 19: Bethesda Md, USA (1975).
4. Dendy PP, Goldstone KE. Radiation Measurements and Doses to Patients in Physics for Diagnostic Radiology. Institute of Physics Publishing, Bristol and Philadelphia, 1999.
5. Dixon RL Patient Dose Determination and Evaluation in Diagnostic Radiology. RSNA Categorical Course in Physics; pp. 135-139, 1996.
6. Dosimetry in Diagnostic Radiology. Radiation Protection Dosimetry Published by Nuclear Technology Publishing. Vol. Nos 1-4, 1992.
7. ICRP (International Commission on Radiological Protection), 1991, Recommendations of the International Commission on Radiological Protection, ICRP Publication 60. Oxford: Pergamon Press.

8. ICRP (International Commission on Radiological Protection), 1996, Radiological Protection and Safety in Medicine, ICRP Publication 73. Oxford: Pergamon Press.
9. EUR 19604-Recommendations for patient dosimetry in diagnostic radiology using TLD. European Commission. . Luxembourg : Office for Official Publications of the European Communities, 2000.
10. Crawley MT, Booth A, Wainwright A. A practical approach to the first iteration in the optimization of radiation dose and image quality in CT: estimates of the collective dose savings achieved. The British Journal of Radiology, 74 (2001), 607-614.
11. EUR 16262-European Guidelines on Quality Criteria for Computed Tomography. European Commission. Luxembourg : Office for Official Publications of the European Communities, May 1999.
12. EUR 16263-European Protocol on Dosimetry in Mammography. European Commission. Luxembourg : Office for Official Publications of the European Communities, 1996.
13. Addendum on Digital Mammography to Chapter 3 of he European Protocol of the Quality Control of the Technical Aspects of Mammography Screening. European Commission, version 1, November 2003.

COLOR SECTION

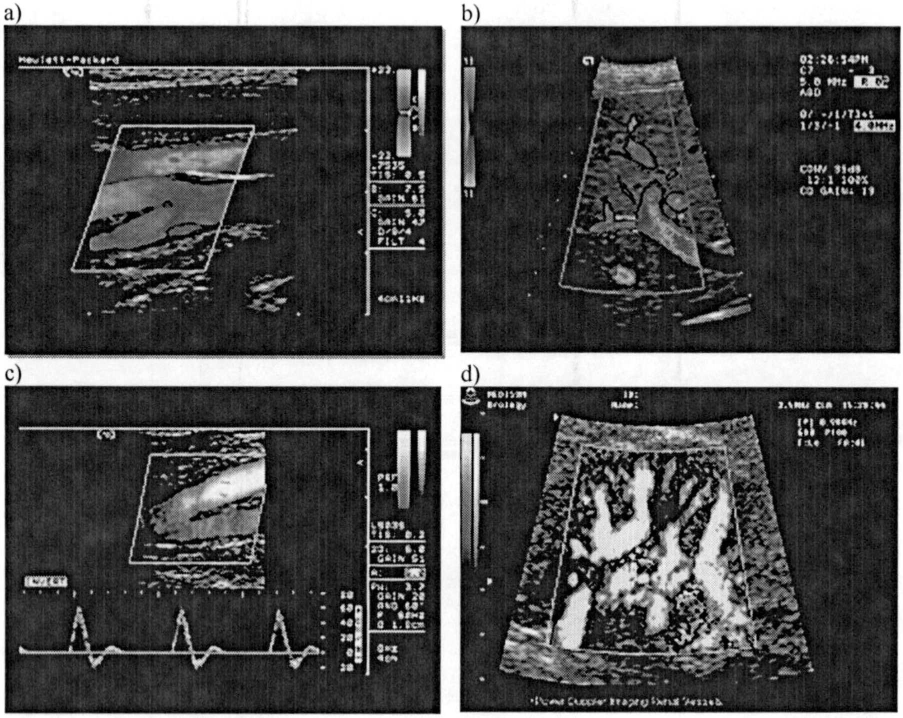

Figure 7. Typical US Doppler images provided by commercial equipment. (a) Carotid artery bifurcation and jugular vein are identified through red and blue colours, respectively, because of the opposite blood flow directions in such vessels; (b) portal vein: sudden colour changes are here due to aliasing; (c) femoral artery bifurcation colour image and spectrogram; (d) renal perfusion estimate through power Doppler.

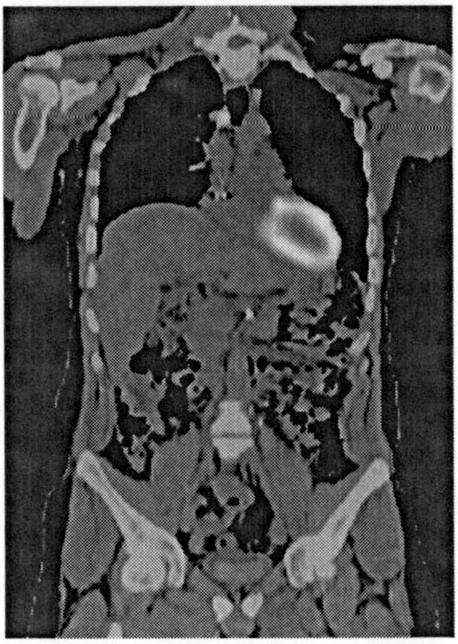

Figure 21. A PET/CT scan of a patient with a lymph node disease. The image shows high uptake of FDG in the lymph node, the heart and the bladder.

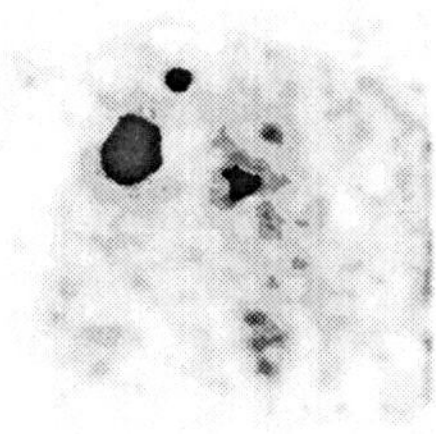

Figure 23. A half-body PETRRA FDG image of the patient shown in figure 19.